U.S. Organic Dairy Politics

Also by the author

India's White Revolution:
Operation Flood, Food Aid and Development

Food and Risk in the US and UK:
Seattle and Newcastle Academics, Firefighters, Motorcyclists and Others Reflect
on Organic and Local Food

Technological and Social Dimensions of the Green Revolution: Connecting
Pasts and Futures
Coedited with Pratyusha Basu

U.S. Organic Dairy Politics

Animals, Pasture, People, and Agribusiness

Bruce A. Scholten

First published in 2014 by PALGRAVE MACMILLAN® in the United States—a division of St. Martin's Press LLC, 175 Fifth Avenue, New York, NY 10010.

Where this book is distributed in the UK, Europe and the rest of the world, this is by Palgrave Macmillan, a division of Macmillan Publishers Limited, registered in England, company number 785998, of Houndmills, Basingstoke, Hampshire RG21 6XS.

Palgrave Macmillan is the global academic imprint of the above companies and has companies and representatives throughout the world.

Palgrave® and Macmillan® are registered trademarks in the United States, the United Kingdom, Europe and other countries.

ISBN: 978-1-137-33060-4

Library of Congress Cataloging-in-Publication Data

Scholten, Bruce A. (Bruce Allen)
    U.S. organic dairy politics : animals, pasture, people, and agribusiness / by Bruce A. Scholten.
        pages cm
    Includes bibliographical references and index.
    ISBN 978-1-137-33060-4 (hardback : alk. paper)
        1. Organic dairy farming—United States. 2. Organic dairy farming—Political aspects—United States. 3. Organic farming—Political aspects—United States. 4. Natural foods industry—United States. 5. Nutrition policy—United States. I. Title. II. Title: US organic dairy politics.
SF246.O74S36 2014
636.2'142—dc23

                                                              2014021497

A catalogue record of the book is available from the British Library.

Design by Amnet.

First edition: November 2014

This book is for my dad, Bastian,
for families in the farming life,
and Martha Clemewell Young-Scholten,
who could have been a splendid dairy wife.

Photo 0.1 An Organic Valley cow greets a man who once farmed this Nooksack River site.

Photo Credit: Bruce Scholten.

Contents

List of Photos and Illustrations

Foreword

Dairy farmers around the globe are politically active and, at the same time, sensitive to political interventions and decisions. The level of activity and vulnerability varies across family farms, cooperative dairy networks, and corporate dairy enterprises in developed and developing worlds. In less developed regions, as in much of India, family-managed dairying is more of a livelihood avenue than a commercial operation, and hence the politics of dairying focuses on protecting the livelihoods of the less privileged. In the developed world, for example in the United States, the focus is on the competitive advantage of production systems and enabling market expansion through branding and product differentiation.

Family farms in both rich and poor countries suffer asymmetrical power dynamics relative to corporate agribusinesses. Industry appropriates on-farm processes through technological innovations in value chains and patents these for profit. Family farms struggle for survival in a vicious cycle of subsistence activity, as corporate farms and agribusinesses tread the virtuous cycle of growth. Are there value chains that can maintain the sustainability of family farms?

Four years ago, Bruce Scholten and I visited a few villages of family-managed dairy farms in and around Bhubaneswar, the capital city of Orissa State in India. The farmers were delighted to have visitors see their dairy activities and ask a few questions. They were appreciative of the cooperative organizations they belonged to, the level of member controls they enjoyed in support of their farming, and confident in the management of these cooperatives. However, they doubted their value chains were robust enough to compete with corporate initiatives, due to the low operational scales and the generic nature of their products. The availability of imported, value-added dairy products on the shelves of supermarkets in urban Bhubaneswar was indicative to them of the evolution the sector is witnessing and the need for their reorientation toward those systems. They do not have the means to reorient. We were apprehensive about the ability of these family dairy farmers to differentiate their products to attract urban consumers in the future. The farmers were also unsure of any political will in Orissa or India to assist such product differentiation.

Regional comparative advantage determines the market competitiveness of farmers and farmer organizations—as well as their influence in politics and governmental regulation. The world's multilateral trade and market environment seeks the political mediation of farming operations and the creation of robust and sustainable value chains to compete and grow. The debate on labeling genetically modified food remains contentious. Meanwhile, differentiating food products by labeling them as "organic," or grown with nonpesticide management, may be a way to market food as safe and healthy. The degree to which small farmers in countries such as China, India, and the United States share success in organics depends on the politics of global trade.

Professor C. S. Sundaresan
Chennai, Tamil Nadu, India
President
Alliance for Advanced Research and Development Initiatives
(AARDI: http://www.aardi.org.in/)

Preface

Memoir: Bleak Winter to Spring Break

The radio droned on. A newsman said world trade talks were underway. Who cared? Farm kids wanted more excitement on the weekend, and the cows did too. A brother asked Dad: "When are the cows going on pasture? They're tired of the barn!" The hoarfrost that had made the windows slates of icy whorls had melted days ago. February sun shone through the windows. It got so hot on the east side of the barn that we opened a few windows for air circulation. The grass past the silos looked green, roused from winter torpor. Cows shuffled their hooves and bellowed. They wanted to leave their stanchions to graze on pasture.

Dad said, "Not yet. The pasture is soggy from rain. Wait another week, or their hooves will tear up the ground so bad they'll ruin the grass." He was right. Seasonal rain and snow left the water table so high the herd would turn the ground to muck, with the sucking sound of hooves stuck in muddy, unready pasture.

After a long winter, the girls seemed wearier of the barn than we were of school by Friday afternoon. Occasionally a cat or dog visited their mangers to sniff the feed. When a cow breathed steam at a cat's face, you doubted they were bosom buddies, but watching each other eased the tedium of bleak midwinter.

To compensate for the enclosure, we made sure the cows exercised on the concrete pad outside the barn when it wasn't snowing. They took turns at the water tank and urea salt block. A couple butted heads. Cows are herd animals, and this was how they established or reinforced a hierarchy they were comfortable with.

Bathing in the sunshine, they licked each other's ribs. An old apple tree trunk was hauled there for their grooming pleasure. They rubbed forelocks and withers on it. Tree branches scratched body parts that everyday social grooming could not reach.

Feeding grain was a popular chore. Kids scooped generous measures from the wheelbarrow onto hay or silage in their mangers, about 20 cows in each of the east and west rows. Dessert was also served—buckets of warm molasses were dripped onto their hay or silage. Kids didn't love chocolate ice cream any more than these girls loved molasses.

The big day finally came. It was a Saturday, so we wouldn't miss the land rush while we were stuck at school. The cows would first graze the field closest to the barn. The big doors on the south were rolled wide open on their trolleys. The girls were excited. The sun was out, and the scents of dry grass and freedom were in the air. Cats watched from the wings as Dutchy, our black-and-white shepherd dog, rushed through the barn onto the dry concrete cow yard outside and set about barking. Dad told him to shush. The girls knew something was up.

"Let 'em out!" called Dad. The two oldest boys began loosening the steel and wood stanchions in both rows. The cows backed out onto the concrete walkway, snorting and bopping the cows ahead of them to walk faster toward the open door. Younger siblings waited at the far north end of the barn, brandishing sticks and ready to yell if any of the less experienced cows went the wrong way, toward the calf pens. In a few minutes, all the girls were in the exercise yard in the sunny south. They nosed around the water tank, salt block, and tree before they were sure this was the special day. Dad had opened the gate of the electric fence.

"*Ka-boss! Ka-boss!*" he called, backing into the field. It's a Dutch-American immigrant call for "Come along!" Cows don't have the sharpest eyesight, so a few squinted where the wire fence normally hung. It wasn't there! The queen of the herd, a ten-year-old named Daisy, shouldered her way through underlings into the field. Other big cows followed, focusing on Dad as he yelled, "*Ka-boss!*"

As Daisy's hooves rustled though the drying spring grass, she betrayed a seldom seen girlishness, bucking and kicking her heels. The rest of the herd hurried off the concrete onto the grass and ran toward the center of the field. One or two of the younger cows actually lowered their girths onto their forelegs before turning on their sides, happily rubbing the sides of their heads on the spring grass. The oldest, slowest cows had new spring in their steps. The rest danced. Those who imagine cows only as grand matrons haven't seen them cavorting on spring pasture.

A junior cow named Hilly, who'd just had one calf, pranced around longer than senior cows of Daisy's vintage. Next year Hilly would weigh more, produce more milk, and need to pay closer attention to the 40 pounds or so of grass a mature cow needs each day. But today was like recess at school. Hilly ran free, a little like she'd done in her first two years as a heifer, with no more responsibility than putting on weight. Eventually the herd settled down to the serious business of grazing. The forage was a mix of timothy grass and other annual and perennial grasses, along with brassicas and legumes like clover (AGA 2001). As a bonus, a few oat volunteers had drifted from an adjacent field.

On Monday morning, as we walked to the school bus, the cows were headed from the milking barn to a fresh paddock, with a few trees for shade. Hearing their satisfied bellows, we knew the new field looked as tasty to them as ice cream to us. Facing another long day of study before chores at evening milking, we moaned, "The cows have it better than us!"

Four or five decades later, it's worth asking whether school kids still envy cows. Are the lives of today's bovines more brutish and short than those of their great-grandmothers? How have the political and economic realities of globalization altered dairy farming's impact on animals, the environment, and the livelihoods of farm families since the introduction of inputs such as antibiotics and genetically modified organisms to the land?

In the mid-2000s, a British family organic firm and an American organic cooperative made videos resonant with the scene above, showing bovine delight at the debut of spring. By this time pasture grazing was the exception in Anglo-American dairy systems driven by market demands. The British family firm was bought by a U.S. corporation that, after winning contracts with hypermarkets, kept thousands of organically fed cows on its "organic" megadairies. Consumers learned this company confined animals in feedlots, unlike the cartoon images of jolly cows grazing on milk labels. They learned that other supposedly organic farms did not pasture their cows, either. Many were confined to feedlots during lactation and milked two to four times daily, for 11 months or longer after birthing their last calf.

In the 1950s it was normal for dairy cows to live seven years and not unusual for them to last past ten. In the confined-animal feeding operations of the United States in the twenty-first century, conventional dairy cows live barely four years; many are injected with genetically modified hormones for maximum milk production; most eat acidosis-prone diets of corn, grain, and other commodities, before aching feet, acidic stomachs, and reproductive problems consign them to slaughter. Some cows in what Michael Pollan (2001) calls the "organic-industrial complex" last little longer than their sisters in intensive conventional operations. Both cases are comparable to cooping up cows like battery chickens, with little chance to express natural behaviors, such as grazing or sampling the odd plants, weeds, and thistles that sprout along the fences of biodiverse pastures. Cow confinement has led many shoppers to abandon milk produced on what they call "factory farms" and reach for organic labels in the dairy case. Trust is what consumers want, and, now that the nation's hypermarkets plan to sell organics at lower prices, the challenge is to maintain an organic program based on ethics and values, not just price.

In the 1950s kids trudging to school envied cows on pasture. Today? Not so much. Most cows are confined to concrete barns and boring feedlots. How can these animals enjoy the longevity and well-being experienced by their ancestors unless today's family farmers are compensated enough to keep them in the style that encourages better health?

Acknowledgments

Not all of the people who helped with this book share all of my opinions, so I begin by taking responsibility for any errors. Most would agree that by the 1990s an exodus was underway by family dairy farmers in Washington State and beyond. These were not "bad farmers," but respected managers who found that pluriactivity—such as giving piano lessons or raising ostriches—did not adequately supplement the income from one hundred cows on one hundred acres, given the politics in Washington, DC. This book is mostly about politics, but also about people who manage change ethically. It is also about animals, cows that seemed like family on pasture in the 1950s, but in contemporary facilities are more remote and shorter lived.

This author was lucky to grow up in the 1950s–1960s, in a family of grandparents, uncles, aunts, and cousins shipping to the Darigold cooperative. We shared labor and equipment in a machinery ring called the Scholten Fox Chopper Co. Some of us remain in co-op dairying, some converted to berry farming or raising organic herbs, fruits, and vegetables. Others work in construction, aviation, education, fishing, or software. All like seeing cows on pasture.

Thanks to my linguist wife, Martha, for editing. Respect to my dad, Bastian Scholten, who, working with the Conservation District and Washington State Dairymen's Federation, kick-started my political interests over radio news during morning and evening milkings. Thanks to my mom, Delores, for faith. "*Tot ziens*" to my siblings and their digitally savvy kids. Credit to my relatives who shared the fun of silo filling. It is astonishing, recalling all the farms I've visited, that we never drove a tractor into a ditch. (Don't mention the cow tipping.)

The University of Washington, in Seattle, spurred my interest in the global agricultural trade. Joyce LeCompte-Mastenbrook offers anthropological insights on Whatcom County in this text. Durham University (UK) provides my base as an Honorary Research Fellow, and Prof. Peter J. Atkins, who supervised my master's and doctoral work on India, the United Kingdom, and the United States, shares my milky interests.

Much obliged to Ms. Jan Chandler for professional advice on survey techniques.

Regards to periodicals in Germany, India, the United Kingdom, and the United States where Steve Larson, my occasional editor at *Hoard's Dairyman*, continues to share his knowledge. Researching articles on genetically modified dairy hormones, animal welfare, and milk quotas amid surpluses, beginning in 1988, taught me much.

Gratitude to Paul Biagiotti, DVM of Basin Bovine Practice, in Jerome, Idaho, who improved my understanding of culling and longevity. Likewise, thanks to Ladd Muirbrook of DHI Computing Service, Inc. in Provo, Utah, for nation-wide records on cows leaving herds.

Thanks for lore to Jerry and Karen Scholten, Joe and Arlene Verdoes, Hans and Colleen Wolfisberg, and anonymous sources. To Goldie Caughlan, formerly Puget Consumers Co-op, now on the board of the Cornucopia Institute—thanks for the photos and more. Kirk Peffers and Cousin Arlene—thanks for assistance in Colorado. Thanks for Wall St. nous, Libby Hawkins. Johanna Young and Lindley Rankine kindly facilitated research in New England.

Appreciation to Farideh Koohi and Robin Curtis of Palgrave Macmillan in New York, whose chats at Association of American Geographers conferences crystallized the project. Thanks to editors Matthew Kopel, Brian O'Connor, as well as Scarlet Neath, Sarah Lawrence, Nicole Hitner, Rachel Taenzler, and Jamie Armstrong. Reviewers earned kudos, such as Dr. Hubert J. Karreman, VMD, veterinarian at Rodale Institute in Kutztown, Pennsylvania. Prof. C. S. Sundaresan, of AARDI India, noted how politics determines benefits to family farmers in global trade. UK nutritionist Paul Robinson assessed bovine diets worldwide. Michigan State University Prof. Phil H. Howard's newest graphic answers the question, "Who owns organic?"

Topics from antibiotics and appropriationism to sentience thread this book on farm and public health policy. Colleagues in Africa and India join me in repeating a lesson from poor countries that applies to U.S. markets: it's better to be a price maker than a price taker. Stakeholders with concerns for animals, the environment, and rural livelihoods can better rely on farmers' cooperatives and organizations that market their own produce than stockholder-driven firms that happen to sell organic foods among nonorganic lines. Better government policies can help U.S. family-scale farmers reclaim their share of power in food systems.

Acronyms and Glossary

AFN: Alternative food networks, including box schemes, farmers' markets, and short food chains. See Colin Sage (2003) or Scholten (2007, 2011) comparisons to global alternative agrofood networks (AAFNs).

AMS: Agricultural Marketing Service oversees the NOP.

Agribusiness: Global systems in production, distribution, and consumption of food and fiber.

Animal welfare: "The sum or integration of an animal's past and present states of well-being as it attempts to cope with its environment; and human values concerning the social or ethical aspects of providing that environment" (USDA NLA 2013).

Animal well-being: "The current state of an animal living in reasonable harmony with its environment" (USDA NLA 2013).

APHIS: Animal and Plant Health Inspection Service in USDA.

Appropriationism: Exchanging industrial processes or synthetics for traditional methods (Fine, Heasman, and Wright 1996; Guthman 2004).

BST: Monsanto's bovine somatotropin brand (Posilac) for bovine growth hormone (BGH); rBGH denotes recombinant (GMO or synthetic) BGH. This text refers to rBGH/rBST.

Bt (*Bacillus thuringiensis*): A soil bacterium producing toxins that are deadly to some pests, engineered into some GMO crops (USDA 2013).

CAFO: Confined (or Concentrated) Animal Feeding Operation.

CAP: Common Agricultural Policy of the EEC and EU.

Conventional farming: In the twentieth century, agribusiness accelerated the appropriation of organic agricultural components and substituted industrialized materials, such as chemical fertilizers, pesticides, and finally synthetic and GMO hormones.

DMI: Dry matter intake, for example, forage on pasture or fodder in feedlots, such as hay and silage.

ECM: Energy corrected milk.

EPA: U.S. Environmental Protection Agency.

EU: European Union; superseded European Economic Community (EEC).

Fodder: Hay, silage, and so forth.

FOOD: Federation of Organic Dairy farmers.

Food system: Everything from farm to plate and back—that is, raw materials, inputs, technology, food processing, distribution, retailing, and waste reuse, recycling, or disposal.

Forage: Grass and so forth grazed by livestock in fields and pastures.

GE/GM: Genetically engineered/genetically modified.

GHG: Greenhouse gas.

GMO: Genetically modified organism, produced in the biotechnology industry.

HMO: Health Management Organization.

IFOAM: International Federation of Organic Agriculture Movements.

LCA: Life cycle assessment or analysis.

MIRG: Managed intensive rotational grazing.

MODPA: Midwest Organic Dairy Producers Alliance.

MRSA: Methicillin-resistant *Staphylococcus aureus*.

NAHMS: National Animal Health Monitoring System.

NODPA: Northeast Organic Dairy Producers Alliance.

NOP: National Organic Program.

NPDES: National Pollutant Discharge Elimination System.

Organic farming: Agriculture without industrialized chemical inputs or GMOs. Until the twentieth century, organic farming was the norm, or conventional way, of farming.

Pastoralism: Dairy systems based on grazing pasture.

PPM: Parts per million.

PMPD: Pounds of milk per day.

USDA: United States Department of Agriculture.

WODPA: Western Organic Dairy Producers Alliance.

Zero-grazing: Confinement systems in which lactating cows seldom or never graze on pasture.

1

Introduction

Photo 1.1 A farmer introduces a child to a cow from a herd. Note another cow silhouetted under the trees.

Photo Credit: Bruce Scholten.

1

Introduction:
What Is Organic Dairying?

What is organic dairying? Idyllic visions of cows chewing wildflowers in verdant meadows vie with factory farms in answering that question. It is a tough question because the priorities of people engaged in dairying vary along food chains according to the focus of stockholders on financial profit, and amid the mix of concerns of stakeholders (including family farmers, rural communities, and urban consumers) for social justice, environmental sustainability, and animal welfare. With so much at stake, metaphors of pasture wars are apt.

The words of a Dutch American uncle ring in my memory. Over coffee one day in the 1980s, he laughed ruefully, saying, "Once a farmer could make a living with twenty cows. Now you can't make it with two hundred." The decline in real milk prices has not been kind to farmers or to cows, whose longevity has fallen. Certified organic dairying was planned to fix those problems, but it is a tale of mixed success.

This book chronicles clashes in organic politics in the context of its past and its possible futures. In his book *Liquid Materialities: A History of Milk, Science and the Law* (2010), Peter Atkins relates how geographers and other social scientists have followed political struggles over quality and safety definitions of milk, which in its natural variability of fats, solids, and microorganisms manifests a sort of material resistance to easy classification, sanitation, or preservation. Atkins (159) notes Zygmunt Baumann's view of the classification of foods such as milk as important to modernity, and he notes that Michel Foucault (1975) goes further, deeming measurement a part of government. Thus have politicians and the media joined battle over organic rules at home and abroad. Our focus is the United States, which exports only about 12 percent of its dairy output but can greatly influence best practices around the world. Therefore, as America harmonizes rules with its transatlantic and transpacific trading partners, it is vital to optimize them to ensure fairness to farmers, processors, traders, and the biosphere.

Reviewing 30 years of organics, Garth Youngberg and Suzanne P. DeMuth (2013: 30) observe, "While many conventional agriculturists continue to reject and disparage organic farming, distorting its image and limiting its broader application, the American consumer has enthusiastically embraced organic products and much of its ideology." Consumers cling to the nebulous but comforting notion of "nature" in organic farming. But business seeks to control or appropriate natural forces in time and space, and substitute more predictable industrial processes when practical, in order to bolster stockholders' profits (Fine, Heasman, and Wright 1996: 150).

Competing priorities confront people from farm to plate, with consequences for animals, the environment, and society. Multiple perspectives are illustrated by competition between family-scale farmers and agribusiness. The U.S. Department of Agriculture (USDA 1999; 2007) defines agribusiness as "industries involved with manufacture, processing, and distribution of farm products" and describes an "escalating concentration of agribusiness . . . into fewer and fewer hands." Farmers' control of markets and their share of consumer spending on food products have diminished. Consolidation brought the loss of over 155,000 farms from 1987 to 1997, while 30 million acres of farmland were lost to urban and suburban sprawl from 1970 to 1997. The trend continues.

The term *agribusiness* overlaps—and is about as old as—our era of intensive dairying (Davis and Goldberg 1957). In his abstract for "Case Studies in Agribusiness: An Interview with Ray Goldberg," Grandon Gill (2013: 203–12) writes, "Agribusiness refers to the collection of global systems involved in the production, distribution and consumption of food and fiber . . . The term was first coined by Harvard Business School (HBS) professors Ray Goldberg and John Davis in the 1950s." Gill's abstract (203, 204) describes agribusiness as "the complex relationships between agricultural products, trade, technology, and public policy." In the interview, Goldberg, who still teaches at Harvard, explains that "the farmer was just as much a businessman as anybody else," and they "really should encompass the whole value-added chain. So we . . . called it agribusiness." Davis and Goldberg "included the subsistence farm as well as the commercial farmer because they were an integral part of the whole system." Unlike many econometrists, they considered the social effects of agribusiness, shown in the following example of the world's largest dairy development program, which increased smallholder incomes in India.

Through junior colleague Michael Halse, Goldberg acted as a guru to the Food and Agricultural Organization of the United Nations (FAO-UN) and farmers' cooperative leader Verghese Kurien in planning India's "White Revolution" in milk (Kurien 2005: 107; Scholten 2010: 185, 221). Despite massive demand for high-protein milk, Indian production in the 1960s dipped as low as 20 million tons per year. This was linked to the sporadic donations of dairy aid from Europe's surplus "Butter Mountain," which amounted to dumping and disincentivized India's farmers. Dumping can harm food security in postconflict environments, such as India's postcolonial transition, or after the Bangladesh war of independence in 1971, and Uganda's conflicts of the 1970s and 1980s (Scholten and Dugdill 2012: 148, 156, 253). India's landless, smallholder, and marginal cooperative farmers averted further dumping in a program called "Operation Flood" (1970–96), which monetized European butter oil and milk powder to fund India's own dairy infrastructure. When domestic processors and traders attempted to substitute cheap dairy aid, the 1984 Jha Committee Report mandated that they pay more for domestic milk, which incentivized smallholders and eventually boosted India's output past that of America's with 81.4 million tons in 1998 (Scholten 1999: 287).

This illustrates Goldberg's understanding of the unequal power relationship between subsistence farmers in India and domestic dairy processors and retailers, who were willing to ignore domestic farmers if they could profit a few more rupees by using cheaper, dumped European commodities. It also parallels, as C. S. Sundaresan suggests in the foreword to this book, the *asymmetrical* power relations between family-scale organic farmers and major American processors and traders who have access to cheaper organic products on world markets.

Technological advances in the Second World War spurred postwar trade with improved command and control of global food chains and transport tracked by, eventually, digital satellite communications. Just as profound were supranational political changes. Since the Uruguay Round Agricultural Agreement (URAA 1994) brought farm trade under the aegis of the embryonic World Trade Organization (WTO 1995), U.S. family-scale organic farmers have faced greater pressure from global agribusiness. Freer trade under the WTO puts farmers in developed countries, such as the United States, at a disadvantage to processors and retailers who can access cheaper foreign food sources.

On one hand, small organic farmers fear free riders taking unfair shortcuts (such as not pasturing cows) that besmirch the integrity of time-honored practices. On the other hand, they fear *appropriationism* via agribusiness's lobbying for regulations that allow for the industrialization of traditional processes, such as substituting feedlots for pasture or permitting cheap synthetic ingredients such as chemically-derived omega-3 fatty acids to be added during processing of milk products certified as USDA organic. Due to the many forms of appropriationism, purists worry that organics are being adulterated around the edges.

Chapter 3 notes a report by The Cornucopia Institute (2006a; c) titled *Maintaining the Integrity of Organic Milk* and its accompanying "Dairy Scorecard." Along with that chapter's "Organic Timeline," they help readers assess the relative positions or competitors in what may be called "pasture wars" (hearkening to the 'range wars' of the nineteenth century). While any farm may be said to be part of agribusiness, the use of that term in this book generally connotes large-scale intensive operations, some with transnational power. In Chapter 7 this book offers anonymized, multiscale case studies, and opinions from respondents to a survey designed and conducted expressly for it. These are supplemented by information from farm visits around the country, some of them pictured herein. Lifting the veil on much of U.S. organics is Michigan State University's Philip Howard (2014), whose flow chart "Organic Industry Structures" maps some of the acquisitions, mergers, and spin-offs documented in this book. Considering the varying priorities and motivations driving actors across scales, Julie Guthman articulates the paradoxes of appropriationism and substitutionism that mark organic dreams and commercial realities in California's organics in her book *Agrarian Dreams* (2004: 209; see also Goodman, Sorj, and Wilkinson 1987). This is why, as Youngberg and DeMuth explain, there has also been tension between advocates of "certified organic" and "sustainable" farming.

There are also ongoing battles between advocates of strictly certified-organic foods and manufacturers and retailers who market products euphemistically labeled "natural" even though they may contain synthetic ingredients and genetically modified organisms (GMOs). So-called natural products are cheaper than organic to make, so there are financial motivations to persuade shoppers they are just as good as or better than organics. The nonprofit Anglo-American publication *Academics Review* (2014: 1–2) compares the relative sizes of the organics and natural sectors: "The global market for organic foods has reached $63 billion while the extended 'natural' products marketplace exceeds $290 billion in the U.S. alone."

At the USDA (2013; 2014) economist Catherine Greene notes that organics already accounted for more than 4 percent of total U.S. food sales in 2012 and were expected to reach $35 billion in 2014 (compare to USDA 2002a by Dimitri and Greene). Greene explains that the government does not disaggregate national organic sales, but that industry estimates of $31 billion in organic sales in 2013 were led by produce (fruit and vegetable) and dairy representing about 43 and 15 percent of total organic sales, respectively, in a pattern that has lasted years.

Background

Social scientists might say the present author's positionality is embedded in years of experience with small-, medium-, and large-scale farms. The author grew up on a family dairy farm in Washington State, and this text is based on decades of observation of dairy politics. Much has changed since the 1960s when the cover picture was taken and when such a barn, plus capital investments in equipment, 40 to 80 acres of land, 40 cows (a bull, plus heifers), with a house sufficient for a typical farm family was estimated at around $250,000. Today that figure is in the millions, due to the greater investments needed to maintain net incomes against rising input costs and lower conventional milk prices.

The classic response to such conditions is to increase economies of scale. In the twentieth century, *conventional* U.S. farmers and agribusiness suppliers tried to control biological cycles, improve yield, and fight disease with off-farm chemical inputs. But some American dairy families choose the *organic* way. In individual contracts or via cooperatives, they sell certified-organic milk for higher prices than conventional. But the premium they receive for organic milk has been whittled down by competitors, whom Michael Pollan, a professor of journalism at the University of California, Berkeley and popular writer on food and farming, dubbed the "organic-industrial complex" in a 2001 article in the *New York Times Magazine*.

This book builds on the author's research conducted for various freelance journalism articles on the global farm trade since 1988 and on graduate studies, including a doctoral dissertation comparing consumer risk reflections on conventional, organic, and local food in Seattle to those in Newcastle upon Tyne in the United Kingdom. Most consumers in both localities stated a preference for local food but were open to organic experimentation, especially when weaning children from human breast milk. Mixed research methods included surveys, focus groups with academics, firefighters, motorcyclists, and others, and interviews with people along food chains.

The dissertation featured a case study of an "organic Pasture War" (Scholten 2007: iii), including discussion of a mid-2000s boycott of organic-industrial milk brands by consumers who feared that factory farms' violations of "access to pasture" grazing rules contradicted a rural idyll falsely suggested in their advertising. It was submitted three years before the U.S. Department of Agriculture (USDA) issued a long-delayed final "Pasture Rule" in 2010 (*Sustainable Food News* 2007). But research continued with participant observation in electronic dairy forums, farm visits, USDA meetings, and a 2013 survey on dairy politics.

Chapters

This book's objective is to explore the views of agribusiness, consumers, farmers, and other actors on salient issues, such as the Pasture Rule and the use of antibiotics, in the hope of optimal outcomes for all involved.

Chapter 1: Introduction: What Is Organic Dairying?

This chapter introduces the "wars" thread in the narrative. From the perspective of human geography, competition for land and resources is a problematic— that is, it is a problem that never goes away. Such struggles, political or by other means, over the distribution of economic wealth recur in time and space according to local conditions and—as this book details—to government rules on organic dairying.

Chapter 2: Agricultural Revolutions: Winter Was Bleak before Haymaking

This chapter summarizes agricultural revolutions since prehistory. The history of farming and food systems is traced in the United States, which, as in Europe, accelerated an intensive *productionist* paradigm after the Second World War. Negative externalities of productionism, such as environmental degradation and food scares, drive current policy choices, between biotechnologically medicalized approaches (using GMOs and nanotechnology) in agribusiness and Big Pharma–driven food systems, versus organic systems from family-scale farms to consumers' tables.

*Chapter 3: USDA Organic Pasture War: Where
Have All the Cow Herds Gone?*

This chapter examines the origins of the USDA Pasture War, as extensive pasture grazing gave way to intensive confinement systems in the twentieth century. Benchmarks include the Organic Foods Production Act in 1990, debate over synthetic ingredients permitted in organic foods, and a consumer boycott of megadairy brands in the 2000s (which shoppers suspected of violating the letter and spirit of organic dairying by denying their cows access to pasture—an egregious error that negatively affected animals, the environment, and the livelihoods of family-scale farmers skilled in keeping them). Chapter 3 runs through the final Pasture Rule of 2010. Despite being an apparent victory for grazers, hostilities continue amid charges of underfunded monitoring of megadairies and lax enforcement by the National Organic Program (NOP).

*Chapter 4: Animal Welfare: From Rudolf Steiner to
the St. Paul Declaration*

The focus is on cow longevity marked by their age at culling and slaughter. Organic dairying is traced from the Austrian Rudolf Steiner, to the British organic movement, the Rodale Institute in America, and the International Federation of Organic Agricultural Movements, whose St. Paul Declaration demands that sentient animals be allowed to perform natural behaviors, such as grazing (IFOAM 2006).

*Chapter 5: Stewardship in the Northwest: Dutch Stewards, Vets, and
Researchers Discuss U.S., Canadian, and European Rules*

Anthropologist Joyce LeCompte-Mastenbrook's ethnography of Whatcom County dairying examines care for cattle and land from dynamic perspectives of Christian stewardship.

*Chapter 6: Antibiotics and Health in the Northeast and Beyond: Experts On
U.S., Canadian, and European Rules*

A visit to a University of New Hampshire organic farm prompts questions on organics rules. Rodale veterinarian Dr. Hubert Karreman parses the advantages and disadvantages of U.S., Canadian, and European rules on antibiotics and alternatives he has helped develop.

*Chapter 7: Family Farms and Megadairies: Effects
on Cows, Land, and Society*

Agribusiness's appropriation of farmers and the vulnerability of USDA organics to GMO adulteration and synthetic ingredients, competition from so-called natural products, and ambiguous marketing are discussed. Cooperatives give cause for optimism, in that their farmer members can control not just production and processing, but also marketing—that is, where the money is.

Organic and GMO Politics

The focus of this book is organic dairying, not genetically modified organisms. But there are sound reasons why the binary of organics/GMOs surfaces in organic politics. Despite attempts by GMO advocates to portray the technology as a natural extension of animal husbandry or plant breeding, which have been conducted for thousands of years, the reality is far from popular understandings of the word "natural." While animal breeding can take place in farmyards and plant breeding in greenhouses, GMOs are impossible to create without the apparatuses of modern laboratories. The transgenic sharing of DNA from one kingdom to another—such as the accomplished insertions of an Arctic fish gene into tomatoes and ice cream—is hardly "natural" in the traditional sense (Smith 2013; 2003: 37, 137–40). GMO opponents claim that the transgenic gene-splicing in what they call "Frankenfoods" is less precise than the biotech industry pretends and that eating these could have serious unintended health consequences in consumers, while the modifications themselves could pollute the genetic inheritance of heirloom breeds.

A dearth of credible testing was slated in an editorial in *Nature* (2013: 5–6) titled "Fields of gold: Research on Transgenic Crops Must Be Done outside Industry if It Is to Fulfill Its Early Promise." It hails the potential of biotechnology but admits that GMOs have not earned public trust because people fear that insufficient safety testing is done, independently of the commercial firms that fund R & D. The editorial acknowledges widespread public "fears of the unfamiliar and 'unnatural'" aspects of GMOs and "concerns about health or environmental impacts" which generate "calls . . . for foods with GM ingredients to be clearly labelled."

Voters on the political left and right see bipartisan support for GMOs. Democrat presidents from Jimmy Carter to Obama and Republican presidents from Ronald Reagan to George W. Bush tout the promise of biotechnology, as the product of U.S. science—and a linchpin of U.S. future export earnings. GMO doubters on the left take notice when former President Jimmy Carter (the peanut farmer from Georgia who is widely credited for strong morals, if not political acumen) questions the morality of GMO opponents who would deny GMO "fortified Golden Rice" that could save the eyesight of vitamin A–deficient people in developing countries. Cynics scoff that vitamin A deficiencies can be addressed by better health interventions. They question the profit motives of companies that bioengineer "terminator genes" into seeds to prevent farmers from using them for the following year's crop—or take farmers to court for allegedly trying to do so.

However, when the Bill and Melinda Gates Foundation joins the venerable Rockefeller Foundation (which helped fund the first Green Revolution beginning in the 1940s) in funding new "Gene Revolution" crop research, and Gates invests $25 million in studies of drought-tolerant GM corn (known as "maize" in much of the world; see AGRA 2013), citizens of all political stripes may muse that it is worth a try. After all, Gates is an intelligent billionaire. His Microsoft cofounder Paul Allen details his investments in biotechnology in his 2009 autobiography. It seems likely that when they appear at the annual World Economic Forums in Davos, they share information beyond public knowledge. Perhaps U.S. presidents and Microsoft alumni trust that if early GMOs are faulty, they can be improved over heirloom varieties as the twenty-first century unfolds.

Food scares can catalyze such placid uncertainty into reasonable fear and permanently alter consumption. Recurrent alarms over Alar (Daminozide), antibiotics, *E. coli*, mad cow disease, salmonella, and the like—as well as the unmasking of food additives such as aspartame, trans fats, and high fructose corn syrup (HFCS) as health concerns—drive shoppers toward the organic food aisle, where provenance is perceived as more traditional and natural.

As this introductory chapter was reworked in late 2013, early in the second term of President Barrack Obama, food politics were discordant. The Organic Consumers Association in America (OCA 2013) relayed warnings by Julian Assange, the WikiLeaks founder who found asylum in the Ecuadorian embassy in London (under the eye of British police seeking to extradite him to Sweden on rape charges), that the United States was quietly negotiating trade agreements against consumers' interests. The OCA and Assange expressed fear that the talks could render citizens voiceless on trade-related intellectual property rights (TRIPs) regarding food and other essential concerns (*Guardian* 2013c). Not even U.S. senators have free access to the negotiations. Assange said talks were being conducted with Europe on the draft of a Trans-Pacific Partnership (TPP) and claimed that "if you read, write, publish, think, listen, dance, sing or invent; if you farm or consume food; if you're ill now or might one day be ill, the TPP has you in its crosshairs."

In the same article journalist George Monbiot said a related Trans-Atlantic Trade and Investment Partnership (TTIP), linking regulatory practices in the United States and European Union (EU), is a "monstrous assault on democracy. Countering that view was Ken Clarke, a senior Tory minister for UK prime minister David Cameron, who defended the regional negotiations, claiming the outcome "would see our economy grow by an extra £10 bn per annum" (*Guardian* 2013c). But if principles of the multilateral World Trade Organization (WTO) are any precedent, these new agreements would treat environmental, animal welfare, labor, and small farmer concerns as nontariff barriers (NTBs) to be swept aside in the interests of aggregate efficiency (Anderson and Tyers 1991). Harmonizing standards could involve the EU abandoning its precautionary principle and accepting imports of milk produced with GMO hormones.

Such a dystopian future is portrayed in the 1984 science-fiction film *Blade Runner*, directed by Ridley Scott, based on a novel by prescient science fiction writer Phillip K. Dick. The bleak cityscape is dominated by the headquarters and robotized factories of transnational corporations, while an underclass of menials provides producer services and security to the TNCs that ran them off the global commons. The doyen of human geography, David Harvey (2002: 310), praises how the film pits "overwhelming corporate power" against a multicultural "street level scene of seething small scale production." Harvey's concern for the urban underdogs is shared by the Organic Consumers Association (2013f), which worries that the Food and Drug Administration (FDA, which oversees fruit and vegetable farms) and USDA (which oversees meat and dairy farms) are implicated in transoceanic trade agreements that could erode local food chains. The OCA warns that the FDA's proposed Food Safety and Modernization Act (FSMA) "would likely hurt small, local producers" and stop them from the traditional spreading of "natural fertilizers, like compost and manure," from "raising vegetables and animals on the same farm," and from using "natural water sources for irrigation" without burdensome Hazard Analysis and Critical Control Point (HACCP) tests for nitrates and *E. coli*. This galls small organic farmers who see conventional agribusiness and organic industrial-scale growers as the progenitors of health threats, such as the 2006 *E. coli* outbreak that affected two hundred consumers in 26 U.S. states. The OCA notes the *E. coli* stemmed from spinach sourced from an industrial grower, Natural Selection Foods in California's Salinas Valley, also called Earthbound Farms. The OCA credits writer Michael Pollan for his rueful prediction that a food panic created by industrial-scale firms would precipitate onerous safety regulation of family-scale farms.

The USDA has been loath to prohibit nontraditional processing of organic soy with the neurotoxin hexane, which is likened to toxic rocket fuel by The Cornucopia Institute (2009a; 2011). The USDA has acceded to demands of major corporations to continue use of carrageenan, a seaweed-based food additive that is banned in Europe. An emulsifier used in dairy foods and infant formula, carrageenan has been linked to gastrointestinal upsets and possibly cancer. It continues to be used despite the fact that organic pioneers disdain it and alternatives exist (*The New York Times* 2012a). One organic dairy cooperative, Organic Valley, plans to switch (personal email, February 2014). Complicating matters is that OV has supplied milk from Midwest and Northeast farmers to Stonyfield, owned by French corporation Groupe Danone. Reportedly, Danone are now removing carrageenan from their products. WhiteWave-Silk-Horizon (owned by $12-billion Dean Foods until 2013) continues to defend its safety, but in response to customer concerns, announced plans to phase it out in 2015 (Odairy 2014e).

USDA policies suggest its boosterism for agribusiness. Although many greens (environmentalists, not necessarily members of the Green Party) also voted for President Obama, Mark Kastel of The Cornucopia Institute expected Obama's secretary of agriculture, Tom Vilsack, to embrace GMOs: "It doesn't matter who's in charge in Washington, there is a pro-agribusiness bias" (personal communication, July 31, 2013).

The fear has provenance. In 1997, when more people assumed biotechnology promised unalloyed progress, USDA secretary Dan Glickman proposed rules allowing GMO ingredients in organically certified foods. But, even then, so many consumers saw this as a cynical attempt by agribusiness to co-opt traditional organics that the proposal withered under a hail of one-third of a million protesting messages from the public.

Outside the United States, observers have mused at the hold biotech companies seem to exert on top politicians. A Swedish radio news program and website called Red Ice Creations (accessed November 23, 2013), hosted by filmmaker and researcher Henrik Palmgren, quoted a Monsanto in-house newsletter of 2000 predicting "agricultural biotechnology will find a supporter occupying the White House next year, regardless of which candidate wins the election in November." As it happened, 2000 was the year Supreme Court Judge Clarence Thomas swung crucial Electoral College votes to Bush (despite Al Gore winning more popular votes). Previously, Thomas was an attorney for Monsanto.

Since the dawn of the Gene Revolution some three decades ago, perceptions have grown that GMO seeds, inputs, and contracts with farmers are contrived to benefit primarily the corporations that sell them—compared to high-yield seeds in the Green Revolution of Norman Borlaug's era, fostered by governments and charitable institutions, such as the Ford and Rockefeller Foundations. Corporate claims that the Gene Revolution will feed the world prompt questions of how the poor will fare in such a world (FAO 2005). Reports from India that traditional seeds have disappeared as the prices of biotech seeds and inputs multiply, while hundreds of thousands of indebted Bt cotton farmers commit suicide, are a disturbing portent of what is to come (*The New York Times* 2012b; *Economist* 2007).

The Economist weekly newspaper has been a staunch global media advocate for GMOs as part of agricultural innovation, since the eighties and nineties, when its editors promoted the doctrine of *substantial equivalence* between traditional plant breeding and laboratory gene manipulation. The logic of their support is that free trade in GMO foods resonates with the newspaper's founding mission to repeal the Corn Laws in 1846. (The tariffs on food imports—that is, the Corn Laws—had helped Britain's landed gentry retain their "economic rents" on food production in Britain. These were contested by the new capitalist manufacturing class, which sought cheap food from abroad to enable lower wages for factory workers.) *The Economist* (2013c) regularly mocks those who worry that GMOs can harm animals, the environment, or human health, and the headline for its article on Washington State Initiative 522 to label GMO food, in the November 2013 election, was "Warning Labels for Safe Stuff." When the journal *Food and Chemical Toxicology* retracted an article published in September 2012 (*FCT* 50 (11): 4221–31) by Gilles-Eric Séralini, Emilie Claira, Robin Mesnagea, Steeve Gressa, Nicolas Defargea, Manuela Malatestab, Didier Hennequinc, and Joël Spiroux de Vendômoisa, *The Economist* (2013e: 16, 86) chortled that greens were guilty of the same resistance to science that afflicts climate-change deniers. The original *FCT* (September 2012: 4221–31) article had concluded that GM NK103 maize (corn) resistant to Monsanto's glyphosate Roundup caused tumors in mice. The political consequence was that France, Kenya, and Russia put holds on GM initiatives. The biotech industry attacked the methodology of the study, claiming Seralini et al. had used too few rats of a type prone to tumors. *FCT* announced that it retracted the article not for fraud or misrepresentation of data, but because the editors later found it inconclusive. Séralini, of the University of Caen in France, then demanded the journal retract a 2004 article supporting Monsanto's position but with fewer of the same rats, and threatened further litigation. *Organic Bytes*, the newsletter of the Organic Consumers Association (2013g) and claiming over 400,000 subscribers, identified a case of the revolving door when "*FCT* created a new editorial position—Associate Editor for Biotechnology—and appointed none other than a former Monsanto employee, Richard E. Goodman, to the post." Goodman is affiliated with the International Life Sciences Institute, which is linked to global agribusinesses, such as Coca-Cola, Kraft, Monsanto, Syngenta, and Unilever.

To its credit, *The Economist* recognizes that trust in privately funded biotech research is low—especially in studies funded by Life Sciences companies. In an editorial castigating the veracity of scientific papers funded by private companies that were supposedly peer-reviewed before publishing in journals, *The Economist* (2013b: 16, 86) warns: "A rule of thumb among biotechnology venture capitalists is that half of published research cannot be replicated."

GMOs in the Obama Era

First Lady Michelle Obama gladdened green hearts when she planted an organic garden on the White House lawn in her husband's first term. But official Obama policies seem calculated to bolster American exports of trade-related intellectual property epitomized by GMOs, while continuing to favor cheap organic imports from countries including China, whose dairy exports have been marred by toxic adulteration (*Hoard's Dairyman* 2012c).

During the 2008 presidential campaign, Obama told fellow Democrats that GMO foods should be labeled to promote consumer choice. But Obama policy appears as pro-GMO as it was under George W. Bush, Bill Clinton, George H. W. Bush, and back to Ronald Reagan when policy became avowedly probiotechnology, abandoning the old-school *precautionary principle* for *risk/benefit* analysis and promises of drought-salt-pest-resistant plants along with higher-yielding dairy cows (Scholten 1990b; c). Sadly, the unintended consequences of GMOs often eclipsed early claims.

Proponents of organic agriculture hope the Obama administration pays more than lip service to organic integrity, but the Republican-controlled House of Representatives, which is the part of Congress that can introduce spending bills, has tightened its purse strings on programs for organics or small farmers. Chapter 7 details congressional support programs for commodity export crops, mocked as "corporate welfare" by greens. About 90 percent of U.S. canola, corn, cotton, soy, and sugar beets were GMO in 2010. Yet biotech penetration of dairy cows has not been straightforward, suggesting sentient beings are different from crops.

Under the Obama administration, the USDA has streamlined certification of GMO crops, whose pollen, claim critics, can contaminate non-GMO or organic fodder and endanger valuable export crops. A rogue strain of GM wheat that escaped trials was found in Oregon in early 2013, whereupon Japan and South Korea put holds on imports of that year's crop—a potential economic blow to Oregon's farmers, who export 80 percent of their grain (*Oregonian* 2013).

The politics of GMOs are fascinating, especially in the United States and United Kingdom, which perceive biotech as an industrial champion for the twenty-first century, just as the automotive, aviation, and electronics industries grew Anglo-American economies in the twentieth. In early 2013, Monsanto spokespersons made comments suggesting it was ceasing promotion of its products in Europe due to antipathy by farmers and consumers. But the ink on the press releases had barely dried before biotech companies assembled UK government ministers in a briefing that warned children in developing countries would wake up blind if they were denied genetically modified Golden Rice with increased vitamin A. Alternative interventions, such as twice-annual vitamin treatments by optical health practitioners, were not aired, but one inference is that milk is another target for alterations. *The Guardian* (2013a) quoted UK environment minister Owen Paterson urging adoption of biotechnologies and GMOs to reduce reliance on pesticides and "use cultivated land more efficiently, [to] free up space for biodiversity, nature and wilderness." Paterson went so far as to accuse protesters of heartlessness when GMOs could feed the world. His opponents suspect him of naiveté.

Emma Watson, deputy editor of *The Economist* (2013a) makes similar arguments in a survey titled "Biodiversity: All Creatures Great and Small." Watson claims her favorite species, the Hainan gibbon, stands a better chance of avoiding extinction if more use of high-yield GMO crops saves space for wildlife sanctuaries.

Critics say GMO advocates promised drought and salt-resistant high-yielding plants decades ago, but the main result has been massive sales of pesticides, such as glyphosate, which transgenic plants are designed to resist. About 90 percent of soybean, corn, and cotton grown in the United States today are GMO varieties designed to be resistant to glyphosate-based pesticides marketed as Roundup or Buccaneer. All this matters to dairying because critics fear the DNA composition of GM plants potentially could alter cows' systems over time and affect livestock products consumed by humans. This could snare agriculture in old-fashioned monocultures that spawn super-weeds resilient under increasingly massive doses of chemicals. Too late—that's already happened, according to organicists. (This book refers to organic advocates as organicists, discussed further in chapter 2.) GMO advocates respond that the adoption of no-till or low-till farming, to save fuel and topsoil while maximizing carbon capture, could be the reason for the appearance of more weeds.

A Dearth of GMO Health Checks

Evidence of direct dangers to human health from biotechnology is scant in the two decades since their inception. But proclamations by advocates that GMOs are proven safe is lacking for the long run. Further, it is not surprising that the scientific establishment of the United Kingdom (standing on the shoulders of James Watson and Francis Crick, who unraveled the genetics of DNA) rallies to defend and extend research in this lucrative area that attracts so much private investment. GMO foes carp that private funding has the corrupting effect of private control, suppressing negative research results on the principle that they were proprietary information.

Biotechnology critics were buoyed by comments by veteran scientist Dr. Arpad Pusztai, who experimented with feeding GMO potatoes modified with a lectin from snowdrops to ward off insects. The potatoes were fed to rats at the publicly funded Rowett Institute in Scotland. As a world expert in lectins, Pusztai was a self-described enthusiast for GMOs, who expected experiments to reflect positively on transgenic technology. In a brief interview on the Granada Television program *World in Action* in 1998, Pusztai divulged that, based on stomach inflammation and growth retardation observed in rats fed on GM potatoes, he would not personally eat such food. A media storm ensued. It is documented in Jeffrey M. Smith's book *Seeds of Deception* (2003: 11–14, 25, 263), along with details of approximately $210,000 (£140,000) in funding received by the Rowett Institute from Monsanto, which one might surmise to have potentially affected results.

The biotech industry immediately joined the UK Labour Party government's chief science advisor, Sir Robert May, in rubbishing Pusztai's remarks as premature. Prime Minister Tony Blair (who touted biotech as ardently as U.S. presidents from Jimmy Carter to Barack Obama) allowed cabinet strongman Jack Cunningham to call Pusztai's research discredited because it had not been published in peer-reviewed journals, but that point was also debatable. The impression was that Pusztai's reputation was sacrificed to agripolitical expediency.

In October 1999, *The Lancet*, a prestigious journal established in 1823 by surgeon Thomas Wakley and acquired by scientific publishing group Reed-Elsevier, did print Pusztai's work in the form of a letter cowritten with his colleague Stanley Ewen (Stanley and Pusztai 1999). The biotech lobby attacked.

A headline in *The Guardian* (1999) read: "Britain's Scientific Elite Continue to Try to Suppress Dr. Pusztai's Research on Dangers of GE Foods: Pro-GM Food Scientist 'Threatened Editor.'" Reporters Laurie Flynn and Michael Sean Gillard wrote that a probiotech rebuttal group in the Royal Society, headed by a Professor Lachmann, warned *The Lancet* editor Dr. Richard Horton not to publish Pusztai's potato results. Horton deplored his impertinence. *The Guardian* writers noted that Lachmann was a former official of the UK Royal Society and on the scientific advisory board of SmithKline Beecham, the company linked to the cloning of Dolly the sheep. Lachmann denied conflicts of interest, defending his actions on the basis of his own capability.

According to *The Guardian* (1999) Pusztai said criticism of his scientific work was unfairly based on preliminary results of the GMO rat-feeding experiments. He explained that this was because later data were confiscated by the Rowett Institute (which allowed the TV interview), after it succumbed to attacks, by seizing his data and forcing him into retirement. In an article in *ActionBioScience*, Pusztai (2001) agreed with an article by J. L. Domingo in *Science* (2000) that the field of genetic modification was marked by many opinions but few data. Pusztai criticized the crude state of the science, the limited utility of the doctrine of substantial equivalence, and the dearth of testing on health risks (*Science* 2000), including allergies and the danger that bacteria may become resistant to antibiotics. (It was a difficult danger to rule out because, as became understood in the commercial use of recombinant dairy hormones, DNA is not always destroyed in the stomach.) In 2003, member of parliament Michael Meacher defended Pusztai. Meacher echoed Pusztai, decrying the lack of GMO safety studies due to the uncooperativeness of the biotechnology industry with independent scientists. Meacher claimed the U.S. Centers for Disease Control had documented a doubling of food-borne illnesses since the introduction of GMOs and that the United Kingdom had seen a doubling of soy allergies since the importation of GM soy began (Smith 2003: 263).

In 2008 Dr. Pusztai told *The Guardian* (2008) he was unrepentant. When reporter James Randerson asked Pusztai if he endorsed supposedly conventional plant-breeding techniques involving radiation and mutation-causing chemicals, Pusztai refused to be sidetracked, replying, "Two negatives don't make a positive . . . It doesn't mean that I agree with those techniques." But Pusztai fingered a political agenda in biotechnology. "Ninety-five per cent of GM is coming from America, so naturally it is in their interests to push it," he says. "I have no ideological grounds against Monsanto. For me it's a scientific argument. They have not done a proper job [of testing], and they are just using their political and economic muscle to foist it on us."

In 2013 Peter Melchett, former Labour Party politician and official at Greenpeace and the UK Soil Association, lambasted the GM lobby along with governments, the Royal Society, and biotechnology firms for trying to conflate opposition to GMOs with antiscientific Luddism. Melchett's counterattack accused GMO zealots of commiting "seven sins against science." The fifth sin, he claimed was that "not once anywhere in the world" had the GM lobby "responded to a scientific study they do not like, by doing what anyone who cared about science should do—repeating the experiment." Melchett noted that the best known such study was conducted by Dr. Arpad Pusztai: "His study, and others that have been conducted since, suggest that some adverse impact was being caused to multiple organ systems in the test animals" (*Ecologist* 2013).

A few years previous, *The Economist* (2010) magazine held an online debate on biotechnology, on the motion "This house believes that biotechnology and sustainable agriculture are complementary, not contradictory." Tom Standage, the moderator, noted that Pamela Ronald, professor of plant pathology at University of California, Davis, claimed GMOs eroded less soil, cut use of insecticides, produced a "halo effect" by protecting neighboring farms from pests, and increased yields and profits for farmers. At Washington State University, Professor Charles Benbrook, chief scientist at the Organic Center in the "other Washington," in the District of Columbia, answered that alternative systems can often increase yields more than GE/GM seeds. Benbrook cited FAO studies concluding that yields were increased by an average of 79 percent across eight systems of agriculture, compared with conventional best practices. Defenders of conventional and GMO agribusiness sometimes grant that organic farmers can boost yields but dismiss their viability on the basis that the cost of labor is too high.

Proponents of organics argue that with funding, promotion, and government facilitation half as good as that enjoyed by conventional farming and biotechnology, the multiple advantages of organics (starting with cutting pollution at the source) would make organics the preferred technology in food systems.

Many people who once fancied themselves GMO fans have become disillusioned, not so much in the *ultimate potential* of biotechnology, but in the dearth of its achievements to date and the lack of long-term testing for the safety of consumers, animals, and the environment. Biotech companies won public support with prospects of plants that needed less chemical fertilizer and pesticides to thrive, but evidence accumulates to the contrary (Binimelis, Pengue, and Monterroso 2009; Benbrook et al. 2013).

Misgivings about the effects of glyphosate weed killer on soil microorganisms were raised in a *New York Times* article (2013). Yet, some of the biggest farmers find biotech profitable. Brad Vermeer, who raises traited—that is, biotech—corn and soy on about 1,500 acres claims his yield would fall by 20 percent if he switched to conventional crops and stopped using glyphosate.

Shoppers could be forgiven for confusion over GMOs. The Organic Consumers Association (OCA) trumpeted *Failure to Yield* (2009), a report by the Union of Concerned Scientists claiming GMO crops did not generate greater yields. The OCA also cited peer-reviewed papers in the *International Journal of Agricultural Sustainability* claiming conventional breeding, not GMOs (contrary to claims), were behind yield increases in major U.S. crops (*IJAS* 2013).

Biotech firms have spent many millions of dollars in congressional lobbying and enjoyed the good will of presidents and cabinet members. But they have not always gotten their way. The Farmers Assurance Provision, called the "Monsanto Protection Act" by greens, began as a rider to a short-term government funding bill in 2012 and then disappeared. The provision would, in a bold assault by the legislative branch of the U.S. government upon the judicial branch, prohibit local and state courts from imposing moratoria on GMO crops, even amid litigation from established conventional or organic farmers whose customers refused GMO-contaminated products.

In one of the latest rounds, the Center for Food Safety (2013) announced, "Victory!! The Monsanto Protection Act is finally dead!" Appropriations

Committee chairwoman Senator Barbara Mikulski (Dem. MD) declared its deletion from the Continuing Resolution budget act (H. J. Res. 59). If the obituary was not premature, it was progress for those supporting the labeling of new GMO fish, and an amendment to the farm bill allowing states to require GMO labeling of other foods, as well as cloned animal meat. But, like the Hydra of Greek mythology, whenever greens cut off one head, the biotech industry grew another (and sometimes regrew the original too). Green campaigns seemed doomed. In California in 2012, Proposition 37 (to label GMO foods) was defeated by a $45-million campaign masterminded by the Grocery Manufacturers Association. Derided by greens as Monsanto's evil twin, the GMA succeeded again in Washington State against Initiative 522 in 2013, even after the attorney general sued the GMA to reveal its secret funders, whose anonymity the GMA had sought to preserve. These turned out to be not only the biotechnology giants Monsanto, DuPont, Pioneer, Bayer Cropscience, Dow Agrosciences, BASF, Dow, and Syngenta, but also venerable brands including General Mills, Kellogg's, and Nestlé.

The story of how I-522, the Washington State initiative to label GMOs, was vastly outspent and defeated is told in chapter 8, along with prognoses for future labeling schemes favored by agribusiness and organics supporters. When General Mills' traditional Cheerios breakfast cereal was advertised as eschewing GMOs, critics pointed out that the brand's other cereals kept using GMOs.

Dairy GMOs and Organics

If we look at the history of GMOs in farming, dairying was the first commercial exercise of genetic modification, mimicking the growth hormone produced naturally by the pituitary glands of animals and humans. After U.S. government certification in 1994, millions of conventional—that is, nonorganic—cows became objects of the GMO known as recombinant bovine growth hormone or recombinant bovine somatotropin (rBGH/rBST). It is synthesized from a gene linked to lactation in natural bovine DNA, inserted in *E. coli* bacteria, and manufactured in biochemical processes not unlike the one used for the production of synthetic insulin (used to treat diabetes in humans). Its original maker, Monsanto, rebranded rBGH/rBST as "BST," or simply "bovine somatotropin," in order to mute negative science fiction connotations that had become popularly attached to growth hormones, and then branded it as "Posilac," to put a positive spin on the synthetic lactic product. From some perspectives rBGH/rBST is the antithesis of organic dairying, and the GMO has been an agripolitical football for two decades.

Marion Nestle (no relation to the Swiss transnational corporation) is a nutritionist who trained as a molecular biologist and edited the 1988 *Surgeon General's Report on Nutrition and Health*. She identifies conflicts between agribusiness and public health in her book *Food Politics* (2003: 101), such as the "revolving door adventures" of lawyers and lobbyists between private firms and posts in the U.S. Food and Drug Administration (FDA), which approved use of Monsanto's GMO dairy hormone. As chapter 8 details, so many Monsanto employees have done stints in government service that small farmers' activists in The Cornucopia Institute ask whether the FDA and USDA are, rather than arms of government, wholly owned subsidiaries of the $12-billion corporation that grew by producing DDT, PCB, Agent Orange, and lately GMOs resistant to their Roundup brand of the pesticide glyphosate (Cornucopia Institute 2013a).

As mentioned above, the Republican administrations of Ronald Reagan and George H. W. Bush boosted the biotechnology industry when they abandoned the precautionary principle in risk assessment, favored by traditional organicists, and adopted the risk/benefit analysis that gave more weight to the economic benefits of innovation (Scholten 1990b; c). President George W. Bush appointed Ann Veneman as secretary of agriculture after her time on the Board of Directors of Monsanto's Calgene Corporation. But there was plenty of bipartisan support for GMOs before words like "terminator genes" and "superweeds" entered the public lexicon. Presidential administrations have had links with Monsanto for decades. President Jimmy Carter recently portrayed critics of genetic technology as extremists in a newspaper piece, and his former chief of staff went to work as a Monsanto lawyer. Former First Lady Hillary Clinton was a legal counsel for Monsanto before her husband Bill's presidency and her turns as a senator from New York and secretary of state for President Obama. Mickey Kantor, global trade representative for Clinton, and Clarence Thomas, a sitting Supreme Court justice, also worked for Monsanto. *Food and Water Watch* (Cornucopia Institute 2013d) reports that board members from the $12-billion company have worked with the USDA, the Environmental Protection Agency (EPA), and served on President Obama's Advisory Committee for Trade Policy and Negotiations.

A more recent example is Michael R. Taylor, who earned degrees in law and politics before becoming FDA deputy commissioner for policy in the 1990s, when he was involved with FDA labeling guidelines on rBGH/rBST. In a switch to the private sector, Taylor then became Monsanto vice president of public policy. In 2009, early in the Obama era, Taylor returned to the Food and Drug Administration. The FDA Office of Foods and Veterinary Medicine (FDA 2013) announced Taylor as the first deputy commissioner for Foods from January 2010, saying that he would lead efforts to "develop and carry out a prevention-based strategy for food safety. . . plan for new food safety legislation [and] ensure that food labels contain clear and accurate information on nutrition." Taylor declares his commitment to making "changes necessary to ensure the safety of America's food supply from farm to table."

Skeptics at the Organic Consumers Association expected some improvements on nutrition labeling but feared the revolving door between Monsanto and the government boded ill for the strong labeling of GMO foods. Industry-government collusion was suspected in quiet moves by the Environmental Protection Agency (established by President Richard Nixon in 1970). These were moves in 2013 to raise tolerances in exposure to glyphosate, the main chemical in Monsanto's GMO pesticide Roundup (OCA 2013b; EPA 2013). This concerned those at the Organic Consumers Association who were aware of studies finding glyphosate in mothers' milk, contradicting claims by Monsanto and the EPA that it is not bioaccumulative.

Thomas MacMillan (2002), who did doctoral work on Monsanto's recombinant dairy hormone at the University of Manchester in the United Kingdom (before leading the Food Ethics Council and then heading the innovation arm of the Soil Association) found fundamental disagreements between U.S. regulators and those in Canada, as well as in Europe, where rBGH/rBST has never been approved. In the late 1980s, the author of this book conducted interviews with German Farmers Association officials and agricultural researchers at the University of Hohenheim for *Hoard's Dairyman* (Scholten 1989a; b; c). Farmers and academics were skeptical of any benefit from the drug when agricultural budgets were strained by the massive surplus Butter Mountain. The only practical use for rBGH/rBST appeared to be as a stimulant for the occasional heifer that underproduced after her first calving. But there were already serious human health questions about the effects of the recombinant hormone on humans. Monsanto public-relations officers based in Belgium admitted to me that rBGH-enhanced milk contained above-normal levels of insulin-like growth factor 1 (IGF-1); however, these Monsanto representatives assured me that IGF-1 was destroyed in the human stomach and thus would not enter the system and affect human health. Several years later, IGF-1 would be subject to speculation as American obesity increased after 1994 when it first appeared in supermarket milk in the United States. The effects on humans seemed less pronounced than predicted when one study found that bovine IGF-1 did not affect human dwarfism. However, GMO gadfly Jeffrey L. Smith (2003: 95) noted that Monsanto's claim that our stomachs stop IGF-1 is contradicted by the *Canadian Gaps Analysis Report* (Health Canada 1998), which states that IGF-1 "can survive" the gastrointestinal tract and is "absorbed intact"; moreover, "the full significance of this finding also was not investigated [by FDA]." Other studies found that IGF-1 can enter the intestines attached to casein in milk.

Risks to cows seemed more apparent in Canadian studies. A 1999 Health Canada (Parliament of Canada) study found Posilac increased a cow's risk of mastitis about 25 percent, infertility by 18 percent, and lameness by up to 50 percent; these conditions in turn increased culling for slaughter and ultimately influenced Canada's ban of the drug. (Smith's book is one source that also expresses doubt that the formula of rBGH/rBST approved by the USDA was the same version on which most safety tests were actually performed.)

By 2003, a decade after it was first sold, the USDA estimated 17 percent of U.S. dairy operations—about 32 percent of all cows, mostly on megadairies—used rBGH/rBST. The drug raised milk production about 15 percent on average but truncated longevity.

Rural sociologist Fred Buttel (1999: abstract, paragraphs 6.5, 10.9) wrote on dairy GMOs that "social resistance to biotechnology" was unlikely to "derail the industry, [but] public opposition will shape corporate strategy and could possibly shape research priorities in public biotechnology research" toward "public-goods goals such as development of non-chemical-dependent and salt-tolerant crop varieties."

Consumers increasingly avoided GMO milk in the latter half of the 1990s. Processors that could do so responded with labels such as "No BST," to differentiate their milk from the GMO variety. The largest milk processor in Maine, Oakhurst Dairy, labeled its milk cartons: "Our Farmers' Pledge: No Artificial Growth Hormones." But the USDA Agricultural Marketing Service (AMS, which administers the National Organic Program, NOP) certifies *processes* but not *products*. Thus, organic makers are not permitted to make claims versus competing products in terms of health or safety. On this basis Monsanto sued Oakhurst, claiming the label disparaged its GMO and unfairly implied Oakhurst's milk was superior to rBGH/rBST-enhanced milk. (It was a strategy that greens suspected Michael R. Taylor developed with FDA labeling guidelines in the early 1990s.) The success of the lawsuit would have serious financial implications for the drug's manufacturers because, *The New York Times* (2003) estimated in the same article that, about one-third of the country's nine million cows had been injected with the drug.

Monsanto spokesman Lee Quarles said: "The purpose of organic standards is to establish a set of production and processing criteria to market foods labeled as organic, not to suggest organic foods are 'healthier,' 'safer' or of 'higher quality' than other foods currently available on supermarket shelves" (*Wired* 2003). After initial defiance, Oakhurst bowed to pressure from Monsanto and the USDA. All the terms of the settlement were not made public, but Oakhurst agreed to a wording change for a less emotive message: "Our farmer's pledge: No artificial hormones used." Additional lettering was added: "FDA states: No significant difference in milk from cows treated with artificial growth hormones." Oakhurst lost its legal case, but GMO milk was already losing in supermarket dairy cases. Darigold farmers' cooperative, headquartered in Seattle, asked its members to stop using the drug in their conventional milk. Supermarket chains, such as Safeway, promoted GMO-free conventional milk brands, and Ben and Jerry's ice cream prominently eschewed synthetic hormones (*Hoard's Dairyman* 2007b; also FDA 1999, 2009).

As the timeline shows in chapter 3, USDA organic regulations had prohibited rBGH/rBST since 2002. Today there is a trio of major U.S. organic dairy lines: the Organic Valley/ CROPP cooperative founded by Midwest farmers, including George Siemon, in 1988; the Horizon Organic company founded by Mark Retzloff and Paul Repetto, veterans of the natural foods sector, in 1991; and Aurora Organic company begun by Mark Retzloff and Marc Peperzak in 2003. These organizations have enjoyed double-digit sales increases, year after year. Organic farmers quipped that their market was stimulated by Monsanto's synthetic hormone.

Negative publicity on GMO milk overshadowed Monsanto's efforts to market GMO plant seeds and the Roundup glyphosate pesticide they were modified to resist. Sales of rBGH/rBST were slowed by a manufacturing problem at its Austrian factory in 2003. (Ironically, European Union regulators allowed the manufacture of the drug but not its use in EU dairying.) Supervisors of some of the cow megadairies (with herds numbering in the multiple thousands) in California were among those who stopped using it upon realizing they already managed the cows' nutrition so closely that further stimulating the cows was not worth the cost of the drug and increased the risk of cow burnout and early culling.

The National Animal Health Monitoring System (NAHMS), as part of the USDA Animal and Plant Health Inspection Service (APHIS), is an important data source in dairy politics. The NAHMS (2007a; b) publication *Dairy 2007: A Report on Changes in the U.S. Dairy Cattle Industry, 1991–2007* related the rise and plateau, if not fall, of rBGH/rBST in America. In 2008 Monsanto's recombinant dairy technology was sold to Eli Lilly's Elanco. To this day arguments simmer over the lack of transparency in rBGH/rBST regulation, certification for commercial sale, and official endorsements of the hormone. Biotech firms routinely portray themselves as agents of scientific progress and their critics as superstitious Luddites. But detractors counter that biotech firms practice sloppy science. Consider that Monsanto's critics include diabetics who are grateful for the GMO insulin that regulates their blood sugar. But the same people expect GMO plants and livestock products to be safety tested as rigorously as synthetic insulin. Industry promises of reduced pesticide use were disappointing. Washington State University scientist Charles Benbrook, of The Organic Center, published an article (2012) in *Environmental Sciences Europe* showing that, rather than the predicted reduction in herbicide use in GMO crops, herbicide use on GMO crops such as cotton, soybeans, and corn actually increased substantially in the United States.

Superweeds, predicted by greens skeptical of GMO crops, have emerged and are spreading in the Corn Belt. Overall, Benbrook found 7 percent more glyphosate use in six GMO crops in the first 16 years of their use in America, with the prospect of more pesticide cocktails to counter evolving superweeds, sometimes resorting to older pesticides to bolster increased applications of glyphosate. Outside the United States, such a "transgenic treadmill" was identified in the evolving "emergence and spread of glyphosate-resistant johnsongrass in Argentina" in an article in *Geoforum* by Rosa Binimelis, Walter Pengue, and Iliana Monterroso (2009).

In order to relaunch the livestock hormone purchased from Monsanto, new owners Elanco (2009) had its public-relations firm Porter-Novelli commission a paper titled "Recombinant Bovine Somatotropin (rBST[/rBGH]): A Safety Assessment," by eight authors led by Richard Raymond, a former undersecretary for food safety at the USDA. The paper was presented at a major dairy conference in Montreal in 2009 and subsequently publicized by Elanco in industry and lay publications.

In a refutation titled "Sponsored Academics Admit Falsely Claiming Safety for Monsanto/ Eli Lilly's Bovine Growth Hormone," Jonathan Latham and Allison Wilson (2013; both are PhDs and cofounders of the Bioscience Resource Project) contested the Elanco paper's claims that rBGH was endorsed by the American Cancer Society and the American Pediatrics

Society. One of Raymond's coauthors commented that the silence of these two bodies had been interpreted as an endorsement. Raymond admitted that the ACS and APS did not explicitly endorse the GMO dairy hormone, but he claimed support from several entities, including the White House. Latham and Wilson ended their report with a riposte from Rick North of Oregon Physicians for Social Responsibility: "Elanco's numerous false statements and misrepresentations on endorsing organizations are only the tip of the iceberg. The entire report is riddled with similar inaccurate, misleading claims about rBGH itself."

Ironically, uncertainty and fear of risks associated with GMOs, as well as food scares, such as mad cow disease (BSE/nvCJD; see Scholten 2007), presaged public preference for organic milk, since GMO inputs are expressly prohibited in it. But GMOs are just one front in agriwars pitting corporate agribusiness against small farms. In the context of this agriwar, we now focus on the farmer.

Rural Livelihoods in the United States

Do farmers in the Global North, let alone the rich United States merit our concern? Concern is regularly accorded poor farmers in the Global South (the Third World), as in Susan George's books *How the Other Half Dies* (1977) and *Ill Fares the Land* (1984). In the same political genealogy, in 2003, the doyen of human geography, David Harvey, published *The New Imperialism*, a leftist analysis of "accumulation by dispossession." His book describes conflicts ranging from the neoliberal quest for oil in the Middle East to repression, if not genocide, of mestizo and indigenous smallholders by indigenous U.S.-backed Latin American elites. In a similar vein, UK writer Fred Pearce titles his book *The Landgrabbers* (2012), which argues that if Paul Collier's prescriptions in his 2008 book *The Bottom Billion* (urging a merger of Africa's commons into big farms growing GMO crops) were followed, these measures would actually serve to increase northern wealth rather than decrease hunger in the Global South. In Pearce's view, past promises by transnational corporations to increase local employment, training, nutrition, and prosperity via the application of science are distractions uttered to cover landgrabs.

America's own amber waves of grain conceal much poverty from sea to shining sea. Many rural dwellers have inadequate food, clothing, and shelter, and such poverty deserves attention. Like their Global South counterparts, northern farmers, including U.S. dairy farmers, are not immune to accumulation by dispossession and the attendant threat of proletarianization. This can be the result of agribusiness's hegemonization of a sector, especially if poor governance allows unfair competition in it (Glassman 2006). Arguably, this has been the case in the U.S. poultry industry, in which few independent family operations exist outside integrated structures controlled by corporations, such as Tysons. This book argues the same hegemonization processes have been underway in organic dairying.

The USDA ERS 2013 report finds that nonmetropolitan (i.e., rural) poverty has worsened in every census since 1959. After the current Great Recession began in 2007 (the worst since the Great Depression of 1929–39), an additional 700,000 rural residents returned to poverty by 2011. Differences in metro/nonmetro poverty rates are highest in the U.S. South and lowest in the Northeast and Northwest regions, where there happen to be strong markets for organic milk.

Rural economies vary in time and space. Many Okies who fled from the Midwest Dust Bowl of the 1930s were descended from peasants expelled from previously common lands in the English Enclosure Movements, centuries before (Dyer 2007). The percentage of rural dwellers in the U.S. population has plunged in the last century to a current level of about 16 percent. USDA secretary Tom Vilsack admits that absolute numbers of rural folk began falling in 2012, as noted by an editorial by organic farmer and educator Joel Salatin (2013) titled "Why Do We Need More Farmers?" Rural depopulation contributes to urban homelessness and erases rural communities. Every time a family-scale dairy farm is abandoned or merged with a neighbor, jobs are lost, depleting local economic activity. *Hoard's Dairyman* (2014c: 151) reports that most dairy farms have exited the business in the last couple decades. Country towns that had their movie houses shuttered years ago are now losing their last cafes and gas stations. Salatin, featured in Michael Pollan's 2006 book *The Omnivore's Dilemma*, was angered by a speech given by Secretary Vilsack in Virginia, when he claimed the country needed more farmers not to increase the number of earthworms in soil or care better for animals or plants, but because "although rural America only has 16 percent of the population, it gives 40 percent of the personnel to the military." No fan of prevailing foreign and energy policies, Salatin seethed: "The whole reason for increasing farms is to provide cannon fodder for American imperial might."

Such talk comes a decade after rhetoric of a rural revival in the 2000s when the biofuel boom spurred land investment and raised many farmers' net worth and credit ratings. In economist Kym Anderson's 2010 edited volume *The Political Economy of Agricultural Price Distortions*, David Orden, David Blandford, and Timothy Josling (2010: 177) write: "One of the proximate causes of the 2007–08 boom in commodity markets was the U.S. ethanol fuel tax credit and ethanol use mandates designed to promote corn-based fuel production." Analysis shows corn ethanol is little more environmentally sustainable than coal, and U.S. tariffs on cane fuel from countries such as Brazil were fiscally imprudent. Granted, the biofuel program did ease dependence on Middle East oil imports in the post-9/11 years, but it did not increase the U.S. rural population in the long run. Higher commodities prices rocked conventional and organic dairy structures, along with the entire food industry.

In the current decade, farming journals have cautioned against investing in land in the expectation of profits in the biofuel business. The new driver of U.S. land prices is fracking—that is, the fracturing of underground geological structures to release natural gas. Although fracking has increased supplies of natural gas and driven down energy prices, it is opposed by many greens. It appeals to policy makers concerned with energy costs and dependence on imports. There is international apprehension at the extent that high-pressure chemical fracking can impact geological stability and water purity. There will continue to be winners and losers in the fight for land and the mineral rights that come with property in America compared to, for instance, Britain, where most mineral rights are held by the Crown.

Dairy families often ignore opportunity costs in income and social mobility, in their 24-hour, seven-days-a-week struggles to retain land ownership. Reasons for such sacrifice include an affinity for animals, the environment, and the lifestyle, the desire to maintain family structures in a certain community, and an atavistic belief that land ownership connotes economic security.

Organic pasture wars are the latest iteration of centuries-old land wars. They are not as physically violent in the United States as they were during the English enclosure movements (beginning in the thirteenth century, resurging in the sixteenth century, and lasting until the nineteenth century) (Dyer 2007), or the Scottish clearing of the eighteenth and nineteenth centuries, which were conducted to increase rents to the aristocratic and ecclesiastical castes by seizing land for the lucrative sheep and wool industries and provide cheap displaced labor to urban industrialists. These, and numerous cases in contemporary Africa, Asia, Latin America, and South America, are called "primitive accumulation" (Glassman 2006: 610–12; Pearce 2012). But the impetus to maximize capital with land is accelerating, due to food safety and hunger scares (Scholten 2007; 2011), demand for livestock products in the emerging BRIC countries (Brazil, Russia, India, China), and the targeting of food commodities by global institutional investors after the Internet bubble burst around 1999, and the housing bubbles that followed suit, along with the Wall Street crash marked by the fall of Lehman Brothers in 2008.

Casualties of the American pasture wars seldom die. Most are priced off their land, leaving lifeless farmscapes as cows are confined in worse conditions than their bovine forebears. How could organic dairying improve this picture?

Twenty-First Century Organics

Around the turn of the century, organic pioneers sounded modest. Organizers and vendors at farmers' markets or others in alternative food networks (AFNs), such as box schemes and community supported agriculture (CSA), had few delusions of grandeur. Typically they demurred, "Organic is just a drop in the bucket compared to conventional farming. We're not even 1 percent of total food sales. We're here to serve our network, not change the world."

But organics have steadily grown in significance. At one time, aggregate organic sales of $10 billion per annum seemed an unreasonable dream. It became reality sooner than many organic pioneers thought possible. The Organic Trade Association (OTA 2012) reports that total sales in the year

2000 amounted to $6.1 billion, cracked $10 billion in 2003, and reached $29.22 billion on 2011. The $30-billion mark was exceeded in 2012.

The Census Bureau, Statistical Abstract of the United States (2012 Table 823: 535) estimated the market value of all dairy cattle and milk production at $34,754,031,000 in 2007.

The Economic Research Service (USDA ERS 2013) estimated that by 2006 organic milk sales of 1,062 million pounds were 1.92 percent of total milk sales of 55,251 million pounds—nearly two percent of all (organic and conventional) milk sales. By 2011, ERS estimated organic milk sales of 2,073 million pounds were 3.86 percent of total milk sales of 53,723 million pounds.

Although organic milk sales dropped slightly in 2009, the trend was upward. In fact, absolute quantities of conventional milk sales in 2011 were trending down. Organic milk sales were gaining in absolute quantities and as a percentage of total milk sales. Organic milk was a bright spot while organic fruit, vegetables, and related sales, such as box schemes, contracted in the recession. Organic dairy remains the best hope for economic sustainability for many small family-scale farmers.

Since the 1974 census, the USDA has defined a farm as any place from which $1,000 or more of agricultural products are sold. This is not much income, just one-third of the U.S. government's poverty line in the mid-1960s. In fact, self-described commercial farmers (on operations functioning as a family's main or sole source of income, capable of replacing their own capital) scorned such small operations as "hobby farms" for retirees or amateurs playing at farming after their regular workdays.

Unfortunately, many of the larger dairy farmers (those with perhaps two hundred cows in the mid-1990s) who talked tough about marginal "hobby farmers" eventually succumbed to the same economic push-pull factors that drive enlargement or merger with other farms. They were pushed on a technology treadmill of rising inputs prices and pulled by declining farm gate prices to increase their economies of scale (Geisler and Lyson 1991) or sell the farm.

Another way to put this was, "Get big, or get out!" in the words of Earl Butz, USDA secretary to President Richard Nixon in the 1970s. A British newspaper report on the disappearance of single-family dairy farmers included statistics so shocking that the editors titled it "Bye-Bye This American Guy" (Scholten 1997). But the 2000–08 Bush administration regarded the dairy industry as ripe for even further consolidation.

It is no surprise that a report for the USDA (2007b; also USDA 2007a: 5) by James MacDonald, Erik O'Donoghue, William McBride, Richard Nehring, Carmen Sandretto, and Roberto Mosheim found "the number of dairy farms with fewer than 200 cows is shrinking, while the number of very large operations, with 2,000 or more cows, doubled between 2000 and 2006." The study by MacDonald and his coauthors did not treat costs and farm sizes of organic dairy farms, but it fulsomely illustrates the rush to scale in conventional intensive dairying, from which so many organic dairy farmers fled, in order to preserve their livelihoods on family-scale farms.

In the middle of the last century, big intensive dairy farms emerged on the U.S. West Coast when overall industry structures were still based on extensive midwestern farms with mixed farming models, featuring pigs and chickens sharing the land with dairy cattle; many of these dairy farms raised all their "replacement" cows, heifers and bulls in closed herds, on the farm. Around 1958 Dutch immigrants to America set up farms in California and quickly specialized in intensive milk production for local consumers tired of mixing powdered milk. The dairies were so successful in suburbs such as Bellflower that relatives in distant Seattle could be enticed to relocate and set up unprecedentedly large dairy farms near Los Angeles. Over decades, dairy farms appeared and disappeared in a series of real estate transactions, in concentric circles around LA, as orange groves turned into dairies, housing, industry, and strip malls, and finally were drawn into expanding business districts (Gilbert and Wehr 2003). Local demand for fresh liquid milk was high, and consumers preferred fresh local milk to milk reconstituted from powder made in Seattle or Wisconsin. But California farmland was beset by commercial pressure that drove land values (and taxes at best economic use) that frightened farmers in states like Washington, where many enjoyed tax relief on land zoned for agricultural production. In her book *Agrarian Dreams: The Paradox of Organic Farming in California*, Julie Guthman (2004: 184) argues that California demonstrates that "the worse aspects of the industrialization of agriculture are in part driven by land values." So there were strong economic incentives to elicit as much profit as possible from each acre of a Californian dairy farm. That is why in the 1960s most farms in Washington grazed cows on pasture, while increasingly more farms in California confined cows in feedlots.

In the 1950s Whatcom County, about 1,200 miles north of Los Angeles, was a nationally acknowledged leader in dairy management and yield per cow. Farms with more than one hundred cows were virtually unknown. At the half-century mark, the county counted over three thousand dairy farms averaging 11 cows apiece. The county's biggest city, Bellingham (population 35,000), hosted a four-year college, and local jobs in logging and plywood manufacture competed with dairy and crops (fruit, vegetables, hops) for employment. Almost every county road revealed dairy or beef cattle on pasture, amid the green canopy of second- or third-growth forest. Most farms had at least a few cattle to fertilize crops, before the switch to chemical inputs was complete after the Second World War. But as specialization superseded mixed farming, cows agglomerated on fewer, larger dairy units, in a restructuration different from, but as dramatic as, the disappearance of horses after the advent of tractors.

Whatcom Farm Friends (2002), a nonprofit farm advocacy service based in Lynden, the county's dairy hub five miles from the international boundary with Canada, records jumps in scale. The period 1962–2001 saw a 77 percent cut in the number of dairy herds, a 390 percent rise in average herd size, a 27 percent rise in aggregate cows, a 50 percent rise in milk per cow (to 21,500 lb./9,752 kg. in 1999), a 992 percent average rise in milk from farms, and a 151 percent rise in milk from the county as a whole. Whatcom County's experience was a harbinger for national restructuring, driven by the spur of California's intensive model: from 3,020 milk-producing farms in 1951 (the birth year of this author) barely 200 remain today.

Work for the U.S. Department of Agriculture by MacDonald and his coauthors (USDA 2007: 1; also USDA 2007b) notes that the USDA reports the national number of all (organic & non-organic) farms with dairy cows fell 88 percent, from 648,000 operations in 1970 to 75,000 in 2006, amounting to a decrease in farm numbers of 88 percent (MacDonald et al. 2007: 1; USDA ERS c.2001 "Milk Production: 1950–2000, Appendix table 1"). Intensified technologies and consolidation of family farms in the dairy industry continued inexorably, while massive increases in productivity per cow allowed decreases in the numbers of cows in the aggregate national herd. (Meanwhile the average age of intensively farmed cows at slaughter gradually decreased.) The national total of dairy cows decreased from about 21.99 million in 1950, to 17.51m in 1960, to 12.00 million in 1970, to 10.79m in 1980, to 9.99m in 1990, to 9.21m in 2000, to 9.1 million in 2006, which is about the same level in 2013 (MacDonald et al. 2007: 1; USDA Milk Production: 1950–2000, App. table 1). The average herd size rose over six times from 1970 to 2006, from only 19 cows per farm to 120.

Hoard's Dairyman staff (2014a: 151) reported, based on USDA data: "Since 1992, the drop in licensed or so-called commercial, dairy farms has declined 84,549 from 131,509 to 46,960. That's a 64 percent drop during that time." Over two-thirds of farms that quit did so between 1992 and 2002. Herd size, which averaged 74 nationally in 1992, shot up to 196 in 2013, a stupendous increase of 167 percent. Much can be learned about conventional U.S. dairy structures from these figures for regional average herd sizes: Midwest 1992/51 cows, 2013/128 cows; Northeast 1992/61 cows, 2013/98 cows; Southeast 1992/104 cows, 2013/188 cows; and West 1992/263 cows, 2013/998 cows. Nationally, total cow numbers fell 0.1 percent 2012–13 to 9.221 million, and milk production fell 0.3 percent to 201,218 million pounds (*Hoard's Dairyman* 2014b: 155).

In this shift from small, family dairy farms to megadairies, human geographers might say human/nonhuman intimacy declined. Another way to express it is that, in a couple generations, it got much harder to remember each cow's name, if she had one besides a number on an ear tag. It was not just larger herds that caused this. New, high-volume milking parlors elevated cows to nearly a meter off the floor. This eased the strain on a farmer's back when lifting steel pneumatic milking machines, but blocked most face-to-face contact with cows during milking (compared to earlier belt-and-bucket methods in which cows and farmers stood on the same level).

Upscaling to higher-volume (typically double-herringbone) milking parlors of stainless steel and glass has speeded milking and can help improve hygiene. But these expensive investments have also affected gender, generational, social, and ethnic relations in rural communities. Social scientists, such as Karl Kautsky (1854–1938), have pondered why families persist in single-family farming when studies show their income is insufficient to maintain the capital invested in them. More recently, Ian Drummond and Terry Marsden (1999) found in a study of Queensland, Australia, that much is explained by a family's determination to retain land ownership by exploiting its own family labor to stay on the land, even when individual members could earn more income in off-farm jobs. The experiences of family dairy farmers there resonate with counterparts around the world. Farmers resisted falling commodity prices by adopting irrigation, mechanization, and chemical inputs after the Second World War to raise yields. But they eventually returned to what Kautsky dubbed "self-exploitation"—working harder, longer hours and foregoing opportunity costs of higher education, holidays, and other rewards. New technologies seemed to usher in a golden era of family farming, but by the 1990s in Australia, the loss of soil productivity, pollution of river waters, and even damage to coral in the Great Barrier Reef showed the unsustainable effects of such intensive agriculture on the environment, as more family farms merged with others or exited from farming.

Julie Guthman (2004: 11) describes the plucky ethos behind family farms in California and perhaps universally: "The agrarian ideal is also an owner-operated farm, self-sufficient to the extent that family members provide all the necessary labor . . . Hiring outside labor is considered a sort of moral failing."

Inevitably in the United States, as in Australia, many small farms depend on income from spouses, usually wives, with part- or full-time jobs off the farm. Wives may keep records on the farm but also work in town in business, education, health care, or retail. But as farms have enlarged and intensified and the twentieth century neared its end in America, wives, sons, and daughters who previously shared milking duties were increasingly channeled out of the barn to the farmhouse to concentrate on school activities in preparation for nonfarm careers. As waves of low-paid Latin American migrants replaced family members as milkers, herders, and general farm labor, U.S. dairying—once a quintessential structure of nuclear family farming—began to assume the two-tier social class system that a generation earlier had characterized crop farming with Latino labor in California.

Farm by farm, U.S. dairy operations have graduated from a scale where family members recognized and knew the name of each cow to a scale that discourages naming. The shift has quickened because yield per cow doubled in just a few generations. As mentioned above, the number of all U.S. farms with dairy cows fell 88 percent, from 648,000 operations in 1970 to 75,000 in 2006, and fewer today.

Amid this process of agri-industrialization, even bottom-line business-oriented dairymen betrayed affection for the beasts. One of the author's most valued informants over the last three decades is Joe Verdoes, a son of Dutch immigrants who grew up farming near Mount Vernon, in northwest Washington. He has earned university degrees in business and history, in a career including dairy management on progressively larger farms in North America, Australia, and New Zealand, before turning to commercial fishing. Asked to describe the intelligences of animals in dairy ecologies, Verdoes was blunt: "I don't make great claims for cow's intelligence but they easily beat out sheep.

Although I do remember a few cow Einsteins (you notice genetic variation when you've had contact with thousands of them) who could open gates, and heifers who had senses of humor." He explained that one jet-black heifer played a game of mock stalking him: "She wasn't tame but she wasn't really afraid of me either." The heifer tagged behind closely, two feet back, stopping when the farmer stopped before proceeding when he did. "It turned into a game to see if I could catch her unawares by stopping suddenly and get her to bump into me. Never did. Later I turned suddenly on her and stepped towards her, and she lifted her tail high in the air and galloped off in mock fear. I started walking again and sure enough she was behind me again," recalled Verdoes.

Cow advocates might respond that intelligence has been bred out of bovines since the times when their ancestors, the larger aurochs, eluded or fought saber-toothed tigers. But few dispute the aesthetic value of pastoral farmscapes or the sociability cows demonstrate by sniffing humans who enter their pasture. When a farm is working well, it is almost as though farmers share intelligence with the herd, in the sense of Hillary Clinton's book claiming "it takes a village" of cooperative stakeholders to raise children. On days when a farm is humming, humans and nonhumans play their parts so that every creature succeeds. All get from each other the food, shelter, and comfort they need. In cold weather many a miserable farm kid learns that after herding the cows from pasture into the milking barn at five o'clock in the morning, the best place on the farm to warm cold hands is between a cow's leg and udder.

As old-timers retired, some were heard muttering that they didn't much like the direction dairying had gone. Confining cows in feedlots year-round devalorized those who prided themselves on the nuances of rotational grazing. Knowing when to put cows on grass and exactly when to and how frequently to switch them among paddocks requires the timing of an orchestra conductor. But old and young dairy farmers realized the real price of milk was increasingly unremunerative with small herds. Many farmers, especially young ones impatient to make their mark, saw intensification and scaling up as necessary to the economic sustainability of the family farm, and to provide for retiring parents. It was common to hear explanations like, "Sure, I'd like to see the cows on pasture, like in the old days. But the family has to make a living too." Before long, a family that had 20 cows in one generation had 700 in the next. Small wonder the downward spiral in real farm gate prices per hundredweight or liter, adjusted for inflation.

But a small minority of dairy farmers refused to accept this was inevitable. They pointed to the intensification of poultry farming. The mixed farms that typified American agriculture until the mid-twentieth century included dairy and beef cattle, pigs (fed partly from cows' milk), and often a chicken coop with open access to large, grassy chicken yards, where hens dug for worms and insects. Chickens might number a few dozen to hundreds. But in the 1960s, intensive battery chicken operations began to dominate the industry. It was retail magic when supermarket reader boards advertised three dozen eggs for a dollar. Today such loss-leader supermarket retailing still depresses prices in the poultry industry. Milk is used similarly, placed in the back of the store as a frugal loss leader, luring customers through aisles of processed food.

Consumers welcomed cheap eggs, in tune with President Lyndon Johnson's so-called War on Poverty. But the reality of battery conditions leaked to the public: seared beaks for multiple hens confined to wire cages with insufficient room to move freely or flap their wings, let alone scratch for grit or bugs outdoors. Apologists for battery farming claimed, "Chickens are so dumb, they don't mind."

But, in later decades, the marketing concept of *free range* foods came into vogue, as farmers and consumers questioned the ethics of factory farming. Organic dairying waxed as demand for conventional milk waned, partly due to consumer concern over mad cow disease and GMO dairy hormones (as noted above) and concerns about antibiotics and animal welfare (as we will see in the rest of this book).

According to USDA data, total certified-organic operations of all kinds increased from 3,587 in 1992 to 12,880 in 2011. In 2008 there were 2,012 organic dairies that reported selling milk (*Dairy Business* c. 2009). The number of organic milk cows grew one hundredfold from 2,265 in 1992 to 254,771 in 2011 (see USDA 2013). Below are some details on how it happened.

CROPP Organic Cooperative

The following chapters in this book offer more details on the development of family-scale and megadairy-scale organic dairying in America. Before returning to other themes of this book, it is worth mentioning the establishment in Southwestern Wisconsin, in 1988, of the Coulee Region Organic Produce Pool. In 2013, one-quarter century after its founding, CROPP (nowadays the Cooperative Regions of Organic Producer Pools) was the biggest *organic* dairy-farming cooperative in North America, with upward of 1,800 farmer-owners in at least 33 states and four Canadian provinces, including a network of organic forage growers. Cooperatives such as these can provide an alternative business model to many family farmers, but, of course, not all dairy cooperatives involve organic farms, and not all organic dairy farms are members of cooperatives. In 2009 the USDA reported 53,300 members of dairy co-ops (of a total 2.2 million members for all farming co-ops), representing nearly $31 billion in total conventional and organic dairy sales (USDA 1996 [2011]).

In 2010 when CROPP's Organic Valley label had 1,652 members, it was only about 3 percent of all dairy co-op members—but growing. Organic Valley includes cheeses, dry and fluid milk products, yogurt, cream, cottage cheese, vegetables, juice, and eggs, plus an organic soy beverage added in 2004. The Organic Prairie label represents meat (beef, pork, and poultry) products. Organic Valley's main noncooperative competitors in U.S. dairy cases have included the Horizon Organic Dairy "happy cow" brand founded in 1991, marketed by WhiteWave (owned by Dean Foods 2003–13). Another significant competitor is Aurora Organic Dairy (AOD, established in 2003), which supplies "own brand" milk to supermarkets.

Debates on the USDA National Organic Program should not obscure the fact that many family farmers express satisfaction with the corporate buyers of their milk. For example, Horizon Organic Dairy's website claims its smallest farm in Vermont has just 12 cows in an apparently cozy relationship. Correspondence with Horizon Organic Dairy since early 2013 elicited emails, but permission to include them was not received by the publication deadline. Therefore, information on Horizon in this book is based on previously published material in the public domain, which is plentiful.

Correspondence with Aurora Organic Dairy brought little new information, so this text relies on information in the public domain.

Queries were also made to the CROPP/Organic Valley cooperative. A response was received to a question on the use of carrageenan: OV replied that it was working to remove that synthetic ingredient from its organic products.

This book references several studies from The Organic Center which has incorporated data from Aurora, Horizon, CROPP/Organic Valley, and other organizations that cooperated in the production of valuable documents, such as *A Dairy Farm's Footprint* (2010).

Transatlantic Comparisons

For decades, a point of pride for conventional American dairy farmers was that so many of them belonged to cooperatives, and about 90 percent of milk was processed via co-ops nationally. Looking across the pond, some U.S. cooperators sniffed that UK dairy farmers seemed relatively unable to learn the cooperative maxim: "Hang together or hang separately." U.S. dairy farmers fancied that cooperatives gave them more power than British farmers, who seemed more likely to sign individual contracts with milk processors, rather than form cooperatives with their neighbors. (Chapter 8 notes how the British government scuttled the cooperative body called Milk Marque when farmers did try to organize.) But pride goeth before a fall: U.S. conventional dairy cooperatives have not stopped the precipitous drop in conventional dairy farm numbers.

After the Second World War the policy of many European governments was to stabilize rural towns, partly by subsidizing crops and dairy farming. The subsidies restored food security, but eventually resulted in costly surpluses. After years of debate about the high storage costs of its well-publicized surplus Butter Mountain and "Milk Lakes," the European Economic Community (EEC, later European Union, or EU) finally instituted milk quotas in 1984 (Scholten 1989a; 1989; 1990d). But in the United States, it became apparent that the Republican Reagan-Bush administrations (1980–92) were hostile to requests for supply management, even by lobbying groups such as the National Milk Producers Federation (normally part of their political constituency). What these farmers desired were production controls akin to those in Canadian and European milk quotas, though they wisely avoided the taboo *q*-word, which smacked of socialism to the laissez-faire sensibility of Reaganites. Reagan and Bush gave the farmers a polite "no" on production controls. The Grand Old Party's (GOP's) only concessions to worried farmers were government efforts to raise farm gate milk prices with a herd buyout in 1986, which entailed culling and sales of cattle to China. The buyout had a limited, temporary effect, although it helped kick-start China's dairy production from very little to 30 billion metric tons in a few decades. Why did U.S. dairy farmers not turn to the Democratic Party for relief in production management? After all, the party had brought New Deal "parity" policies to aid farmers in the Great Depression. Democratic President Jimmy Carter, 1977–81, was a peanut farmer, whose churchgoing may have impressed some farmers. But Carter's grain embargo against the Soviet Union, after its 1979 invasion of Afghanistan, hurt farmers financially and turned them against him. Further, since the 1960s, the Democratic Party was perceived as an urban party with a liberal social agenda that nonplussed many rural Republicans. Dairy farmers were left to work out their own salvation. Fortunately, some useful strategies were developed by cooperatives to slow—if not stop—the technology treadmill that was throwing so many farmers off their livelihoods.

The late 1980s saw the establishment of CROPP. According to the Organic Valley (n.d.) website history headed "Our Story," when a handful of farmers met to organize CROPP in January 1988, "family farms were on the brink of extinction". But they had the tacit support of consumers, many prompted by food scares, the herd buyout, and increasingly aware of the revolving door between agribusiness and government regulation. Pro-organic sentiment had built since the 1960s. Apprehension was also starting to mount over the introduction of dairy GMOs, and rumors were beginning to circulate in Europe and America about carnivorous cattle-feeding practices in the UK dairy industry that eventually emerged as mad cow disease (BSE/nvCJD; see Scholten 2007a, b).

Around the world it is joked that organic movements began with a bunch of hippies. That is only part of a longer story, but many well-respected organic leaders did emerge from the 1960-70-80s counterculture. In the case of CROPP, a lanky, long-haired farmer named George Siemon was a founder-member and the original chief executive officer (or, as he prefers, "C-E-I-E-I-O," rhyming with the "Old MacDonald" children's song). His family had farmed Wisconsin's Bad Axe River valley since 1977. Siemon helped establish strict pasture standards in Organic Valley and has helped lead the fight for strong national standards for organic certification. CROPP has also opted to go dairying without antibiotics. Harriet Behar (personal communication, Portland, Oregon, NOSB meeting, April 8, 2013) recalled: "I worked with Organic Valley when it was just seven farmers. They wondered what they'd do without antibiotics, but now there are about 1,800 farm members."

Siemon remains a CROPP leader today, articulating the message that consumers demand a more sustainable food system and that supplying their needs can sustain family farms and rural communities. When CROPP headquarters suffered a major fire in early 2013, messages from longtime leaders such as Siemon reassured members and consumers that Organic Valley products would appear uninterrupted in dairy cases.

From the 1960s, organic pioneers in several states made grassroots attempts to regulate the sector. They wanted to clear their market of what economists call "free riders," those who called their products organic but neglected the investments in time and money required for true organic production, processing, or sale. Efforts to establish national organic standards accelerated about the time CROPP was established in the late 1980s. A major political achievement was congressional passage of the Organic Foods Production Act (OFPA) of 1990, which brought U.S. state programs under a federal umbrella.

However, agreeing on standards and permitted inputs was difficult. Compromises were required, some lasting decades. Former secretary of agriculture Dan Glickman said in 2000: "Let me be clear about one thing. The organic label is a marketing tool. It is not a statement about food safety. Nor is 'organic' a value judgment about nutrition or quality" (*Academics Review* 2014: 2). That helps explain why, when he was farm secretary under President Clinton in 1997, some surprising items were proposed by USDA for inclusion in the organics ingredients list: genetically modified organisms (GMOs), sewage sludge containing heavy metals, and ionizing radiation. An upsurge of 275,000 consumer messages to the USDA showed widespread opposition to this "Big 3." Purists won a major victory when the USDA published the National Organic Program (NOP) rules in 2002, excluding the Big 3 from the National List of Allowed and Prohibited Substances (National List).

The guardians of organic integrity, by policing against prohibited materials or practices, were USDA-accredited certifying agents, including California Certified Organic Farmers (CCOF) and Quality Assurance International (QAI). (For more information see "Instruction: Accreditation Policies and Procedures. NOP 2000," effective, Feb. 28, 2014).

But the devil is in the details. According to the USDA Agricultural Marketing Service (AMS 2013a) website, last updated April 4, 2013, accredited agents of the National Organic Program (NOP) certify that organic operations "demonstrate that they are protecting natural resources, conserving biodiversity, and using only approved substances." In organic crops, "the USDA organic seal verifies that irradiation, sewage sludge, synthetic fertilizers, prohibited pesticides, and genetically modified organisms were not used." In organic livestock, "producers met animal health and welfare standards, did not use antibiotics or growth hormones, used 100 percent organic feed, and provided animals with access to the outdoors." And organic multi-ingredient food "has 95 percent or more certified organic content. If the label claims that it was made with specified organic ingredients, you can be sure that those specific ingredients are certified organic."

If these standards were as simple as they sound, there would be no need for this book. The U.S. Environmental Protection Agency (EPA) notes pesticides derived from natural sources (such as biological pesticides) may be used in organic production. But the National Center for Food and Agricultural Policy, a group linked to the chemical industry, complains that two fungicides allowed in organics, copper and sulfur, have been used at higher rates per acre than synthetic fungicides (*Scientific American* 2011). The NCFAR portrays this to the public as a chink in the armor of organics.

Less well known is that U.S. tree fruit crops, such as apples and pears, have been allowed to use antibiotics until a 2014 "sunset"—that is, phaseout—was agreed to by the National Organic Standards Board (NOSB) at a meeting in Portland, Oregon, in April 2013 (Granatstein 2011; 2013; WSU 2014). The original rationale for allowing antibiotics had political aspects—that is, to include as much acreage as possible under the organic aegis, after the passage of the Organic Foods Production Act (OFPA 1990)—with the understanding that antibiotics use would soon cease. But extensions on streptomycin and tetracycline were passionately demanded by some veteran fruit growers at meetings of the NOSB standards board, including Seattle in 2011 and in Portland in 2013. They maintained that USDA rules did not in fact prohibit antibiotics in crops or livestock production. Finally, after two decades, an industry consensus including the National Organic Coalition (NOC 2013), Organic Trade Association (OTA), and The Cornucopia Institute agreed on a 2014 endpoint at the Portland gatherings. But organic politics is seldom "over." See chapter 8 for later developments.

At a higher level, the International Federation of Organic Agricultural Federation's (IFOAM 2006–13) *Basic Standards for Organic Production and Processing* is a "broad church" for varying interpretations of organic philosophy. Antibiotics mark the biggest difference between U.S. and EU organic standards (USDA 2002c; 2004). The antibiotics ban in U.S. organic dairying is virtually absolute, compared to Europe, where antibiotics are allowed in organic dairying, with a period of milk withholding from the market after use. Antibiotics are prohibited in fruit in the European Union. In the United States, some use of antibiotics against staphylococcus and other microbes is only now being phased out. Harmonizing U.S.-EU rules on antibiotics could have far-reaching benefits for animal and human health, a critical policy juncture in the effort to maintain an arsenal of the modern variants of penicillin. Discovered by Scottish biologist Sir Alexander Fleming around 1928, it has been a wonder drug against fatal infections since its mass production by the United States during the Second World War. (Fleming was reputedly a poor public speaker with little imagination for commercial development. Fortunately the drug was channeled to use by Britons aware that more soldiers died from infection than initial wounds in previous wars.)

At the Rodale Institute, Dr. Hue Karreman (2004; 2011) is a veterinarian who has led the development of alternatives to antibiotics in cattle. Karreman sees both sides of the issue but notes that worry about antibiotic resistance is fostering interest among those in audiences where he has given presentations in North America, Britain, Germany, Holland, South Korea, and Turkey (personal communication, August 31, 2013).

The range of political philosophies among actors in the organic movement invites comparisons to the German Green Party, whose socioenvironmental wings fractured between *Realos* (realists) and *Fundis* (idealists). Idealists want no compromise with organic principles on the family-farm or corporate-agribusiness scale. Realists are more apt to accept any improvement on the status quo, in hopes it nudges more of world agriculture toward sustainability. One realist in Washington State is Gene Kahn, a founder of Cascadian Organic Farms before it was sold to conventional agribusiness giant General Mills. Kahn fought for the inclusion of some synthetic chemicals on the organic permitted list. Kahn, who famously argued (Pollan 2001) that it should be possible to make organically certified Twinkies (a sugar and preservative-rich confection in its conventional form, manufactured by Hostess Brands), claims the organic sector would have been dead in the water without such compromises.

The West Coast—sometimes called the "Left Coast"—has been prominent in organics since the 1960s. In those less than politically correct days, TV talk show hosts referred to the state of California as "the land of fruits and nuts," an allusion to the stereotypical appetite of celebrity spouses for blended fruit juices and organic and alternative foods. Oregon and Washington States had important forest and farming industries but relatively less powerful economies than California, although Oregon was developing its sport shoe industry around Portland, and Washington was buoyed by the high technology of Boeing around Seattle, which frequently trumped California's own post–Second World War aerospace companies. Nevertheless, Washington had a high preponderance of family farms and was an early hotbed of alternative food networks. Washington's organic certification program was nationally admired, and it reassured many pioneers when former Washington State Department of Agriculture organics official Miles McEvoy became deputy administrator of the National Organic Program (NOP) at USDA in the "other Washington," in the District of Columbia on the East Coast.

Observers of a suspicious nature question to what extent respected organic stalwart McEvoy might be green window dressing for USDA secretary Tom Vilsack. As governor of Iowa before his cabinet appointment by President Obama, Vilsack won the Governor of the Year Award from the Biotechnology Industry Organization, with members including Monsanto. Since then, he has expedited certification of genetically modified crops, such as alfalfa, which organic and many conventional growers claim endangers the integrity and commercial sale of $4 billion of annual alfalfa exports to Japan. They claim this is just one recent example of government preference for biotechnology as an industrial champion, successor to the automotive, aviation, and computer technologies that matured in the United States before their manufacture was relegated to emerging economies, including China. The logic is that, as a font of scientific research, development, and innovation, the United States can maintain global revenue streams from GMOs and biotechnology via patents in trade-related intellectual property (TRIPs; see Anderson and

Tyers) under the World Trade Organization (WTO), a policy shared by every presidential administration since Ronald Reagan.

Black, White, and Grey

This book is not a morality tale in black and white. While some animal charities paint modern livestock agriculture in terms of pure profit and cruelty, the reality of conditions for animals may be better than a hundred years ago in some instances. One example is the improvement of slaughterhouse design prompted by the insights of animal psychologist Temple Grandin (1989; 2005; 2006). She challenged America's huge meat industry, saying that since people were plainly committed to eat animals, the least they morally owed animals was to diminish fear and pain in their conditions. Thus Grandin helped design feedlots and abattoirs that saw floors, runways, and lights from the perspective of cattle, eliminating disturbing reflections, dangling chains, and high-pitched noises. Around half of North American industrial livestock processing plants have been improved according to her prescriptions. Grandin identified two types of animal abuse. The first was obvious, sins of commission, such as neglect of feed and shelter, overuse of cattle prods, or worse. The second was less obvious, sins of omission, such as consigning animals to boredom in overcrowded facilities (http://www.grandin.com/welfare/welfare. issues.html). Grandin observed that providing animals with toys improved welfare. The same principle was shown in Europe, where veterinarians found that pigs enjoyed playing with mobiles hanging over their pens (Scholten Feb. 25, 1990).

Overseeing a U.S. population of dairy cows exceeding nine million, there are legions of animal lovers on both conventional and organic megadairies owned by corporate agribusiness. U.S. TV networks have broadcast incidents of cruelty on big and small farms, usually by untrained, low-paid workers coping poorly with too many animals. Such incidents are also possible in the family-scale organic sector. A contemporary of mine, a veteran of the dairy industry, confides, "On every farm there's always something . . . that inspectors would question." Iowa and other farm states have passed legislation making it illegal for job seekers to deny links to welfare groups, such as Compassion in World Farming (CWF) or People for the Ethical Treatment of Animals (PETA). *Hoard's Dairyman* (2012a) has editorialized that this is the wrong approach and that training is the best strategy to stop abuse and to discontinue practices, such as tail docking, that concern the public. Public pressure can shift attitudes to promote calf and cow welfare in everyday best practices if cattle keepers are properly trained and paid.

Environmental protection is another fundamental concern on any farm with more than a few cows. Manure and urine runoff can alter creek chemistry, spawn growth of reed canary grass, and wreck an ecosystem previously good for fish and other aquatic species. Farmers, organic or otherwise, may be tempted to overlook what they see as a minor problem—until that distant day when they have time to manage such pollution better.

Social justice is a sticky area. Corporations often play the baddies in tales of dairy morality, such as the popular 2008 film *Food, Inc.*, narrated by prominent writers Michael Pollan and Eric Schlosser. But even the most trusted farmers' cooperatives have, on occasion, been accused of sharp practice in their business operations and of sometimes leaving their own farmer members in the lurch. The struggle for dairy market share has not been pretty, and, like a military retreat, it is almost every man, woman, and cow for themselves.

Even before 2007 when the world economy began to unravel, a trusted source in the organic dairy community told me, "There are no good guys and no bad guys." Since then, widespread drought and soaring feed and petroleum costs have made economic survival more precarious. Yet, per capita sales of organic milk are still increasing in the United States.

With family-scale dairy farms at the center, this book is an account of the complex, varying priorities placed in social and economic goods in a competitive market economy drawing on finite natural resources. Some actors prioritize the supply of affordable organic milk for consumers valuing dairy products produced without antibiotics, synthetic hormones, or chemical fertilizers and pesticides in their production. Other players prioritize a system of milk production linked to the socioeconomic tradition of farmers' cooperatives as one that delivers the best combination of animal welfare (with cows grazing on grass for as much of the year as possible), low environmental impact, and social justice for family farmers, many of whom combine their resources and political economic power in cooperatives.

This begs the question of whether standards for animal welfare should be raised and strictly enforced in both conventional and organic dairy sectors. From ancient times until at least the era of Thomas Jefferson and his ideas based on agrarian democracy, it was common to think people's health and even national wealth were based on the quality of soil, water, and other natural endowments. But the switch of the chemical industry from munitions to farm inputs after the world wars of the twentieth century was less benign than just knocking swords into plowshares. By many tokens it resulted in more food but of poorer nutritional quality and at the cost of negative externalities, such as poorer soil and water quality, not to mention less demand for rural labor, resulting in faster rural-to-urban migration.

The introduction of chemicals in farming smacks of Gresham's Law, named after British Tudor financier Sir Thomas Gresham, on whose watch fiat money drove intrinsically more valuable specie, such as gold and silver, out of the system. Likewise, government promotion of intensive, chemical farming made organic farming economically unsustainable for most farmers, until the second half of the twentieth century when some organic pioneers fought back with special certification, marketing schemes, and cooperatives.

Organic farmers urge others to adopt their techniques to cut greenhouse gas (GHG) emissions and reduce petroleum use. These include fertilizing soil with animal manure, lime, and other applications; practicing crop rotation and fallow, cover crops; sanitation; early season weed control (blind cultivation); and using good non-GMO seeds.

Conventional farmers might reply that they already follow some of these practices, such as low-till or no-till crop cultivation, and mapping areas that need lime or other inputs with precision GPS technology. Looking at U.S. Dairy Industry Statistics, 2007 to 2012 (*Hoard's Dairyman* 2013a: 206–07) conventional dairy farmers might also contend that U.S. herds comprising 9.233 million cows, each producing an average 21,697 pounds milk in 2012, had less negative environmental impact than the fewer 9.189 million cows producing 20,024 pounds of milk in 2007, because of annually rising yields per cow. Summarizing this view on the relative efficiency of organic milk versus conventional milk, a veteran dairy writer who also manages a dairy farm states flatly (personal communication): "Organic milk production is less efficient because of the lower levels of milk production. Maintenance cost is spread over fewer pounds of milk."

To that claim of conventional efficiencies, traditional organic dairy farmers might reply that a holistic analysis of all inputs sourced off farm for conventional farms would reveal unsustainable dependence on finite supplies of petroleum and environmentally harmful chemicals, which produce more local pollution and climate-destabilizing GHGs than closed-system organic farms and that, fundamentally, there are social costs when megadairies drive family-scale dairies out of business and deplete rural communities. Catching their collective breath, organic farmers might add that green consumers expect animal welfare and longevity on organic farms to be better than on conventional farms.

The Organic Center's Chuck Benbrook (2012: 1, 6) pinpoints pasturing, cow longevity, and production as three factors distinguishing organic production from conventional, concluding that "the first two are generally greater on organic farms, and the third is greater on most conventional farms." Benbrook makes counterarguments on efficiency based on life-cycle analysis (LCA; see chapter 8), claiming that cow health and longevity are low priorities for the conventional dairy scientists who are intent on developing a cow for the future, to reduce methane emissions. But that could change as evidence of a culling crisis mounts within the conventional sector (Biagiotti 2014; DHI-Provo 2013).

Organic purists go further, claiming the debate is more complicated than a simple organic-versus-conventional faceoff. Just as insidious, according to such idealists, are organic-industrial megadairies with a patina of USDA organic certification masking a focus on ingredients instead of organic *process* (Pollan 2001; Odairy Apr. 27, 2014). These are simulacra, which outsource great amounts of organic inputs but operate according to environmentally unsustainable, conventional business plans (Scholten 2010b: 149). True organic food systems, according to idealists, are those in which farms are networked in more regional food systems, rather than the present globalized system in which many components, including highly processed milk, travel an average 1,500 miles between cow and consumer.

Before returning to these contemporary debates, the next chapter's walk into prehistory will be useful. It helps us imagine how the domestication of ruminants, such as dairy and beef cattle, sheep, and goats, assisted the development of commerce, politics, science, and the arts, spreading civilization from the equator to the poles.

2

Agricultural Revolutions

Photo 2.1   Tractors appropriate for small farms can last 50 years.

Photo Credit: Bruce Scholten.

2

Agricultural Revolutions: Winter Was Bleak before Haymaking

The earth warmed, and glaciers retreated some 10,000 years ago in the early Holocene epoch. Sheet ice that had reached as far south as present-day Long Island or Yorkshire gradually withdrew. Wherever they were located, the nearer the herds of nomadic pastoralists were to the ice, the more animal numbers dwindled in winter as grazing turned sparse. Those keepers who had herded animals to lush summer highland pastures moved them down to the flatlands and warmer temperatures. There the chances were better of finding grasses and sedges to keep cattle alive, producing milk into autumn and the coldest months, till spring when most would calve.

Forage conditions varied from winter to winter, but inevitably animals' ribs began to show. Some cattle were slaughtered, and their meat consumed or hides traded with people encountered in permanent settlements along the way. In like manner nomads in the time of the Biblical Abraham might have pastured herds on the Golan Heights, which benefit from occasional snow, en route from Ur of the Chaldees through arid deserts to the cities of Egypt.

This chapter evokes some of what we know about prehistory and follows dairying through a series of agricultural revolutions to the present day, when peremptory applications of biotechnology in recombinant bovine growth hormone (rBGH) and food scares, such as mad cow disease (BSE/nvCJD), helped kick-start a modern organic dairy industry in the United States and elsewhere.

For millennia cattle keeping was nomadic, and in many parts of the world, it still is nomadic or seminomadic. The International Fund for Agricultural Development (IFAD n.d.: 1) estimates nearly two hundred million pastoralists still traverse the world. IFAD terms them "nomadic" when their seasonal paths are irregular and "transhumant" when they are consigned to annual progressions form lowlands to mountain meadows, as is typical of cow herds in Switzerland or of reindeer herds of the Sami in Scandinavia.

Bovine migrations encourage upland vegetation because, as biologists know, animals are a rare means of transporting organic nutrients upstream, where geologically eroded loess may combine with animal waste to encourage plant growth and, over the seasons, produce humus and tilthy soil conducive to plant growth and further grazing.

Recurring droughts, climate change, and urbanization are constricting the movements of the Masai in Africa, some of whom have recently been attracted to settled dairying in the East Africa Dairy Development project (EADD; Scholten 2013b) project. Similarly in India, encroaching urbanization has induced Advivasi castes of hill tribes in the state of Gujarat to join farmers' cooperatives in the country's successful Anand Pattern of low-input/low-output dairying (Scholten 2010; 2011). A 2013 policy document from the UN Food and Agricultural Organization details how dairy-industry development programs assist smallholders, women, and children in countries including Bangladesh, Thailand, and Uganda. Milk production improves nutrition and incomes for dairy families, who then purchase better nutrition, education, and health care for themselves. This stimulates local businesses in a virtuous circle that, in a demographic transition, prepares some of the children for jobs in manufacturing or service sectors of the economy and slows rural-to-urban migration. Dairy activity also saves surprising amounts of foreign exchange for developing countries that previously had to import billions of dollars' worth of dairy products (FAO-UN 2013).

Pastoralists are too often seen as relics of prehistory, but the settled world should honor their perseverance against encroaching urbanity (*Nairobi Star* 2011). Their sensitivity to natural ecosystems imparts the knowledge to herd cattle in seemingly peripatetic patterns that, for example in East Africa, may be more productive in agricultural products and income than more settled ranching in the United States. Contrary to a common prejudice, pastoralists seek to avoid overgrazing. Their goal is to forage long enough in an area to feed cattle, but there is an unintended but happy consequence for the environment: the stimulation of plant regrowth by the physical action of hooves on plant roots and soil is synergistic with the deposit of wastes. The International Livestock Research Institute finds that nomadic pastoralism is often not just an ecologically efficient use of arid land, but is also economically efficient (ILRI 2011). In Africa, for example, the International Union for Conservation of Nature (IUCN 2013) estimates that mobile pastoral farming generates 50 percent of Kenya's gross domestic product (GDP)—an astounding figure when considering that Nairobi is a hub for continental and multinational trade.

Settled dairying had little chance in climates with long winters. New technology was needed, a technology so basic it is often overlooked—hay. Cattle herds have always been more than mobile meat and milk. As ancient symbols of wealth and prestige, cows are also surprisingly practical, fluid wealth similar to modern cash cards from which value can be deducted and traded inside a tribe or auctioned to others. Keepers are perennially tempted to keep more animals than they can adequately feed. But livestock can be exchanged for other goods—foods, gems, hides, spices, and weapons—on short notice. Yet, animals looked leaner and meaner as the darker months progressed. Seasonal forces apply to the few thousand remaining Sami reindeer herders in Arctic regions, as they did to nomads seeking pasture as the ice receded after the last major glaciations.

In this warming period, it wasn't so much that it snowed less but that more melted every year, leaving less ice and more open ground for more days of the year, as the ice retreated and the climate gradually warmed in northern latitudes. Taiga emerged, filled with trees. Methane- and carbon-rich peat remained as permafrost, but the steppes above warmed in the sun, sprouting grasses from horizon to horizon. It tempted herders with seasonal livestock, with one eye open for signs of prey animals, like deer and aurochs, ancestors of present-day cattle with shoulders as high as a hunter's eye (extinct perhaps three hundred years). The travelers kept their eyes open for their own predators, bears and saber-toothed cats.

In northern climes the mean temperature rose, and human population gradually increased. Birds followed camps and supped on the remains of game discarded by hunters. They also picked seeds from fruit and vegetables discarded from temporary camps. Birds passed these seeds from one place to another. British geographer Brian K. Roberts (Atkins, Simmons, and Roberts 1998) reckons the camps' cooks were generally women and acted as protofarmers when they returned to annual encampments to find volunteer plants sprouting in garbage areas from previous visits. By weeding out plants they disliked and encouraging those they remembered as tasty or useful in a medicinal way, they practiced an early form of agricultural intensification.

Sometimes campfires got away, especially in dryer months, becoming wildfires that cleared brambles and bushes between trees. More often, lightning from late-summer storms struck vegetation and blackened it to the ground. Frightened animals and herders fled from such desolation and sought safety. Imagine their joy, a season or two later, finding the earth covered in verdant raiment of grasses that fed their herds and attracted deer, elk, moose, and other prey. Before long people understood the pattern of fire and pasture and began to use fire as a tool to manage the landscape. Such a tool could be more bludgeon than scalpel. Greed surfaced in prehistory, long before eighteenth-century European explorers slaughtered the dodo bird into extinction on Mauritius or nineteenth-century American thrill seekers nearly did the same with buffalo, shooting them from moving trains. Atkins, Simmons, and Roberts (1998) note evidence that the environmental footprints of some early North Americans were deeper than necessary for their own self-preservation. Archeological and anthropological remains show fire was used to drive many more buffalo to their deaths off cliffs than the hunters could practically eat or even preserve as meat, bone, and hide. Like a fox in a henhouse, people sometimes just get carried away.

By chance and reckoning, they discovered ways to improve their lives by making animals thrive, with technology fundamental to human well-being. Fast forward to the mid-twentieth century. The transatlantic Allied powers have just been victorious over the German-Italian-Japanese Axis in the Second World War. Moral victories associated with the Allied powers became conflated with the science and technology that produced the atomic bomb. In dairy advertising, laboratory images replaced bucolic scenes of milkmaids and cows as the guarantors of health (DuPuis 2002). Progress was men in white lab coats. In the 1950s, nuclear fission was declared the future of energy production, supplying energy in volumes so enormous that it would be too cheap to meter. Reductive science taught elementary school pupils that "man" was the only creature to make and use tools. But civilization remains dependent upon carbon fuels, and, as it happens, birds and nonhuman mammals are among creatures that use tools to feed themselves. Granted, humans' tools are more complicated than animals' sticks or levers. But humans' tools go awry in bigger ways, as did nuclear reactors in Pennsylvania in 1979 and Chernobyl in 1986. Supposedly fail-safe systems can fall foul to natural forces. Emergency cooling systems failed to protect the core of Japan's Fukushima Daiichi reactor when tsunami waves hit the site in 2011. The loss of firefighters' lives and of regional electricity production was compounded by an embargo on agricultural goods inside the quarantine area and an exodus of the local population to safer areas. Radioactive contamination spread from the damaged facility and surrounding area. BBC News (2013a; see also Huffington Post 2012) quoted Mycle Schneider, a consultant who has advised French and German governments, saying: "[Fukushima] is much worse than we have been led to believe, much worse."

Was the invention of nuclear power a net mistake? Some proponents of the precautionary principle claim the scope of damage from nuclear accidents is so immense that no benefit can justify fission at its present state of development, or at least involving processes that produce plutonium. However, if Chernobyl, Three Mile Island, and Fukushima are not polluted for all time, advocates of a risk/benefit approach (or risk analysis) to policy might view risks linked to fission as tolerable, since its outcomes are beneficial (for example 80 percent of French electricity is generated by nukes) if we ignore the persistent problem of radioactive waste disposal and the concomitant production of plutonium, which can be diverted to weapons of mass destruction. Events in the twentieth century led to the suspicion of human invention and of "men in white lab coats."

Bill Gates (1999), cofounder of Microsoft, notes that one hundred thought leaders were asked by writer John Brockman, "What is the most important invention in the past 2,000 years?" Freeman Dyson, renowned physicist at Princeton University, answered: "Hay." Dyson explained that before some unknown genius invented hay in the so-called dark ages, Greek and Roman civilization existed only in climates warm enough to support winter grazing. Hay made it possible for empires to move north of the Alps and to build Vienna, Paris, London, Berlin, and eventually Moscow and New York.

Before hay, in agricultural hearths such as Mesopotamia, farmers were able to produce enough surpluses to support new formal sectors of complex economies, such as religion, mathematics, and science. Turning surplus grasses into hay, to sustain cattle and horses through the winter, allowed civilization to advance intellectually while physically following retreating glaciers toward the poles. Humans forged a series of innovations that kept hunger at bay. But human invention also increased world population and pressure on nature.

Agricultural Revolutions

The *first agricultural revolution* followed millennia of hunting and gathering and nomadic pastoralism that began in prehistory (Atkins and Bowler 2001: 25–35). It deserves the appellation of "revolution" because it produced significant changes that increased productivity and improved human well-being in ways recognizable as progress—compared to our widespread contemporary ennui, fearful that technology rebounds in unfortunate ways, exemplified by anthropogenic contributions to climate change. But in simpler times, seed agriculture offered hunter-gatherers the option of settling down, a sedentary life and landscape that could be defended against marauders in dwellings amounting to capital that succeeding generations might inherit.

As mentioned above, it may have been the women, among the nomadic pastoralists, who developed seed agriculture by selecting and fostering volunteer plants around settlements. This nascent iteration of crops-livestock agriculture intersected with the domestication of sheep about 9,000 years before present (BP), and goats about 7500 BP, utilizing mammals' milk for humans in the hearths of the Fertile Crescent of Mesopotamia and the Nile valley. According to geographer Carl Sauer, South Asia also participated in the first agricultural revolution about this time. But Atkins, Simmons and Roberts (1998: xiv–xv) point out that the succession of human societies from preindustrial to industrial/modern to global/postindustrial conditions was diachronic, occurring at different times in different places and not always in relation to glaciations. For example, farming probably took off in southern Europe about 9000 BP (7000 BC), but not until the nineteenth century in Australia as it was not developed by aboriginal peoples there (Diamond 1997). In North America there is evidence of sophisticated crop farming by Native Americans in the lower Mississippi River valley, but communities in and around what is present-day Maine in the Northeast and Puget Sound in the Northwest might be categorized as hunting-and-gathering societies, rather than agricultural.

Knox and Marston (2007: 310–11) note the domestication of cattle as draft animals about 6500 BP occurred along with the domestication of wheat and rice. Recent research shows nomadic peoples were indeed taller, healthier, and longer-lived than their successors in primitive settlements, including working-class Britons, who were found to be short and undernourished when called to military service in the First World War.

Living cheek by jowl with animals resulted in humans sharing a great many zoonotic diseases but may also have made human immune systems more robust. Farm animals offered traction to relieve the work burden on humans and speeded transport and communications, affording food surpluses, which in the long run contributed to the rise of cities and specialization of labor. Life was better for many people and animals.

The *second agricultural revolution* is disputed as to its timing and origins, but there is consensus that it accelerated in the early nineteenth century, when the Industrial Revolution was centered in Britain (Atkins and Bowler 2001; see these authors for background on how the second agricultural revolution overlapped with the world's "first food regime," marked by Britain's adoption of free trade policies and the repeal of the Corn Laws in 1846). Although there was significant innovation in metallurgy and chemicals in the pan-German areas of Europe, the financial ability to invest and profit from it was greater in Britain. This helps explain why, in 1687, sword makers from Solingen settled in northeast England. Advanced German steel furnace technology was brought to Derwentcote near Durham and Newcastle early in the eighteenth century, by metallurgists who may have felt underappreciated in their homeland. In Britain, capitalist finance exploited new inventions and diffused manufactured farm implements and processes abroad. These included iron tools, water power, pumps, coal-powered steam engines, bridges, and breweries, which fed brewers' wastes to adjacent milk cows in cities, such as London (Atkins 2010). This was an early example of a confined-animal feeding operation (CAFO), and conditions for cows in airless sheds adjacent to breweries were alarming. This was comparable to the abuse suffered by horses drawing London's taxicabs that was exposed in Anna Sewell's bestselling 1877 book, *Black Beauty*. One reform stemming from Sewell's book was that taxi license fees were lowered, to reduce the moral hazard of animal abuse. Life was hard for some animals.

It certainly was not all optimism for humans in Britain either. Scottish economist Thomas Malthus's dismal predictions of famine were rife. American humorist Mark Twain (n.d.) may have borne Malthus in mind when he advised others: "Buy land, they're not making it anymore." But for optimists in the nineteenth century, swords were giving way to plowshares in the spirit of modernism, as new discoveries diffused from Britain in efforts to prove Malthus wrong. These included fertilizers and technologies, such as crop rotation, to improve yields and soil fertility. Ditches were dug to drain swamps or fens in areas such as East Anglia, effectively creating arable land. This second agricultural revolution leaped the Atlantic Ocean. Massive landscape modifications were made possible by innovations in capitalist finance and government fiscal capacity. The Bridgewater Canal, in northwest England (finished in 1769), and the Erie Canal, joining the Great Lakes with the Atlantic (completed in 1825), improved the distribution of commodities and lowered transport costs for farmers far from population centers and seaports. It didn't hurt that Prince Albert, the consort of Queen Victoria, who reigned 1837–1901, was interested in agriculture (resonant with present-day Prince Charles, known for his passion for organic agriculture). The Victorian legacy was better crop and livestock yields, with improved yokes for oxen, new uses for horses, and an acceleration of road and canal building.

New technology brought winners and losers. New transport routes were zero-sum technologies for some. Today, New England, in the United States, is more forested than a century ago, with nearly 87 percent of the arboreal cover that existed when the first European settlers arrived (*Hoard's Dairyman* 2013b: 244). One reason is because new transport precipitated farm declines in the early colonial states. Northeast farms that once supplied New York City lost this prime market as they were displaced by farms in the fertile Midwest breadbasket linked by canals and trains. Today we associate rural New England farms with the stones that percolated through springtime mud to furnish the rock in Robert Frost's subtle poem "Mending Wall." But there is evidence that it was not always so, because wooden rails were not superseded by stone walls until the 1800s, when over a century of unsustainable monoculture by the European settlers had so depleted humus and loam in the upper strata of soil that plowing was impossible without removing stones from the fields.

U.S. government support for the ideology of Manifest Destiny promoted western expansion, along with apocryphal exhortations of newspaper writers to "Go West." The McCormick reaper, developed near Chicago, decreased labor and increased arable productivity in the Midwest, where the plains were amenable to mechanization that gave breadbasket farms greater economies of scale than New England.

Before transport and cold-chain improvements, such as refrigeration, in the late nineteenth century, most food and dairy consumption (except for hard cheese) depended on local production. In her delightful 2002 book, *Nature's Perfect Food: How Milk Became America's Drink*, Melanie DuPuis details how milk trains, on a growing rail network from New York City, northward through the Hudson River valley, linked remote farmers to urban consumers. Such vertical linkages and upscaling weakened the practice of colonial-era families to keep a milk cow in the backyard. DuPuis records growing reverence for nature amid nineteenth-century industrialization. In America as in Britain, a trend developed in which the growing middle class sought summer respite from the smoky city in raw nature. Milk advertising offered images of these rural idylls, milkmaids and cows, and the public accepted the role of milk as part of a healthy child's diet.

Bovine milk is indeed nutritious but, unfortunately, a vector for many diseases humans share with domesticated animals. With tragic irony, milk proved to be an endemic source of bovine tuberculosis among children in families prosperous enough to guard their health with bovine milk. In *Liquid Materialities* (2010) Atkins documents how farmers' opposition to tuberculosis (TB) eradication measures contributed to about 600,000 human deaths in Britain, 1860–1930, before disease identification, eradication, and herd compensation measures were introduced in the mid-twentieth century.

TB remains endemic in the United Kingdom, where cattle and wildlife, such as badgers, can transmit the disease to each other. The BBC (2013b) reported that the Department for Environment Food and Rural Affairs (DEFRA) had authorized a pilot cull of badgers implemented by private companies under license by Natural England. The largely futile cull began in

Somerset in August 2013 and spread to Gloucestershire in September, with plans for marksmen to shoot approximately five thousand of the burrowers in order to improve biosecurity. Animal welfare groups insist a vaccination campaign makes more sense, but the National Farmers Union counters that vaccination would compromise future attempts by parts of the United Kingdom to regain TB-free status in global trading. The problem for livestock inspectors is the difficulty of distinguishing the residue of vaccination in a cow's system from signs of live transmission. If pharmacologists could apply some sort of molecular marker to vaccine, it might be commercially viable, as well as a boon to animal welfare and farmers' economic sustainability. In the United States, better screening has helped lower rates of tuberculosis, but demands for a national cattle identification system are opposed by some small farmers who believe they will have to bear an unfair share of the costs and paperwork. Meanwhile, the rise of drug-resistant TB in other parts of the world makes the monitoring of human and nonhuman immigrants and visitors important to public health.

The *third agricultural revolution* grew out of the second in the late nineteenth century. New technologies and social mobilities facilitated global dairy cold chains, which in the twentieth and twenty-first centuries came to be understood as parts of a general, world-wide "White Revolution" in dairying. Anil Sharma notes, in a Food and Agricultural Organization (FAO 2003) publication on the World Trade Organization (WTO) Agreement on Agriculture, that success in India's Green Revolution in crops and White Revolution in dairying led to cognate efforts called a "Yellow Revolution" in oilseeds and a "Blue Revolution" in aquaculture (see also Scholten forthcoming).

In *Merchants of Grain* (1979), Dan Morgan notes that train and ship transport from the American Midwest enabled the U.S. heartland to rival traditional global breadbaskets, such as the Ukraine, as tractorization, irrigation, and chemical inputs of the third agricultural revolution followed the horses, oxen, and mules from the second.

Refrigeration was a major advance in dairy cold chains globally. Transoceanic telegraphy via undersea cables and a shift from sail to coal to oil power enhanced global command-and-control capabilities.

The biotechnological origins of the twentieth century's Green Revolution are in Mexico in the 1940s, where institutions such as the Rockefeller Foundation funded experimentation to meet Mexico's nutrition needs. Led by Dr. Norman Borlaug, who is seen as the "Father of the Green Revolution," this constitutes a genetic intensification of the third agricultural revolution, but closer to the patient plant-breeding techniques made famous by the Austrian monk Gregor Mendel than the genetically modified organisms (GMOs) discussed in chapter 1. The first example of the Gene Revolution is the 1990s Flavr Savr tomato developed by Calgene for resistance to spoilage in transport. An early version added a fish gene from the animal kingdom to a tomato in the plant kingdom, but Jeffrey M. Smith notes that this was not commercialized (Smith 2003, 2010, 2012). Stomach lesions and the deaths of some rats were recorded in the studies, which Dr. Arpad Pusztai has called poorly designed. Another version of Flavr Savr tomatoes was commercialized but got mushy. Observers said an unsuitable species of tomato was chosen as the base for development. It was later withdrawn from Anglo-American markets, and Calgene sold it to Monsanto.

Back in the mid-1940s in Mexico, classic research in the original Green Revolution spurred development of rust-resistant wheat and high-yield variety (HYV) maize (corn), rice, and wheat seeds in partnership with institutions in the Philippines and Asia. In India, Borlaug, the Rockefeller Foundation, and the Ford Foundation worked with the late Shri C. Subramaniam, known as the "Father of India's Green Revolution." Assisted in field trials by Dr. M. S. Swaminathan, the country enjoyed its own Green Revolution as rice yields tripled from the 1960s to the 1990s.

The ingredients of the Green Revolution were essentially a package of HYV seeds, chemical inputs, such as nitrogen (N), phosphorus (P), and potassium (K) fertilizers in the familiar formula, pesticides, irrigation, electrification, and mechanization. As the Green Revolution multiplied yields in bumper grain harvests around the world, observers questioned why the dairy sector did not adopt similar technologies. The unique properties of milk complicate answers.

Additionally, the ethics of animal welfare have often accompanied discussion of modernization in livestock agriculture. These often focus on the ability, or not, of animals to perform instinctive behaviors in breeding, feeding, and so on. Thus, when artificial insemination (AI) was propagated in hundreds of India's village centers in the 1950s, a national debate ensued among people uneasy about what they perceived as a denial of cows' natural relations with bulls.

Peter Atkins (2010) notes how milk developed from a local by-product of cattle raised for meat or traction to commercial products, such as butter and powdered, that could be marketed as "fresh" products in markets thousands of miles from their origin. Like DuPuis (2002) Atkins observes that, centuries ago in Europe and North America, buffalo, cow, goat, sheep, or other mammalian milk was less often drunk as a liquid than as processed products, such as buttermilk, butter, or cheese, due to its propensity to spoil within hours after milking. It turns out that the twentieth-century American fashion for drinking liquid milk is an anomaly in world history.

Ice was a less-than-ideal way to preserve milk or meat before the advent of refrigerators, in which heat exchangers pumping greenhouse gases, such as Freon, cooled the product. Refrigerated trains and ships transported butter, meat, and fruit from farms to metropolitan centers. Refrigeration helped build the economies of countries around the world. For example, it helped Argentina near U.S. levels of economic development at the end of the nineteenth century and until the 1930s, partly due to its exports of corned beef and grain from the Pampas to Britain and other countries. (Refrigerant gases are contentious: some used in cars are flammable, and others are far worse than carbon dioxide as greenhouse gases. Fortunately, alternative refrigerants are being developed.)

In the North American Midwest, pioneer farm families enthused over prairie loam three feet deep, produced by generations of buffalo dung over the millennia following the glacial retreats. They claimed grain yields that dwarfed those of New England. But not all areas were so deeply fertile. In the territory that was to become the state of Wisconsin, early settlers produced good crops. The law of diminishing returns set in, and yields fell as soil depletion and wind- and water-borne erosion afflicted the area.

Enter W. D. Hoard, a newspaperman from New York State who diagnosed Wisconsin's plight. Hoard found that the glaciers had left just a thin layer of topsoil that, after soil depletion by pioneer farmers, needed replenishment. His background in New York suggested that pasture dairying, in rotation with grain crops, could rebuild soil and add to it. Evidence that W. D. Hoard's prescription worked, according to biographer Loren H. Osman (1985), is that his later career included stints as a governor, 1889–91, and education reformer. Wisconsin became the nation's leading dairy state with many small family-scale dairy farms, until California superseded it with larger megadairies in the late twentieth century. In the journal *Rural Sociology*, Jess Gilbert and Raymond Akor (1988) titled their notable article "Increasing Structural Divergence in U.S. Dairying: California and Wisconsin since 1950." The trend could not be stopped.

Today, Wisconsin also boasts large megadairies amid its traditional mom-and-pop farms. Hoard's demonstration farm remains a test bed for research published in *Hoard's Dairyman*, the eponymously titled magazine, established 1885, that remains a leader in discussions of animal productivity, health, longevity, and farm management—all relevant to this book.

Biodynamics and Organics

W. D. Hoard's leadership in Wisconsin reasserted the principle that healthy agriculture is based on healthy, fragrant soil. Likewise in ancient Rome, Pliny the Elder's *Natural History* (c. 79 AD, or nearly 2000 BP) observed that one test of fertile soil was its sweet aroma after turning with a plow. The lawyer, orator, and farmer Cicero went one better, declaring the taste of good earth to be better than the spice saffron. Contemporary U.S. organic principles are more formally scientific than Pliny's and Cicero's subjective observations, but that is not to say the ancient commentators did not, figuratively, plow a straight furrow.

Rudolf Steiner (1924a), the father of biodynamic farming, addressed the problems of degraded soils, crops, and livestock in lectures on "the spiritual foundations for a renewal of agriculture" in Europe till his death in 1925. He developed farming systems championing biodiversity over monoculture. He also developed nine homeopathic preparations to transform degraded land into healthy soil, often linked with Germanic philosopher Wolfgang von Goethe's (1749–1832) ideas on soil and including practices such as anointing crops under the full moon. Steiner's influences include the ancient Greeks and Romans, as well as Goethe. Goethe's saying "Love does not dominate; it cultivates" is the ethos of biodynamic farming, as opposed to chemical monoculture. Central to biodynamics, and the organics that grew from it, was the concept that each farm should be a self-contained system in tune with the cosmos.

The U.S. Environmental Protection Agency terms farms more prosaically than Steiner as "nutrient management systems" and prohibits them from point pollution of waterways. Conventional farms must not allow chemical fertilizers, such as nitrates, to leach into streams, and the same goes for manure. Organic farms, dependent upon manures and cover crops for fertility, have extra incentive to prevent the pollution of water because it is profitable to pinpoint manure on crops. ("That's the smell of money," explained one old-timer to this author.)

While shunning industrial chemicals, biodynamic farming venerates homeopathic preparations described by Steiner that may be stored in earthen jars and spread on new crops from an animal horn by the light of the moon. As amusing as this may sound to those of a reductive scientific bent, some present-day biodynamic oenologists may also join in chuckling about it— while describing just such rites of spring that are followed in their California vineyards, perhaps in a mixture of belief, tradition, and whimsical notions that there are more things in heaven and earth than can logically be accounted for. The thought also remains that there could be unknown biological worth in such preparations.

Steiner's ideas persist in the twenty-first century, influencing a sustainability approach called "permaculture." In northeast England, in Durham City and at the nearby Abundant Earth cooperative smallholding, permaculture teacher Wilf Richards (2014) instructs people on how to design sustainable systems, such as vegetable gardens and livestock farms. Richards' depiction of permaculture as a secular version of biodynamics, without the rituals, might describe the approach of some other biodynamic practitioners, perhaps including those on the California vineyard above. In a personal communication in 2013, Richards explained to this author:

> Permaculture, unlike Biodynamics, doesn't normally involve working with stars or lunar cycles or working with special preparations such as cow horns filled with special muck and herbs. Permaculture is more about ecological design of any system, not necessarily farming based, whereas Biodynamics is focused on particular agriculture methods. They are not incompatible and there are several cases where they work together quite closely, and they certainly share ethics such as caring for the earth and people.

Going back several decades, Steiner's ideas were absorbed and spread by Lady Eve Balfour and Sir Albert Howard before and after the Second World War (Reed 2001; 2006). They criticized new chemical practices for destroying soil, while extolling the teeming life within soil that was venerated by classical observers, such as Pliny. Among the adherents of Balfour and Howard was a Welsh farm family who managed to avoid the British government's chemical fertilizer directives in the Second World War and maintained their farm's wartime production with judicious use of green vegetable and animal manures.

After the war, Dinah Williams, daughter of academic agriculturalists at the University of Aberystwyth, appeared in public with Lady Eve Balfour, asking farmers not to join the chemical fertilizer and pesticide bandwagon. These efforts led to Williams's family farm's status as Britain's first certified-organic dairy farm and the establishment of the influential Soil Association. This small farm birthed a small family firm: Dinah's daughter Rachel began Rachel's Organic Dairy on the kitchen table, as recounted to this author by her husband Gareth Rowland at a Colloquium of Organic Researchers at Aberystwyth (2002). After years of successful sales in the United Kingdom with supermarket chain Sainsbury, Colorado-based Horizon Organic Dairy bought Rachel's in 1999. Horizon was bought by Texas-based, $10-billion transnational corporation Dean Foods in 2004. Rachel's iterations provide an instructive thread in the annals of global organics. It was an open question as to whether Rachel's brand marketed in America would stimulate the growing organic movement or lose its way.

One definition of *organicist* is someone who believes some systems, for example, social groups and even the universe (in Plato's philosophy) function like biological organisms. This understanding is akin to Steiner's views on the interdependency of soil and animal and human health. Before the Second World War, European and British organicists' ideas drifted across the Atlantic to J. I. Rodale, who founded the Soil and Health Foundation, forerunner to the Rodale Institute, in 1947.

The world wars altered farming in America as dramatically as they had in Britain. By some accounts, the traditional spreading of manure, the planting of cover crops, and multicropping continued in America through the First World War, 1914–18. By other accounts, fertilizer inputs, especially nitrates, were diverted from U.S. farms to munitions production after 1914, and lower yields revealed poorer soil fertility and tilth in much of the once fertile plains—a portent of Dust Bowl days to come. At any rate, the ramp-up of chemical production for munitions in the Great War shifted, after the 1918 Armistice, to production of petroleum-based fertilizers, pesticides, and herbicides, as well as synthetic ingredients, such as preservatives and additives, for industrialized farming in the interwar period. Such soil amendments spread from the United States to Britain.

The chemical trend was less intense in France than in the United States and Britain, whose more laissez-faire attitudes to food and farming have been mutually reinforcing since at least 1846, when Britain's grain tariffs (i.e., the Corn Laws) were repealed, in favor of free trade in food. Curiously, the trend to oil-derived fertilizers was less in Germany than in the United States. This was ironic because the father of chemical farming was arguably Count Justus Liebig, whose essay "Chemistry in Its Application to Agriculture and Physiology" had in 1840 reduced the mysteries of plant growth to mechanistic N-P-K formulae. In his defense, Liebig cautioned that soil fertility was actually not so simplistic and reportedly acknowledged that microorganisms and trace elements play important roles in soil. But once chemical inputs were graphed next to their yields on paper, the association tended to eliminate further analysis or—increasingly passé—philosophical inquiry.

Organicists believe the bacteria, fungi, protozoa, nematodes, and other organisms in soil have their counterparts in the human body. Michael Pollan observes in his popular book *Cooked* (2013: 322–23) that the "microbiota . . . bacteria, fungi, archaea, viruses and protozoa" inhabiting people are far more numerous than previously thought. They have been identified as much more symbiotic than in the days when the work of Louis Pasteur was interpreted to treat microorganisms largely in terms of disease.

In mid-twentieth-century America, the organic ideas of Balfour and Howard were marginalized, and respect for microflora was in short supply. But, beginning in the 1940s, Rodale publications, such as *Organic Gardening* (formerly *Organic Farming and Gardening*), gained traction. Although partisans for conventional industrial agriculture sought to marginalize organicists as kooks, many of their ideas were later validated scientifically. Mainstream society chuckled when organics were espoused by beatniks, hippies, or foodies. But after Rachel Carson pointed out the negative effects of agrichemicals on birds and the food chain in *Silent Spring* (1962), organics won more respect. Although Rodale publications were out of the mainstream, Rachel Carson's television appearances on Johnny Carson's top-rated *Tonight Show* increased recognition and credibility for it too.

There were other voices in the wilderness. Kentucky writer, farmer, and activist Wendell Berry (1972, 1986) absorbed the European ideas of Steiner, Balfour, and Howard, combined them with local knowledge and warnings on chemical inputs from American Cassandra, Rachel Carson, and persuaded generations of farmers and consumers of the long-term wisdom of organic agriculture.

Alar Prompts Organic Politics

With every food scare, mainstream shoppers have given organics more consideration. This phenomenon was epitomized by the Alar scare of 1990 when the chemical Daminozide, which conventional fruit growers sprayed on apples as a ripening retardant, was identified as a possible carcinogen. The CBS network show *60 Minutes* reported on the Alar scare in February 1989, and the ensuing panic depressed the apple industry in Washington State, where apples were a major export to the Pacific Rim and beyond (Scholten 1990). This incited a firestorm of popular concern that such unknown risks were a regular part of U.S. conventional apple production.

Even the USDA National Organic Program has had an issue discussed sub verbo, at least until recently, somewhat like an eccentric aunt flitting behind the upper-story curtains of a family mansion. This is the decades-long exemption allowing antibiotics in some apple and pear orchards against fire blight, referred to in the previous chapter, and further discussed in following chapters. But back in the early 1990s, the spotlight was on the Alar scare, which brought a short- to medium-term fall in exports of conventional apples to key markets, such as Japan. The crisis may also have accelerated a shift by conventional growers to new apple types, dubbed Fuji and Jazz, more piquant than the Washington Red Delicious apple, an industry mainstay since the 1920s, but whose beauty came to be surpassed by its cardboard-like taste.

Most German, British, American, and kindred permutations of biodynamic or organic farming are portrayed by their adherents in terms of biodiversity and living soil. This is fields apart from the inert chemical processes of contemporary industrial agriculture dubbed "green concrete" because huge acreages of GMO monocultures (e.g., GM cotton, rapeseed, or maize) are treated with so much glyphosate that little other flora or fauna complicate the land. But organicists maintain the interconnectedness of living food systems and that biodiversity is healthier than monoculture.

Michael Pollan builds the case against monoculture in his long-running critique of the politics of the U.S. organic movement. In 2001, Pollan published an article in the *New York Times Magazine* titled "Behind the Organic-Industrial Complex," which ignited national debate on agribusiness appropriation of the grassroots organics movement. In that year it was unusual for this author to find a serious organizer in alternative food networks unaware of the piece, and several kept it in their desks.

In a boost to the burgeoning "eat local," or locavore, movement, Pollan argued in a 2003 article, "Getting beyond Organic" in *Orion* online magazine, that the choice for consumers was no longer *conventional or organic* but *local or organic*. Pollan (2003; also Scholten 2011: 129) claimed the original organic dream shunned paradoxes such as the organic factory farm and organic TV dinners because the movement rested on three non-industrial sustainable legs: "(1) harmony with nature . . . treating animals humanely . . . [without] chemical pesticides; (2) [prioritizing] food co-ops, farmers' markets, and community supported agriculture (CSA) . . . and (3) belief [in] biodiversity."

In his earlier 2001 article Pollan used case studies of Cascadian Organic Farms based in Washington, and Horizon Organic Dairy headquartered in Colorado, as examples of organic-industrial corporations (see also Scholten 2010b). In the nicest possible way, Pollan had questioned Cascadian cofounder Gene Kahn about abandoning two of the three legs that Pollan says supported the original organic dream (*Orion* 2003). In his 2006 book, *Omnivore's Dilemma*, Pollan returns to conversations with Kahn and describes Kahn's openness on how organic philosophy is "morphing" into the way the world is. This meant clinging to leg #1 (i.e., farming without synthetic, chemical inputs) but largely abandoning the cooperatives and CSAs of leg #2 in favor of existing conventional food systems and significantly abandoning the biodiversity of leg #3, insofar as global logistics require foods capable of appearing fresh (whatever their taste and nutritional status) after an average 1,500-mile journey from farm to plate (Pirog et al. 2001).

Like W. D. Hoard, who went west from New York to Wisconsin to make his mark, Gene Kahn left a postgraduate program to travel west to Bellingham, Washington, a hotbed of organics for several decades. Kahn and his cofounders began farming a few acres in a remote and stunningly beautiful part of the Upper Skagit Valley in Washington State's North Cascade Mountains in 1972. Early success was accelerated by sales at Seattle's Pike Place Farmers' Market (familiar to viewers of the TV sit-com *Frazier* in which effete sibling psychologists Frasier and Niles shop for organic foods, oblivious to 20-pound fresh salmon lobbed over their heads by fishmongers). Cascadian Farms' success was likely sealed by supply contracts with Puget Consumers Co-op (PCC), which had its roots in 1950s buyers' clubs and continues to influence organic politics. By the turn of the century, PCC was a multifaceted cooperative. One of its initiatives is farmland preservation, which it promotes with farmer-members, including some on the Olympic Peninsula near Sequim. The PCC Farmland Fund changed its name to the PCC Farmland Trust about 2005, acquiring Sunfield Farm, an 83-acre former dairy farm near Port Townsend on the peninsula. Sunfield's organic roots were rhizomes, stretching underground to Britain and Europe, with its philosophic orientation to biodynamics. A farm manager was sought for community supported agriculture (CSA), market gardening, animal husbandry expertise, and general community outreach skills to preserve the historic farm's place in its setting.

Back in the 1970s, as Michael Pollan (2001, 2006) tells it, Gene Kahn's successes with PCC were followed by financial reverses in a market trough. Kahn had to sell his idealistic organic operation to a corporate giant. Kahn accepts the description forthrightly, like many a business survivor resilient enough to keep momentum after failure. He explains that, after overinvesting in the business amid the Alar apple scare (when organic prospects looked limitless), cash flow problems forced him to sell Cascadian Farms to Pillsbury. Eventually, the company joined General Mills, with Kahn as a vice president. Some brushed him off as an ex-hippie with a General Mills logo embroidered on his denim shirt. But an organic pioneer we can call Betty (pseudonym; Scholten 2011: 121), whose smallholding has produced organics for decades and who has known Kahn since the 1970s, told this author she still exchanged waves with Kahn when her old pickup met his Lexus on Skagit County roads.

Pollan did more than identify anomalies in organic horticulture. He ground the meaning of the term "organic-industrial" into the lexicon of many who once assumed every dairy label portraying a pretty cow munching clover in a meadow accurately represented the daily lives of all cows on USDA certified-organic farms. The reality, according to Pollan (2001, 2003), could be closer to battery hens, with cows packed into feedlots known as confined-animal feeding operations (CAFOs). Organic CAFO conditions, wrote Pollan, were little different from conventional confined operations, although the use of antibiotics was prohibited.

It is true that the nominally organic cattle on organic CAFOs were spared the drug-and-hormone-induced synchronization of heat (estrus) that enables artificial insemination (AI) technicians to breed cows en masse. But industrial-organic cattle experienced a life less bucolic than on Old MacDonald's farm. Pollan pointed out that, like conventional cows, such cattle spent much or most of their time on concrete slabs or feedlots and grazed on pasture just a couple months of the year when they weren't lactating. Contact with manure (or slurry) by cows kept in CAFO conditions increased contact with pathogens and led to maladies, such as foot inflammation. There is a fine line between feeding rich grain to maximize milk productivity on one hand and risking cow health on the other. Like their conventional counterparts, the more that organic cows' diets are made up of high-energy grains, such as corn, the more acidic their systems are and the more vulnerable they are to dermatitis, slurry heel, and other infections. Small organic farmers who maximized cattle health resented conventional or organic megadairies that prioritized profit. Feedlot conditions were suspected to be standard on some of the largest suppliers of "own-brand" milk to large grocery stores and giant discounting hypermarkets. Michael Pollan (2001) described Horizon Organic Dairy's main organically certified operation in Idaho as essentially an eight-thousand-head feedlot operation split into halves of four thousand organic cows and four thousand conventional cows.

Consensus was that cows on farms owned by Organic Valley cooperative farmer members or on the Wisconsin-based cooperative's less well-known megadairies in the western United States generally experienced traditional pasture grazing conditions, which were more conducive to expressing cows' natural behaviors than on dour feedlots. The next chapter will elaborate these points.

Bovines have evolved grazing on grass and other natural plants as forage. But since the 1960s, many confined cattle have been fed what are called "totally mixed rations" (TMR) for an optimum combination of roughage, energy, minerals, protein, and vitamins. These are sourced from a variety of ingredients, which may contain fodder, such as local silage, hay, or alfalfa, but are distinguished by the addition of agricultural fodder and commodities, such as barley, corn/maize, cottonseed, millet, oats, rice, rye, sorghum, triticale, wheat, and some ingredients the consumer would not expect to be fed to cows, such as concentrates, proteins, and vitamins from industrialized food commodities. These also include minerals, such as limestone, which may be needed to buffer high-energy components, such as corn, that acidify cows' stomachs, according to the American Grassfed Association (2011). The best TMR managers purchase commodities as carefully as a nutrition-conscious mother on a tight budget, but the pressure to maximize yield is incessant.

A senior veterinary scientist, Dr. Sambraus (Scholten 1990a) from Bavaria, maintains that the ideal herd size is no more than 75 cattle, for their own socialization and well-being as hierarchical animals. Therefore, grouping cows in groups of 50–75, according to their stage of months in lactation, is a sensible practice. Blending feeds together can ensure that cattle get an optimal mix for their stage of lactation and can also reduce costs for farmers. Experts at Penn State (2005: 3) note that "feeds such as urea, limestone, bicarbonate, fats, and by-pass protein sources like blood and fish meal can be added to TMRs in reasonable amounts without significant reduction in feed consumption."

However, Michael Pollan (2006, 2013) zeroes in on the chief ingredient responsible for a paradigm shift in U.S. dairying that has swept the world: corn, or maize, is the high-energy, high-carbohydrate commodity that propels conventional or organic cows' production skyward. But bovines evolved eating grass, not the corn that often "burns them out" (the phrase used by feeding specialists as more cows develop lameness, reproductive problems, or other maladies that consign them to a slaughterhouse before their time).

The result of these practices is that an unexpectedly large number of organic-industrial cows have been so stressed by feedlot conditions and poor diets that health problems consigned them to lameness or early deaths—just like their conventional sisters. There are rumors that some major USDA certified-organic farms burned cows out so fast that they could not even produce enough calves to replace the mothers. Consumers privy to these charges were appalled. Many consumers buy organic milk not just for the health of their children or themselves, but also in the altruistic belief that organic milk is produced in ethical conditions better for the welfare of animals, farmers, and the environment than is often the case in what they spurn as "factory farming." They want more than value for money; they demand humane values for money (Scholten 2010b). This was the origin of the milk boycott of the 2000s at the center of the next chapter.

Small and midsize organic dairy farmers believed that corporate mega-dairies, along with complicit organic certifiers, were manipulating USDA National Organic Program (NOP) rules to save feeding costs. The Cornucopia Institute, a family-farmer-advocacy think tank in Wisconsin, charged that corporate megadairies were "gaming the system" against them (NOFA 2006). In some instances, certifiers interpreted rules to allow 20 percent of the diet of cattle transitioning from conventional to organic herds to be cheaper conventional feed. So, around 2006, farmers in the Northeast Organic Dairy Producers Association (along with Western, Midwest, Food Farmer, and other groups) began asking the government for a strict "origin of livestock" rule to end such practices, which are interlinked with what standards USDA-accredited certifying agents used to manage whole herd conversion. At issue is how long cows may be fed on cheaper conventional feed before gaining organic status (AMS 2006; AMS 2014). Hopes that an origin of livestock rule would be clarified sooner rather than later languished during the Bush era of the 2000s.

Conventionalization of Organics

One landmark outcome prompted by the 1989 Alar scare was congressional passage of the Organic Food Production Act (OFPA) in 1990. For a historical moment, it seemed organic pioneers had gotten what they'd wished for: a legally protected niche for humanely produced foods grown on healthy soil and traded in farmers' markets and short food chains. What Colin Sage (2003) called "alternative food networks" would bolster local economic communities and nutrition, and would be policed to remove imposters and free riders. In retrospect, such a dream seems slightly naïve. Today's reality is that retail behemoth Wal-Mart sells more organics that its competitors in America, and supermarket chain Carrefour does the same in Europe. And there were some organic pioneers, such as Gene Kahn, who might turn the question around and ask idealists, "Do you want the biggest retailers *not* to sell organics?" The implication is that if Wal-Mart did not sell organics, then less of the world's farmland would be certified organic, and fewer people would consume organics.

But by the early 2000s, many actors in alternative food networks (Sage 2003) began to ask if the organic label masked a disturbing reality. Ironically, Michael Pollan's claim (2003)—echoed by many smallholders—is that when the USDA (2002b) published federal standards, small farmers lost control of organics. By 2004 smallholders suspected they were losing control of what was already a $10-billion annual market to agribusiness, whose organic-industrial products might meet the letter of regulations in the National Organic Program, but violated their spirit. It was especially galling to organic pioneers when agribusiness confined cows, instead of letting them graze grass like their grandmothers.

It was not simply a major loss of livelihood for organic smallholders to lose a $10-billion niche (€8.3 billion euros; see Sahota 2004: 21). (Note: the U.S. organic market in 2012 was as high as $31 billion.) Such a loss was more wounding if the market they pioneered was captured by agribusiness, which they saw as an industrialized free rider on their organic dream.

Michael Pollan (2003) deplores a dichotomy between the social goals of pioneers and the profit goals of the organic industry that bodes ill for the sustainability of pioneers. Julie Guthman, whose 2004 book *Agrarian Dreams* subtly follows the evolution of the influential organic sector in the state of California, shows how quickly conventional inputs and monoculture replaced the biodiverse hippie dream of the 1960s. She found that the production of some organic fruits and vegetables was so lucrative that profit-led growers abandoned crop rotations and outsourced humus and other ingredients to maintain monocultures. They were also driven by commercial land pressure in the Golden State, which was becoming one of the world's top seven economies. California does have lower taxes for land zoned as agricultural under schemes such as the Williamson Act (Guthman 2004: 85–86, 184), which explains why some suburban homeowners are taxed as "farmers" amid their apple and orange trees. Thus, it is not so much taxation that motivated the geographical relocation of dairy farms from the suburbs, such as Arcadia and Bellflower, but commerce, which worked against the "kinder and gentler" alternative food systems envisioned in the counterculture. Farmers in Washington and other states observed that once Californian farmland was sold for commercial use, it was taxed at "highest and best use" rates and seldom regained its former status. They dug in their heels to maintain their own status quo of zoning and taxation.

Guthman (2004: 166, 169–71, 173; see also Pollan 2003) discusses the challenging concept *beyond organic* and claims that some hybrid practices may actually be more holistically sustainable than industrialized "organic" systems dependent on organic inputs sourced from far away. The phrase *beyond organic* has spread among organic factions, especially those committed to social justice via local networks, as disgruntlement with aspects of the USDA program have surfaced (Howard and Allen 2006).

Far north of California in Washington State, Pollan found that if the cofounder of Cascadian Farms, Gene Kahn, had ever shared the hippie dream, it had morphed into the capitalist realities of global food systems far from the short food chains of locavores. When Pollan questioned the ethics of producing an organic Twinkie (the epitome of long-shelf-life processed snacks), Kahn forthrightly defended the concept—and organic TV dinners to boot—because they mean more organic acres on the planet than conventional fare. This is a hard argument to trump. But Pollan finds it ultimately unconvincing and notes that Kahn has largely given up farming in Washington's Cascade Mountains to become a manager of contract farmers in California and other states, for Cascadian organics and Small Planet Foods (along with Muir Glen tomatoes) for General Mills, which bought the business from Pillsbury. Cascadian Organic Farm has gone virtual.

Mining the Earth

Pollan's views resonate with those of Tim Lang, professor of food policy at City University London's Centre for Food Policy. Lang in Britain, like Pollan in America, is one of those familiar talking heads, summoned by radio or television broadcasters to discuss every local or global food scare or farming controversy. He earned a doctorate in social psychology at Leeds University, and, when he became a hill farmer in the 1970s, this unusual career step was in tune with the back-to-the-land movement of that decade and embedded him in debates on the environment, health, and social justice. Lang is keenly aware of a nuanced question at the heart of debates between Jeffersonian and Hamiltonian views of democracy—that is, whether people should be seen as citizens or consumers (Bonanno 2000). It is safe to say Pollan, Lang, and family-scale dairy farmers generally rate the role of citizen over consumer. In 2004 Lang and coauthor Michael Heasman published the book *Food Wars: The Global Battle for Mouths, Minds and Markets*. The authors blame the productionist paradigm of postwar Europe's Common Agricultural Policy (CAP) and America's commodity subsidy regime for cotton, corn, soy, and wheat for degrading soils, waterways, the atmosphere, and human health. Lang and Heasman (2004: 34–46) claim this productionist paradigm is wilting and may be replaced by a life-sciences paradigm responding to individual health needs *or* by an ecologically integrated paradigm more in tune with the perspectives of organic agriculture. It is far from clear which paradigm will come to dominate.

In the hungry years after the Second World War, Americans and Europeans assumed human health would automatically proceed from a rich, varied diet. But the proliferation of cardiovascular diseases, diabetes, and obesity brought disillusion with the postwar productionist paradigm by the 1990s.

The life-sciences paradigm is in tune with biotech seed companies, such as Monsanto, Syngenta, Bayer, and Pioneer; this paradigm is just an expansion of the third agricultural revolution (and the Industrial Revolution) as more technologies, such as genetic modified organisms (GMOs), are invented and commercialized. It is also in sync with new health and drug services responding to an individual's unique DNA, offered by health management organizations (HMOs) and Big Pharma (i.e., the agropharmaceutical and chemical industry). This is in the spirit of neoliberalism, ushered in by the administrations of British prime minister Margaret Thatcher and American president Ronald Reagan and carried on by their successors.

Advocates of the ecologically integrated paradigm criticize the life-sciences approach as trying to cure disease when it is wiser to prevent it. Their view of health integrates healthy soil with healthy people—like Steiner, Balfour, and Howard before them, Lang and Heasman, as well as Pollan, favor the ecologically integrated paradigm and are joined by U.S. organic dairy farmers and millions of like-minded people around the world who see the need to supplant the chemically based productionist paradigm with methods linking planet, people, animals, and soil in more benign, biologically linked patterns (Basu and Scholten 2012).

How did we get to the point where the "Dead Zone" in the Gulf of Mexico, the unintended consequence of agricultural runoff in the U.S. Midwest, kills fish in an area larger than some states? Chesapeake Bay fisheries suffer continued pollution from surrounding farms and industry. Across the world, the intricate coral structures of Australia's Great Barrier Reef were already damaged by farm runoff (Drummond and Marsden 1999) and now face dredging and the dumping of millions of tons of sludge for new coal ports. Back on the U.S. West Coast, Puget Sound fisheries face the same threat from coal sludge if coal ports are established there. If biodiversity is one foundation of ecological sustainability, there are disturbing signs, such as bee colony collapse, that industrialization is overwhelming the biosphere. For much of the preceding century, farming was tied to natural ecologies of symbiotic processes, such as biodiverse multicrop rotations and the application of animal wastes to renew soil vigor. In practical terms this meant dairy farms were marked by pluriactivity, relating to what Germans call "multifunctionality," according to Geoff Wilson (2007; see Stolze et al. 2000 and Tuomisto et al. 2012 for comparisons of yield and nitrate leaching in crops), with extensive pasture dairying, mixed with sheep and pig rearing, amid rotational cereals, fiber, and timber production. Properly managed, this was a closed loop that regenerated land as production rotated from one field to another.

In the Green Revolution of the mid-twentieth century, agriculture became an industry that mined the earth for petroleum products to fertilize, protect, and power production. The chemical industry turned from the use of nitrates for munitions in the Second World War to fertilizer. Nerve agents were converted to inputs, such as pesticides. Petroleum-powered tractors displaced oxen and horses, improving labor productivity but increasing greenhouse gases, making farming less environmentally sustainable.

Lang and Heasman (2004: 237; Pirog et al. 2001: 1) cite the *Food, Fuel, and Freeways* study at Iowa State University (2013). Rich Pirog and his coauthors found that conventional food arriving in Chicago from within the continental United States in 1999 averaged 1,518 miles, a 22 percent increase on 1981 when it travelled 1,245 miles. Another finding builds the case for nonconventional, alternative food networks. Farmers supplied institutions, such as conference centers, hospitals, and restaurants, in three Iowa local-food projects, resulting in dramatically reduced food miles: "The food traveled an average of 44.6 miles to reach its destination, compared with an estimated 1,546 miles if these food items had arrived from conventional national sources." Consider also, that average food miles generally increase as the shift continues from local to globalized food systems.

The twentieth century turned agriculture on its head. While slash-and-burn farming has been practiced in various parts of the world in combination with fallow farming (and still is, in parts of the Global South), more sustainable rotations with alternate or cover crops make settled farming sustainable in the long-term. Important lessons were learned on the American frontier, where thousands of square miles of grain thrived above rich prairie loam, a meter deep, from generations of buffalo roaming on prairie grass. Lang and Heasman (2004: 235) note conventional agriculture's ominous demonstration of the law of diminishing returns as the "energy mix" of U.S. corn/maize farming changed over the years 1945–85: "Labor was reduced 5-fold while energy from machinery increased by a factor of 2.5 . . .Energy input from fertilizers and irrigation increased 15- and 18-fold respectively." Disturbingly, Lang and Heasman claim that "the yield/ratio actually declined from 3.4 to 2.9" over these four decades.

Nowadays, critics of petroleum-based agriculture question monoculture as an unsustainable system mining the earth of finite resources including aquifers in an age of Peak Oil (the theory that half of the Earth's extractable petroleum has already been pumped; see Pollan 2006: 45 and Kunstler 2007). Skeptics of the Peak Oil theory take comfort in oil and gas recently sourced via fracking, but that still leaves the climate-linked problems of carbon-related greenhouse gas (GHG) emissions.

Pertinent here is the reliance of dairy confinement systems—conventional confined-animal feeding operations (CAFOs; EPA 2012) or organic feedlot operations following zero-grazing models—upon petroleum to carry fodder and outsourced rations to cows' mangers, and then burning petroleum to carry or pump cattle wastes from feedlots to fields—when, as a Washington State University extension agent remarked to me, cows actually perform those functions quite well themselves (Scholten 2010a,b; 2011). From 1950 to 2000, the average yield of U.S. dairy cows more than doubled, from 8,000 lb./3,636 kg per year to 20,000 lb./9,090 kg. (In 2014, some cows are in the stratospheric 30,000 lb./13,000 kg. range.) But each hundredweight (cwt) of milk yield required greater capital investment and mechanization, which farmers tried to amortize by the economies of scale of larger herds fed in the feedlots of even larger CAFOs. Energy questions are vital to the long-term sustainability of megadairies.

As the result of production stress, many confined cows do burn out young, say dairy workers, who organize their globally sourced totally mixed rations. Such feeding results in significantly more hoof, stomach, and breeding problems among cows in confined, concrete-floored megadairies than pastured cows grazing on grass, according to the *Journal of Dairy Science* (Haskell et al. 2006; 2007). Cows on such megadairies may look brisk, with clean coats, due to washing (sometimes with car wash brushes that activate on contact). Nonexpert visitors may perceive a generally young appearance. Even without the added stress of the GMO hormone rBGH/rBST (discussed in chapter 1), production stress means that few cows survive for very long. Their lives are in confined herds with cull rates around 40–50 percent (DHI-Provo 2013), Instead of living seven years to a decade or more, many are slaughtered before they are four years old. Cows have potential to live 20 years or more, but poor welfare has been identified as the cause for 30 to 40 percent involuntary culling rates in Canada and worse in Turkey and the new industrialized dairy farms in China according to the Government of Canada (DeLaval Longevity Conference Proceedings 2013: 3–4). Culling and longevity will be further discussed below and in succeeding chapters.

Organic Snapshot

The USDA's 2008 Organic Production Survey was the first comprehensive government study of its kind. The results lent gravitas to the new sector, showing organic milk had broken the psychological 1 percent barrier to represent 1.5 percent of total milk volume. Organic dairying brought attention from those who observed its annual double-digit growth figures. Based on a 2007 Census of Agriculture, the 2008 Organic Production Survey listed two categories of "organic" dairy farms: nearly all accredited by state or private certifying agencies and a few dozen exempt from certification because they produced less than $5,000 in annual sales. Altogether the USDA reported 2,065 certified and exempt organic dairy herds in 2008. Organic cows totaled as many as 219,031 in 2008 and fell about 8 percent, to 201,960 cows, by December 31, 2008.

At *Dairy Business*, an information service "supporting, and targeting, larger herd producers across the country and beyond," writer Dave Natzke (*Dairy Business* c. 2009) reported, "Organic dairies represented 3.5% of U.S. herds, 2.1% of cows and 1.5% of milk volume in 2008." The 2,012 organic dairies that reported selling milk during 2008 "produced 2.757 billion lbs. of milk, with a gross value of $750.15 million. At that value, organic milk produced in 2008 carried an average value of $27.21/cwt. (27.57 million hundredweights divided by total value of $750.14 million)."

According to Natzke the organic picture isn't completely clear because the report withholds cow and production data from several states to avoid disclosing information about individual operations. The point is sensitive: some vertically integrated corporate megadairies are so large that a few farms can skew statistics for an entire state. It connotes a monopoly scenario by one corporation big enough to affect prices by itself, and perhaps affect legislation too. Depending on who is buying and selling, such arrangements can be described as monopolies, oligopolies, or oligopsonies. More simply put, when processors and traders have great market power, it may be in their interests to influence the rules making, and the monitoring and enforcement of organic regulations—to the detriment of smaller producers.

One example of potential pressure by processors is addressed by the Farmers Legal Action Group in Minnesota. In her article "Confidentiality Clauses in Organic Milk Contracts" (FLAG 2008a: 3), Jill Krueger reviews contracts for sale of milk with Dairy Marketing Services (DMS), Horizon Organic, and HP Hood, explaining that, although such contracts are not necessarily better or worse than open ones, "this means that the producer agrees not to share the specific contract terms with parties such as family members, lawyers, accountants, lenders, other producers, another processor, or farmer organizations without first seeking approval from the processor." It is thin gruel when a solitary farmer, unable to reveal the farm gate price even to family members, faces drought and rising grain costs (*Seattle Post-Intelligencer* 2007). More reassuringly, Krueger also notes confidentiality clauses in some farm contracts are unenforceable in some states, including Arkansas, Illinois, Iowa, and Minnesota. In a companion article, "When Your Processor Requires More Than Organic Certification" (FLAG 2008b), Krueger notes that HP Hood and Organic Valley cooperative contracts had "detailed requirements regarding access to pasture and the amount of forage in a lactating cow's diet." This would be a relief to pasture advocates. Concern about corporate power by farm activists, such as The Cornucopia Institute, remains strong. Grazing and replacement practices on newly built megadairies, such as those owned by Horizon Organic Dairy farms in New Mexico, may be in accord with the final Pasture Rule, but many consumers would like to know more details. NOP deputy director Miles McEvoy reportedly visited them before a National Organic Standards Board (NOSB) meeting in April 2013, but some attendees worried that he was closemouthed about what he saw.

Geographic Shifts

What the USDA 2008 report made clear was that traditional dairy states with temperate climates had many farms selling organic milk, but their volumes were being exceeded by megadairies in western prairies and irrigated deserts. Wisconsin had 479 organic dairies (i.e., dairy farms) selling milk in 2008, followed by New York with 316 (many owned by Amish); Pennsylvania, 225; and Vermont, 179. By the end of 2008, California counted the most organic cows, with 35,333, followed by Wisconsin, 25,916; Texas, 18,854; New York, 17,431; and Oregon, 16,290. According to Washington Tilth Producers (2012), the

Washington Dairy Products Commission notes that consumer demand for organic milk is relatively high in the state compared with the rest of the nation. There were 46 organic dairies in Washington State in 2008 when organic milk represented about 5 percent of all milk sales in Spokane (ranked eleventh in the United States) and 10.5 percent of all milk sales in Seattle (ranked second). Nicknamed the "Emerald City," Seattle's high organic demand reflects its green environmental credentials and active participation in the milk boycott that marked what this author dubbed the USDA Organic "Pasture War" of the 2000s (Scholten 2007c; Spokane 2010b). However, because the politics of land ownership and governance are dynamic over time, it can also be accurate to refer to various pasture *wars*, akin to power struggles over landgrabs.

Wisconsin has been a premier state for milk production, but the dramatic growth of California into the world's seventh-biggest economy (along with Dutch immigrant intensive dairy technology) had long toppled it by 2008, with 501.8 million pounds, followed by Wisconsin at a still doughty 329.0 million pounds; Texas, 284.2 million pounds; and Oregon, 261.1 million pounds (USDA 2008). Average annual prices for organic milk were: California, $26.60/hundredweight; Wisconsin, $25.87/cwt.; Texas, $28.47/cwt.; New York, $28.46/cwt.; Vermont, $28.21/cwt.; and Oregon, $26.43/cwt.

The USDA (2008) report noted that 1,623 herds sold 39,922 organic cattle during the year, with a total value of $33.47 million. Crucial to the future economic sustainability of any farm are cattle replacements. Thus, if herds are unable to raise enough heifers to replace their milk cows, it may be a sign of poor management or too much stress on cows to produce milk.

Cow Longevity Conference 2013

Leading dairy equipment manufacturer DeLaval supported a Cow Longevity Conference in Tumba, Sweden in 2013, which clarified current knowledge and best practices in dairy welfare. From the University of British Columbia Dairy Education and Research Centre, J. Rushen and A.M. de Passillé (DeLaval 2013: 3–4; Government of Canada 2012) explained "The importance of improving cow longevity" because it is directly related to welfare practices: "While cattle have the potential to live 20 years or longer, on most modern dairy farms few dairy cattle will live longer than 6 years." Rushen and de Passillé stated that dairy farm culling rates in Canada average 30 to 40 percent, and that most culling on intensive farms in North America and Europe is "involuntary," largely due to "reproductive problems, mastitis or poor udder health, lameness and problems with feet and legs, and other forms of illness or injury" (disease such as turberculosis can also hike culling rates). In new intensive dairies in China, Rushen and A.M. de Passillé (DeLaval 2013: 3–4) see "even higher percentages of involuntary culling due to . . . reproductive problems, udder problems and feet and leg problems." Other types of illness or disease were more prevalent in Turkey and Mexico. Rushen and de Passillé laud farms with good longevity practices, which can inform "clear, industry-led animal welfare standards, and greater use of benchmarking of farm performance" to improve longevity.

While USDA organic certified cows often experience better welfare than their conventional sisters, both systems can cause stress by diet or other conditions. A frequent stress trigger is consumption of too much high-energy (i.e., high-carbohydrate) corn, leaving cows lame, with acidic systems unable to achieve natural estrus, conceive or bring healthy calves to term. Cows that cannot conceive face slaughter. Culling can exceed 50 percent in confined herds.

The Cornucopia Institute, which advocates for small organic farmers based in Cornucopia, Wisconsin, points out that one of the major loopholes in USDA National Organic Program (NOP) rules involves replacements and how rules are monitored by USDA-accredited certifying agents, which have varied regionally. The issue is even more complex when it comes to whole herd conversion. At the time of writing, organic dairies were allowed to replace culled stock with pregnant heifers from conventional (i.e., nonorganic) herds as long as the addition spends the final third of her calf's gestation consuming organic forage and fodder. Heifers from conventional farms are cheaper to buy because they are raised on cheaper grain than heifers from the smaller national organic pool. Thus, organic purists believe present cow replacement rules comprise a moral hazard to animal welfare—a disincentive for organic megadairies to refrain from practices that impact organic cows. Purists demand that all organic herd replacements be raised in organic conditions.

The USDA 2008 Organic Production Survey had further insights on milk and crop production. Organic dairy farms are usually major growers of organic crops for their own use. Thus, about 30 percent of the organic acreage in Washington State is represented by pasture, forage, and fodder for organic cattle. For a decade, rising food commodity prices—boosted by demand in China and India, by the U.S. biofuel program—along with transport cost increases (linked to oil embargoes), drove up the cost of outsourced grain on dairy farms in the United States,. In the mid-2000s this author predicted rescaling to larger dairy farms for more on-farm grain production, and that seems to be the case (Scholten 2010a).

The price of corn doubled in the United States, within years of President George W. Bush's 2005 launch of the biofuel program and mandated use of ethanol in people's cars (Orden, Blandford, and Josling 2010). *Hoard's Dairyman* (2007a) reported that ethanol was changing the dairy industry as corn and soybean prices soared. Goals included greater national energy self-sufficiency and, not coincidentally, strengthening farmer support for the Republican Party. (Note: from a personal communication from a Libertarian source with the pseudonym "Urs": "The biofuel program is suspected to be less Pres. G.W. Bush's farm policy toward U.S. energy self-sufficiency than a plot by plutocrats to control the economy.") Increased farm size is evidence that the trend to bigger farms, which grow their own forage (grass, silage, corn, alfalfa, etc.), is underway, as farmers seek to guarantee their economic sustainability against commodities price spikes. Followers of global debates on doctrines of food sovereignty versus food security in developing countries might acknowledge that post-2008 commodity volatility is persuading many rich-country farmers to opt toward microsovereignty.

Overall, U.S. dairying is a mature industry, with the drinking of conventionally produced liquid milk decreasing among children and adults, partly due to competition from the soft drinks industry. Yet, the consumption of organically produced liquid milk was rising in 2013. Of course the conventional dairy business seeks ways to sustain dairy product consumption and credits the expansion of pizza consumption since the 1970s as one savior of the industry. More recently, a bright spot is rising yogurt consumption, especially "Greek" recipes.

But the most dynamic sector of American agriculture for decades has been organics in particular. Why? Because milk is a perennial gateway to organic consumption, even by people not previously concerned with the environment, animal welfare, or health food. The Hartman Group (1997; 2004; 2006), consultants in Bellevue, Washington (near Seattle), find that when new parents (especially highly educated ones) wean babies from human breast milk, they often determine that the safest substitute for their progeny is organic bovine milk. This helps explain the relatively stable sales of organic milk sales in the economic debacle marked by the fall of the Wall Street firm Lehman Brothers, during the presidential election involving John McCain and Barack Obama in 2008. Total organic sales in the United States (and the United Kingdom, whose banking crises already had made headlines in 2007) took hits of around 7 percent, unprecedented in a market that had been booming in double digits for years. Yet, the mainstays of organic sales remained dairy products, such as liquid milk and infant milk replacer, for parents insisting on milk unadulterated by antibiotics, chemicals, and genetically modified organisms (GMOs).

The impelling draw of organic milk for new parents who take this dairy gateway leads to their providing organic fruits and vegetables to their fast-growing children. At the same time, this demographic is largely populated by people in their early careers, with low seniority and little disposable income. Low cost explains why discount supermarket Wal-Mart (branded as Asda in the United Kingdom and Walmex in Mexico) is the largest U.S. retailer of organics. Such discounters source milk from megadairy farms and processors, such as Aurora Organic Dairy, whose economies of scale achieve low wholesale prices. Critics claim an obsession with cost cutting led (especially before the USDA 2010 Pasture Rule) to more confinement of cows, three times or more daily milking, animal stress, and a general cutting of corners on conditions for animals, the environment, and communities by weakening the competitiveness of smaller family-scale organic farms (those averaging herds of 65 milking cows instead of thousands).

UK-U.S. Studies

Concern about cattle confinement is not restricted to the United States. In Britain, animal welfarists and environmentalists scorn plans for an eight-thousand-cow confinement megadairy repeatedly proposed near Lincoln. On a much smaller scale, on the English side of the Scottish border, Newcastle University splits its 725-acre Nafferton Farm into organic and nonorganic halves to test their respective merits, with funding from the UK Department for Environment, Food and Rural Affairs (DEFRA), the European Union (EU), and others. Led by Gillian Butler and Carlo Leifert (Butler et al. 2008), the Nafferton Ecological Farming Group team made global headlines, showing milk from pastured cows has significantly more nutritious conjugated linoleic acid (CLA) and omega-3 fatty acids, and lower levels of omega-6 fatty acid, than milk from cows on conventional cows eating totally mixed rations (TMR). This elicited political hostility in the United Kingdom, where policies of the Conservative, Labour, and Liberal Democrat Parties generally extol genetic modification (GM) while paying lip service to organics and the labeling of GMO foods. The UK Food Standards Administration (FSA) claimed the nutritional advantages of grass-fed organic milk were insignificant—while overlooking the lack of chemical residues, environmental benefits, and the often better welfare standards of food produced organically.

That was added motivation for Newcastle professors Butler and Leifert to replicate the UK nutrition study in a U.S.-wide 18-month study with Chuck Benbrook and Donald Davis of the Organic Center and Washington State University (WSU), and Donald Davis of CROPP/Organic Valley. They published new findings of significant organic benefits in a major paper in *PLOS One* (Benbrook et al. 2013). Although CROPP and Organic Valley helped fund the study, they had no role in its design or analysis, which was supported by the Center for Sustaining Agriculture and Natural Resources at WSU. In this first large-scale, nationwide study of fatty acids in U.S. organic and conventional milk, their findings corroborated those of the UK study. Benbrook and his coauthors (2013: 1) reported:

> Averaged over 12 months, organic milk contained 25 percent less ω-6 [omega-6] fatty acids and 62 percent more ω-3 fatty acids than conventional milk, yielding a 2.5-fold higher ω-6/ω-3 ratio in conventional compared to organic milk (5.77 vs. 2.28). All individual ω-3 fatty acid concentrations were higher in organic milk—α-linolenic acid (by 60 percent), eicosapentaenoic acid (32 percent), and docosapentaenoic acid (19 percent)—as was the concentration of conjugated linoleic acid (18 percent).

Nutrition is a serious consideration, along with productivity and all-around sustainability, because Earth's population is projected to grow from its present 7 billion people to 9.5 billion people by 2050, entailing 50 percent more food production to satisfy rich consumers. Professor Jules Pretty, a member of the Royal Society at Essex University, proposes "sustainable intensification" to achieve these targets without further deforestation (Basu and Scholten 2012). This approach relies on complex biological understandings, which underscores the need for less laboratory-based research and more land-based research, as long advocated by organicists (including Steiner, Balfour, and

Rodale), rather than the productivist chemical paradigm of the twentieth century.

Conclusion

This chapter condensed thousands of years of animal-human relations, from hunting and gathering in prehistory, through a series of three agricultural revolutions (including different forms of dairy, or White Revolutions, and a litany of political economic paradigms), to the present dawn of the Gene Revolution. The struggle of people to ensure their survival amid unruly nature was explained in terms of technology and the amusing but serious observation by Bill Gates and others that human culture's greatest innovation may have been the invention of hay.

Simple, sweet-smelling hay—dried grasses and forage—allowed farmers to maintain more cattle as capital through the winter, rather than having to slaughter them as forage diminished, and diffused civilization far from the equator. The same is true of the more recent innovation of grass or corn silage, which preserves more of the original plant matter's nutrient value than hay does, through an anaerobic storage method akin to pickling (Short, Watkins, and Martin 2007). But the central point is that technology enabled agricultural surpluses, which helped liberate people to develop culture, as individuals and communities rose on a scale from poverty to prosperity toward what psychologist Abraham Maslow called the status of "self-actualization," a state that Aristotle implied was the natural realization of human potential.

But technology can be a two-edged sword. As the last chapter showed with the advent of the Erie Canal, changing transport technologies can also threaten farmers' livelihoods. Jared Diamond seconds this notion in his 2005 book, *Collapse*, which begins with brooding over the stone wall remains of a 165-cow barn, abandoned over five hundred years ago. It was a veritable megadairy of its day, on Gardar Farm, belonging to the Norse bishop of southwest Greenland. Geographical constraints of climate, disadvantageous distance from markets, and dietary inflexibility (the Norse apparently refused to eat fish) doomed the barn, the farm, and the socioeconomic sustainability of Norse culture there.

We cannot ignore the possibility that, even if some dairy practices are commercially sustainable, there may be ethical reasons to resist them.

The next chapter focuses on the first decade of the new millennium, when consumers began to suspect some of the most prominent cartons in dairy cases were cynical simulacra of what they ought to be. Conflict and distrust over organic governance did not involve the barbed wire and guns of the range wars between cowboys and sheep herders in the Old West of the late 1800s. But after consumer, corporate, and farmers' cooperative groups got lawyered up, for years of boycotts and bitter litigation that went all the way to the Supreme Court, the proceedings can aptly be called the USDA organic Pasture War of the 2000s.

A Note on Food Regimes

Atkins and Bowler (2001) detail food regimes as a chronology of political economic configurations. These should not be confused with the (largely technological) agricultural revolutions that they overlapped and embedded. Scholten offers this synopsis in respect to dairying (forthcoming):

> **The First Food Regime**, pre-1914, was marked by repeal of Britain's Corn Laws in 1846, removing tariffs on imported grain. Free trade was resisted by Britain's landed gentry, but welcomed by new manufacturing classes keen to moderate wage demands with cheap wheat and meat from abroad. Although Britain retained lively trade with its empire, the repercussions of Corn Law repeal were global, spurring exports from North and South America, and boosting an unprecedented period of globalization until it was constrained by the First World War. Trade contracted 1918–46, in post-war recession, beggar-they-neighbour economics, and unabated Depression from 1929 until the Second World War.
>
> **The Second Food Regime**, 1947–1970s, was framed by the Cold War, as the US and its Western allies pursued a General Agreement on Tariffs and Trade, but largely ignored farm trade (GATT excluded dairy and sugar which were subsidised in many countries). Western European governments supported farmers with price supports that achieved continental food self-sufficiency about 1960—and problematic grain and dairy surpluses in the 1970s (e.g. Butter Mountains and Milk Lakes). Dan Morgan describes how the USA used PL480 food aid shipments in surplus disposal from the 1950s (Morgan 1979; Scholten 2010, 2011). Communist governments in the Soviet Union relied on collective farms to supply mass markets, but often resorted to barter rather than cash payments. Non-aligned Third World countries picked a policy mix from East and West after independence.
>
> **The Third Food Regime** is dated by Atkins & Bowler (2001: 29) as 1980s–present, beginning with the neoliberal eras of Ronald Reagan in the US and Margaret Thatcher in the UK. These transatlantic soul mates touted market solutions to stagflation spawned by oil crises, US abandonment of the Bretton Woods system in 1971, and the expense of farm supports. Richard LeHeron (1993) describes this Third Regime as consisting of increased global trading of food, consolidation of capital in food manufacturing, government certification of new biotechnology and genetically-modified organisms (GMOs), consumer fragmentation and dietary change, amid declines in farm supports and regulation. Storing Europe's surplus commodities was costly: 70 percent of the EEC budget was allotted to farming and milk support was half of that. But subsidy removal was politically difficult in the face of farmers' lobbies. Finally, on April 1, 1984 milk quotas were imposed to curtail European production (Scholten Jan. 10, 1989; July 1990d. Note: EU milk quotas are due to end in 2015.). In the Uruguay Round Agricultural Agreement (URAA 1994) and establishment of the World Trade Organization (WTO 1995), sugar and dairy were brought under liberalized world trade rules for the first time. This supranational development did anything but reduce pressure on small U.S. farmers. The next chapter relates how attempts by small farmers to carry on pasture dairying in this harsh global environment met competition by corporate players in the USDA National Organic Program.

3

USDA Organic Pasture War

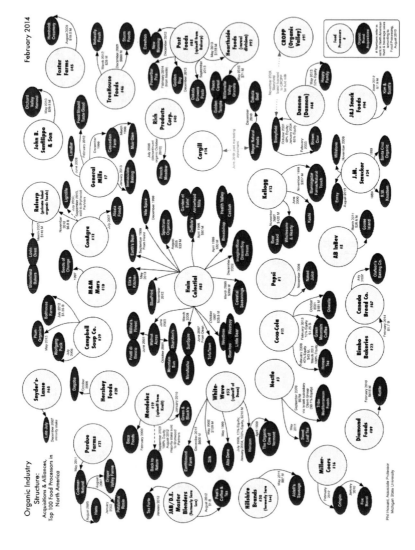

Figure 3.1   Graphics by Prof. Phil Howard (MSU) show corporate ownership in U.S. organics.
Photo Credit: Phil Howard.

USDA Organic Pasture War: Where Have All the Cow Herds Gone?

Today about 80 percent of conventional U.S. dairy cows and many in the United Kingdom, Canada, and other countries have been consigned, with few if any breaks on pasture, to concentrated or confined-animal feeding operations. These CAFOs maximize profit for intensive farmers, who claim their use enables them to deliver affordable food while making efficient use of resources. Critics call CAFOs factory farms for agribusiness and cite the hidden environmental, health, and social costs that bankrupt family farms and deny animals their natural behaviors.

This is the view of the Pew Charitable Trusts, an independent research organization with about $5 billion in assets established by the original owners and scions of the Sun Oil Company. Those who made the Pew fortune in petroleum were politically conservative; for instance they distrusted President Franklin D. Roosevelt's federal-level New Deal policies to combat the Great Depression (1929–39), such as parity between rural and urban incomes, as inimical to small businesses. Recently, the Pew Trusts (2012a: 1) have turned a critical eye on consolidation in livestock agriculture, charging, "Over the past 50 years, the United States has lost more than a million farms, yet more animals than ever are being raised, slaughtered, and processed." Many of today's CAFOs are vertically integrated into the meat industry, which is dominated by a handful of corporations. Pew Trusts (ibid: 1) note these firms are "far larger than those of years ago" and describe the role that the ensuing "lack of competition has played in squeezing out small and midsize farms and ranches and in changing the nature of animal agriculture across the country." Consolidation has soared since 1980. Pew (2012a: 2–3) finds: "The percentage of the market held by the four largest corporations has risen steadily since 1980 for beef, chicken and pork." In 1980 the corporate share of beef was 36 percent; in 1990 it was 72 percent; in 2000 it rose to 81 percent, and by 2010 it was at 85 percent—dominating the market. The consolidation was so dramatic that Secretary of Agriculture Tom Vilsack and Attorney General Eric Holder held hearings on the issue in 2010, but the agribusiness juggernaut carried on. Although beef production has become somewhat segregated from dairying, this corporate consolidation of beef supply financially affects small dairy farmers when they are forced to cull or sell cows, which often enter the meat supply. With so few processors buying beef from so many farmers, the situation is that of oligopsony—that is, a type of group monopoly—with the consequent dangers of squeezed farm incomes and even increasing costs to consumers.

Karen Steuer, director of the Pew Environment Group (2012b) campaign to reform animal agriculture, notes that confined operations are prone to use antibiotics intensively in order to preempt the spread of animal disease and to accelerate animal growth, and they are the main source of nitrate pollution in rivers, according to the U.S. Geological Survey. From the negative consequences of unnecessary antibiotic use and water pollution from huge facilities, it follows that corporate agribusiness squeezes consumers, bankrupts small farmers, risks widespread antibiotic resistance, and pollutes the environment. Agribusinesses are slated by some as accepting welfare. Although retail food prices are lower in the United States than in many other countries, few taxpayers realize how much they subsidize major the commodities from which corporations reap more than family farmers (see Leonard 2014).

In popular understanding, the epitome of CAFOs is a battery chicken facility where multiple birds are confined in cages too small to stretch their wings, where their beaks are seared and blunted to prevent injury to cage mates. CAFO opponents mock them for confining cows like battery chickens. Proponents of large-scale farms respond that contemporary megadairies should not be confused with practices in poultry production. They claim that if their cows do not actually graze on pasture, they do have access to exercise yards. They seek to disarm critics by saying bovine health is prioritized because cow comfort is essential to productivity—that happy cows are more profitable. They may claim that there were occasional lapses in animal welfare on smaller farms in the past. Perhaps cows on muddy paths in northern states envied their cousins in sunny southern feedlots. Empirically, it is difficult to measure cow happiness, but one parameter of cow welfare is longevity, and it generally appears shorter in large confinement operations.

Not many decades ago, practices were different on most family dairy farms. Few things in life are eternal. But a farm kid in the 1950s may have expected that the after-school chore of herding cows from pasture into the barn for evening milking was (if not in bleakest midwinter when cows were in the barn) a chore to be done every day.

It was common for farm children, as soon as they could walk, to accompany parents to inspect the condition of pastures and fences. Kids followed their folks, ducking under electric fences, climbing over old-fashioned woven-wire fences, gaining trust in the docility and good nature of milk cows grazing on grass. Parents kept eyes on their children but worried little for their safety because cows are curious but sociable. They enjoy humans, whom they'll approach and make no moves more threatening than polite nuzzles, seeking carrots or other hidden fare in pockets of raggedy barn clothes.

Normally friendly, cows can be dangerous in exceptional circumstances. Decades ago, when one of the most prominent dairy women in northwest Washington State was killed by one of her favorite cows in a maternity pen, news of her death shocked farmers across the nationwide farming community. But bulls, fraught with testosterone, are powerful and require close control at all times, usually with nose rings and chains; this is true even of friendly bulls who know their keepers. One practice of experienced cow hands is to approach within bovines' field of vision (as prey animals, cows have evolved panoramic vision).

In *Behavioral Principles of Livestock Handling* (1989), renowned animal handler Temple Grandin notes (citing Prince 1977): "Cattle and pigs have a visual field in excess of 300 degrees." Grandin also cautions that shouting stresses animals, thus, "people need to keep their mouths closed except for a gentle 'sshh' or talking softly to their cattle." Cattle are more averse to high-frequency noise than humans, but they can adapt to continuous white noise (such as vacuum pumps in milking barns) and talk and music radio stations.

When approaching cattle in herds from behind and during milking or veterinary procedures, it is wise for herders to broadcast their presence. This can be done by calling familiar phrases, such as a favorite of Dutch Americans, "*Ka-boss!*" Being kicked by 1,200-pound mammals is a weighty matter. If one survives, a bruise is a silent reminder to let cows know one has entered their personal space (whistling or calling her name will suffice).

On family-sized farms where cows graze, a year or two after beginning elementary school, seven- or eight-year-old boys or girls might be assigned to get cows in for the evening milking. This is usually an easy chore because mature cows are conditioned to expect tasty scoops of grain during milking. Nor is it difficult to rouse new mothers, those heifer cows who have just birthed their first calves. These hierarchical creatures soon learn to follow older animals, often led by a dominant cow, identifiable as queen of the herd, to the milking barn.

In 1790 when farmers composed about 90 percent of the U.S. labor force, most children had the opportunity to learn about livestock handling. Material contact with domestic creatures enriched their knowledge of the similarities and differences of humans and nonhumans. The demographics of 1990 showed farmers represented only 2.6 percent of the total population, and urban dwellers no longer had any contact with animals besides pet cats and dogs, since automobiles had long before replaced horses.

Today's U.S. farm population has further shrunk. Few children or adults besides professional farmers have contact with farm animals. Even for children in rural areas, after school there's no need to get the cows home because they are already in a feedlot next to the milking barn. So many more cows are enclosed than even in 1990 that far fewer animals are even *seen* on rural farmscapes by people in passing cars. Yes, there are fewer fence posts. Fields appear more ordered. But they are less pastoral and instead sterile, like factories.

Family-scale farms remain the best setting for kids and extended families to experience livestock on a daily basis. Below we explore consumer and small-farmer actions to preserve the U.S. National Organic Program as one arena where traditional pasture grazing is legally mandated, as a boon to animals and the farmers who care for them.

CAFOs

While observers refer to confined-animal feeding operations variously as "concentrated" or "confined" animal feeding operations, confinement is a primary aspect, connoting the absence of pasture. The U.S. Environmental Protection Agency (EPA 2012) monitors CAFOs because, depending on the number of animals involved and the surrounding environment, they are points of pollution in waste and wastewater. Because manure and wastewater can pollute groundwater and streams by breeches of waste lagoons and storage structures—in accidents, rainstorms, or the unseasonal spreading of wastes on land adjacent to creeks and rivers—they are regulated under the National Pollutant Discharge Elimination System (NPDES).

Testifying for the Congressional Committee on Agriculture (House of Representatives 1999) J. Charles Fox, the assistant administrator for the EPA, observed that the Clean Water Act had done a good job of cleaning up human waste in the nation's rivers, which 27 years previously amounted to open sewers. But by the end of the century, noted Fox, one hundred times more animal than human waste was produced annually in the country, and the Clean Water Act needed to evolve to address pollution from large animal concentrations. About one billion tons of animal wastes were produced annually in the country. The EPA proposed that areas where CAFOs were dense required nutrient management plans and maximum effluent caps to control manure pollution.

Over a decade later, Karen Steuer, leader of the Pew Environment Group's (2012b) efforts to reform animal agriculture, said, "Some of our nation's most prized waters are at risk from CAFO pollution, but the current permitting process is rife with loopholes." She spoke at a conference in Maryland, where citizens worry about enclosed Chesapeake Bay, where saltwater pollution patterns compare with the inland saltwater of Puget Sound in Washington State. Streuer cited EPA estimates that 43 percent of Americans have suffered some level of pathogen contamination linked with livestock operations. Pinpointing manure as the biggest source of nitrogen pollution in rivers, Streuer urged the EPA to bring CAFOs under the aegis of the Clean Water Act, passed in 1972 during the Nixon administration and which, after some neglect in the Reagan-Bush years, began getting more attention during the Clinton-Gore administrations.

CAFOs are defined by the EPA (2012) as lots or other facilities where "animals have been, are, or will be stabled or confined and fed or maintained for a total of 45 days or more in any 12-month period," and in which "crops, vegetation, forage growth, or post-harvest residues are not sustained in the normal growing season over any portion of the lot or facility." Excluded from these definitions are facilities for aquaculture (referred to in the context of the Blue Revolution in chapter 2), such as farmed salmon, shrimp, prawns, and so on, which are also controversial for the pharmaceuticals needed to maintain them in confinement.

While cattle herds of perhaps thousands of cattle on the open range figure in American history, contemporary concentrations of animals are unprecedented, as noted by the Pew Charitable Trust (2012a). A rule of thumb is that the danger of disease transmission increases inversely to the proximity and number of animals. CAFOs are vectors of disease, and this risk has only been surmounted—albeit in the short term—by the use of antibiotics. As antibiotics (like GMOs) are strictly regulated under USDA National Organic Program (NOP) rules, this topic will be elaborated on in following chapters in discussions on animal welfare.

According to the EPA, CAFOs substitute structures, such as corrals, feedlots, and barns, for land and labor and for more extensive conditions in forests or fields. They substitute equipment and temperature controls for feeding, housing, and manure management. CAFOs for mature dairy cattle are categorized as large (700 or more), medium (200–699), or small (fewer than 200). CAFOs with cow-calf pairs may have up to 50 percent animals: large (1,000 or more), medium (300–999), or small (fewer than 300). The largest facilities, those confining more than 1,000 cows, are worrying. Academic papers claim that feedlots, which got their start in the U.S. Midwest, so polluted groundwater in some places that stricter regulations were needed to inhibit pollution. The industry response has been to move feedlots north to Canada, to communities so eager for employment that they ignore the environmental cost.

McDonaldization

By terms of the North American Free Trade Agreement (NAFTA), which came into force in January 1994, there is little if any constraint placed on American retailers and fast-food chains on distributing meat south and north across the borders. At the University of Maryland, George Ritzer critiqued such processes in the modern restaurant industry (which has come to provide about 40 percent of meals in the United States) in his 2007 book, *The McDonaldization of Society.*

Ritzer extended his McDonaldization thesis to other parts of modern life, such as credit cards, throwing light on America's rampant consumerism, and bright, throwaway Styrofoam and paper packaging, which successfully markets billions of fast-food units. The secret of the success of food some might consider unaccountably bland is suggested in a master's thesis in theology submitted by Daniel Martin at Durham University, in the United Kingdom in 2006, when Ritzer gave a lecture to a packed audience. Ritzer (1993:7–13; also Martin 2006) describes the process of McDonaldization as one by which fast-food restaurants influence not only competitors, such as Pizza Hut, but also the procedures of supermarkets, financial services, and university education. The principles of this process are: (1) efficiency, (2) calculability, (3) predictability, and (4) control of human beings via drive-through service and even uncomfortable seats. In other words, time saved by paying a known amount of money for edible burgers and milk shakes compensates consumers for less than ideal quality, while maximizing corporate profits.

The formula works. McDonald's (2013) "Company Profile" boasts it is the "leading global foodservice retailer with more than 34,000 local restaurants serving approximately 69 million people in 118 countries each day. More than 80 percent of McDonald's restaurants worldwide are owned and operated by independent local men and women."

The chain is popular among youth in France and even Italy, despite these countries being hotbeds of the environmentally conscious "slow food" movement. McDonaldization is far from the original organic dream described by Michael Pollan (2003). McDonald's principles, prioritizing time-saving methods over bespoke quality, are contrary to the "back to nature" turn to quality described by Jonathan Murdoch and Mara Miele in their influential 1999 paper in *Sociologia Ruralis*, noting the growth of organic food, farmers' markets, and box schemes.

Backing up Ritzer's assertions of poor health effects from fast-food culture is Marion Nestle. A professor in food studies and public health at New York University, her profile rose as editor of the *Surgeon General's Report on Nutrition and Health* in 1988. Her observations on the high sugar content of breakfast cereals and the amounts of fats and salt in fast foods, such as chicken nuggets, which children find so manageable, tally with Ritzer's McDonaldization thesis. Nestle (2002: 178–79) notes that McDonald's advertising and websites target toddlers, with play facilities, and children aged 8–13, with clever TV commercials. She adds, "The amount of money spent on marketing directed to children and parents rose from $6.9 billion in 1992 to $12.7 billion in 1997." Nestle (2002: 79, 80, 190) is perhaps best known for her battles with food industry lobbyists over the Food Guide Pyramid, a visual system to guide consumption. Nestle said the conventional dairy industry was averse to distinguishing between high- and low-fat labels on pyramids and exaggerated the need for dairy foods in the average diet. The dairy industry took some comfort in Nestle's (2002: 78) critical stance on trans-saturated fats in plant-based margarine, which had whittled away on sales of dairy butter.

The U.S. beef industry is the largest in the world, and the government explains that cattle feeding is most concentrated in the Great Plains, but is also important in parts of the Corn Belt, Southwest, and Pacific Northwest (USDA-ERS 2012a). The regimen for beef cattle is to be fed grain and concentrates for about 140 days, and for most cattle the average weight gain is 2.5–4 pounds per day on about six pounds of dry-weight feed per pound of gain. In the 1960s there was public enthusiasm for a wave of grain-fed beef advertised as tastier than grass-fed beef. Before long, however, health experts soon bemoaned the marbled fat in beef cuts they blamed for higher rates of cardiovascular diseases (CVDs).

The trend continued in the beef industry with ripple effects on dairy. Today the U.S. beef industry is largely separate from the dairy industry, but it has influenced the massive rise of confinement in dairy operations. Beef feedlots with 1,000 head or more of capacity comprise less than 5 percent of total feedlots but market 80–90 percent of fed cattle. Size matters: feedlots with 32,000 head or more *beefies* comprise about 40 percent of fed cattle. Such operations and their attendant slaughterhouses are documented in books almost a century apart, in Upton Sinclair's exposure of the meat industry in *The Jungle* (1906), and Eric Schlosser's exploration of Colorado slaughterhouses in *Fast Food Nation* (2001).

Upscaling

The conventional dairy sector has scaled up dramatically. The USDA (2000) map "Change in Animal Units for Confined Milk Cows: From 1982 to 1997" showed that increases in the number per county, or combined counties, of confined dairy farm herds increased in the Southwest region by over one thousand units in Northern California (as dairying moved away from Los Angeles) and the state's interior, with similar gains in Arizona, New Mexico, Nevada, and parts of Utah and Colorado.

This author's home area of Whatcom County, bordering Canada, and Skagit County, to its south in Washington State—both with a century's tradition of pasture dairying—experienced confinement trends at nearly the same rate as California. East of the Cascade Mountains, an explosion of confinement dairying occurred around Sunnyside and other towns in eastern Washington, irrigated by the Columbia River. One stimulus to dairy growth in eastern Washington was that cities in western Washington had begun to discourage large animal concentrations by imposing stricter water quality regulations. This was partly in response to urban voters in the Seattle and Tacoma metropolitan areas, who looked askance at confinement and perhaps held their noses at large groups of cows. But tougher water regulations were also demanded by Native American groups, who blamed farm runoff for fecal contamination of ancient clam beds. The resulting "red tide" alerts closed them to harvest almost every year.

Southwest Kansas, northwest Florida, central Florida and northwest New York State also saw the growth of confined dairies. But consolidation in some areas weakened dairying in others: the USDA (2000) map shows that in the same period, 1982–97, many counties in traditional dairy states suffered losses of dairy units, most pronounced in southwest Washington, Oregon, coastal California population centers, Minnesota, Wisconsin, Michigan, New York and New England, Florida, and south Texas.

CAFO cynics might say the trend to confinement of cows parallels the trend in criminology that resulted in *One in 100* (Pew 2008), concerning U.S. citizens confined in prison. Michel Foucault (in *Discipline and* Punish, 1975) might observe, were he alive today, that in America, as in Enlightenment-era

France, the sovereign's aim is control of natural or subjective forces inimical to an ordered body politic. Be that as it may, the normalization of grazing ruminants to factory conditions merits skeptical inquiry.

This author's article for the British publication *Dairy Farmer* in 1997 noted that U.S. herd size jumped from an average of 50 to 90 in 20 years (Scholten 1997). Just 15 years later in 2012, the USDA found the average herd size had again almost doubled to 187 (*Hoard's Dairyman* 2013a). The pervasiveness of CAFOs in a very rapid increase in average U.S. herd size in dairy structures is clear.

In Washington State's Whatcom County, a few conventional farms still graze their cows and ship to the region's conventional Darigold farmers' cooperative, according to Henry Bierlink, executive director at Whatcom Farm Friends, an information bureau. Nowadays a handful of farms ship to the Organic Valley cooperative, active in the county just a decade. Sooner or later there may be more organic cows grazing Whatcom pastures than the conventional Darigold cows that for decades dominated its green landscape. Certainly the bucolic images and appeals to nature used by dairy product manufacturers to market products in the nineteenth and twentieth centuries contributed to consumer assumptions that cows should graze on pasture (DuPuis 2002; 2003).

A study by Phil Howard and Patricia Allen (2006: 439; also Howard 2005) based on focus groups and a sampling of one thousand households in the central California coastal region found wide support for animal welfare: "standards for the humane treatment of animals have the highest level of support, followed by a standard for local origin, and for a living wage for workers involved in producing food." A goal of many consumers and small-scale family farmers in the organic movement is to keep animals grazing on grass.

The material conditions of cows producing milk certified organic by the USDA but living in factory-like organic-industrial confinement with extremely limited access to token pasture, if any, is a public relations problem for some of the largest operations (Pollan 2001; Schlosser 2001; Fromartz 2006). The organic dairy pasture grazing debate affects how many animals are visible on farmscapes.

The summer 2006 issue of the Northeast Organic Farming Association newsletter, *The Natural Farmer* (NOFA 2006: 7–8), records massive annoyance among grassroots farmers who once assumed the mandate to graze dairy cows was implicit in the Organic Foods Production Act (OFPA 1990). Why was a sixth major (and expensive) meeting in five years needed to define *pasture* and *grazing*? asked attendees, many of whom had spring planting to attend to at home. The article in this newsletter by NOFA policy coordinator Steve Gilman notes how frustration spilled over at a National Campaign Organic Committee meeting in Pennsylvania, April 17–20, 2006. Former National Organic Standards Board head Jim Riddle (NOFA 2006: 8) suggested, "The NOP had turned around the question of 'how do we implement pasture regulations,' to one of 'what would happen if we regulate pasture?'" Michael Sligh, the organic committee chair, called attention to a report by the government's National Standards Institute that asserted that "the NOP has consistently been way out of line" with International Organization for Standardization (ISO) requirements.

By 2006 it was past time that organic pasture and replacement rules be clarified. Mark Kastel (NOFA 2006) of The Cornucopia Institute charged repeatedly that some of the largest corporate megadairies were "gaming the system" in order to garner market share. One firm had secured about one-third of the market, according to Kastel, who estimated it profited up to $1 million annually by selling all its organic calves soon after birth to avoid feeding these newborn bovines organic milk, which was more profitable to sell on the market. The firm bought milk cow replacements from conventional herds cheaply, as they had not been raised on expensive organic feed until shortly before they calved. (The scarcity and high price of organic feed made it a chronic bottleneck to the system. After years of drought, a *Hoard's Dairyman* [2014b] article described organic feed prices as so high that they made conventional farmers shudder.)

Back in 2006, Gilman's article (NOFA 2006: 7–8) noted that NOFA-Vermont called for a "simple, verifiable pasture standard that is fair to all." Lisa McCrory outlined a definition requiring a minimum 120 days on pasture constituting at least 35 percent of the cow's diet from grazing." McCrory explained this is easily doable even in northern Vermont with its short growing season and that certified organic dairies already kept total annual feed records for easy verification of compliance. In fact the Vermont rules originally called for 150 days minimum on pasture with 50 percent minimum dry matter intake (DMI). All northern dairy farmers this author contacted said these minimums were easily met, but the standard was watered down to 120 days/30 percent DMI, so the companies that owned the new organic-industrial feedlot megadairies built in arid areas would agree to it more easily.

Rise of Cornucopia and Organic Consumers Association

Pasture wars hastened the rise of The Cornucopia Institute, the populist advocacy group for family-scale organic farmers cofounded by Mark Kastel and Will Fantle in Cornucopia, Wisconsin, in 2004. Fantle has a background in environmental journalism and consultancy for public and private groups, and acts as Cornucopia's research director. Kastel's background is conventional agriculture. Kastel and Fantle began Cornucopia to counter new threats to small farmers' livelihoods. Their campaigns include what this book calls "pasture wars," as well as campaigns against synthetic inputs allowed in organic processing, such as carrageenan, soybeans processed with nontraditional chemicals (comparable to rocket fuel), checkoff schemes that benefit processors and retailers more than farmers, lack of GMO labeling of foods, definitions of so-called natural products, which undermine the product differentiation of NOP organics, and poor regulations on antibiotics, which affect animal welfare and longevity as well as human health.

The Cornucopia Institute numbered about ten thousand members in 2014. Kastel is its senior farm policy analyst and directs its Organic Integrity Project. He was an executive with agribusiness technology giants International Harvester and J. I. Case in the 1970s, before health crises triggered by chemical pesticides turned him toward an organic diet and activism. Kastel became a political consultant to the Farmers Union and other family-farm organizations, revealing information on use of GMO dairy hormones, such as its impact on bovine health (Kastel 1995). As hormones became national news, Kastel worked with Will Fantle in the front lines of the pasture wars. The two men cofounded the Cornucopia Institite which they call a watchdog organization for economic justice to the family-scale farming community. They are complementary in their dealings with other actors in the dairy sector, including corporate agribusiness and organizations like the National Organic Standards Board (NOSB) and the Organic Trade Association (OTA). Kastel is an aggressive advocate for small family farmers. Principles seem top priority to Kastel with his spirited Airedale's nose for conflict of interest in organic politics. Fantle has the reassuring presence of an intelligent Labrador, who calms discussion at critical moments. A few years ago this was illustrated by a memorable fracas on Odairy, an email discussion list counting over 750 subscribing members from the Organic Trade Association, National Organic Coalition, Accredited Certifiers Association, farm extension, academic research, and farmers' associations, including the Northeast Organic Dairy Producers Association (NODPA; *Hoard's Dairyman* 2014b), whose executive director, Ed Maltby, has moderated discussion since 2005. Organic pioneers and innovators do not always come from agricultural backgrounds. Maltby's father was an accountant in Brighton, England. Young Maltby worked in conventional farming in the United Kingdom before immigrating to the United States. He began working on New England dairies that practiced managed intensive rotational grazing (MIRG), which was excellent preparation for his post with NODPA.

Considering the uneven scales of the socioeconomic actors in organic food chains, it was probably inevitable that tensions would rise between actors who wanted the organic sector to grow without a hiccup and idealists, such as Kastel. He can be unrelenting when pursuing conflicts of interest, even among pioneers and longtime friends of the organic movement. Industry figures warned that washing dirty laundry in public might confuse consumers, prompting them to ask the dreaded question, "What does organic mean?" Kastel was not afraid to air a little laundry, in order to protect the integrity of the sector, according to the letter and spirit of the Organic Foods Production Act (OFPA 1990) and National Organic Program (NOP). One detractor, who claimed Kastel indulged in self-righteous grandstanding, nevertheless admitted: "I agree that no one individual did more to bring to public awareness the abuses in the organic confined animal CAFO system than Mark Kastel, 2004 to 2006. He certainly deserves recognition and respect for that."

In 2009 Kastel was suspended from the Odairy discussion list for what some members considered undiplomatic language and others called bullying.

A subgroup headed by a New York farmer reportedly set up a new email list to disseminate Kastel's views. The Cornucopia Institute immediately set up an alternative email list called Independent Odairy. The diplomatic Fantle also acted as liaison, relaying Kastel's comments to the NODPA-Odairy list run by Maltby. The Kastel and Fantle double act was effective, and this author had the opportunity to witness it, at the first ever meeting of the National Organic Standards Board and the National Organic Program in Seattle, April 26–29, 2011. Other agenda items included the use of antibiotics to control fruit tree diseases and how the board decides whether an input is natural or synthetic. (This author testified on dimensions for free stalls in loafing sheds, in the same slot as Cornucopia. My view, informed by family farming experience, was that there is a danger point between narrow stalls and those just wide enough to tempt cows to try turning around, when they may get stuck, panic, and injure themselves. See also *Hoard's Dairyman* 2013d.)

Two years later Kastel was unable to attend the spring 2013 National Organic Standards Board (NOSB) meetings held in Portland, Oregon, but Will Fantle attended, with Cornucopia Institute staff Charlotte Vallaeys (senior analyst for food safety and sustainability), and Pamela Coleman (a farm policy analyst and plant pathologist based in East Wenatchee, Washington, who has worked in both conventional and organic production). Coleman initially made Cornucopia's case at a daylong premeeting at Hotel Monaco on Monday, April 8, 2013. Liana Hoodes, executive director of the National Organic Coalition (NOC n.d.), organized pre-NOSB meetings before the semiannual rules events. NOC (2014) members are farmers and farm organizations; nonprofit consumer, environmental, and animal welfare organizations; and businesses dedicated to organic integrity, such as the Center for Food Safety, Food and Water Watch, the Midwest Organic and Sustainable Education Service (MOSES), NODPA dairy farmers, Ohio Ecological Food and Farm Association (OEFFA), Rural Advancement Foundation International (RAFI-USA), and Union of Concerned Scientists.

This author heard the room buzz with formal and informal discussion. It was no secret that emotive issues would be addressed in the coming days. Squabbles on organic certification of hydroponically grown foods would have to wait. More urgent was a rider attached to congressional legislation, dubbed the "Monsanto Protection Act," which would block local judges from imposing moratoria on GMO plantings.

Rumors of an unprecedented application of the Clean Water Act aimed at nitrate pollution threatened the time-honored practice of spreading organic manure. While this seems counterintuitive, composted properly, manure is pathogen-free and "sequestrates carbon" (Logsdon 2010: 43; *Hoard's Dairyman* 2014a). A USDA study found that pasture sequestrates 3,400 pounds more methane per acre annually compared to row crops (Benbrook 2012: 7; Perry 2011; *Hoard's Dairyman* 2014a). Observers conjectured that, if this use of the Water Act to prohibit manure spreading is not naïve, it could be a covert attack on animal agriculture by vegans. At any rate, the FDA appeared to quash the proposals in spring 2014, after protests.

Organic hops, just three hundred acres nationally, were about to enter the brewing sector.

Far more controversial than hops was an issue that could burst into wild-fire within the NOC. This was a checkoff advocated by the Organic Trade Organization (OTA), manufacturers and processors for organic milk. Similar to the conventional beef checkoff, organic milk producers would be obliged to check off a certain percentage of each sale for promotion of the product. While promotional advertising was generally good for sales, organic dairy farmers felt they bore an unfair portion of the burden when, as has happened, processors and traders source cheaper foreign commodities and then adver-tise them with *their* checkoff money.

Traders and processors had surely been allies of farmers when lobbying the USDA to establish regulations for the newly certified market. But the checkoff was bitterly resisted by The Cornucopia Institute and farmers who saw it as another tax when their meager milk checks barely afforded expensive feed in a drought. As mentioned, another reason they resented the prospect of a generic organic dairy checkoff was precedent. They feared some processors and retailers would cut corners with cheap "organic" imports of dubious qual-ity. They claimed that was what Dean/WhiteWave had done when it stopped buying U.S. organic soybeans and secured cheaper beans from sources such as China. There was also danger to America consumers. A decade before, Goldie Caughlan, a nutrition educator and spokesperson for Puget Consumer Co-op (PCC) in Seattle (until her retirement when she joined The Cornu-copia Institute Board of Directors), cited the labeling of product origin as a priority for food safety (personal communications 2002–03). Her prescience was validated when milk products from China were found adulterated with toxic melamine several years ago.

Over April 9–11, 2013, in Portland, it remained for the NOSB to decide on antibiotics in organic fruit trees. It was said that 68 percent of apples and pears, including the new Gala apple favored by consumers, were vulnerable to fire blight. But The Cornucopia Institute's Pamela Coleman said she had previously worked as an inspector in eastern Washington State, where 25 to 30 percent of organic growers have orchards certified to European standards. (The eastern Washington growers who participate in the European program have some of the orchards grown without antibiotics. However, they may also have some orchards not certified to European standards, and others only to organic standards or even conventional. In other words, the growers may have some parts of their operation growing apples without antibiotics and other parts of their operation growing apples with antibiotics.) Coleman sug-gested American growers could rise to the challenge with biological controls, blossom thinning, and planting trees further apart. Passionate arguments for antibiotics extensions were made by apple and pear growers who said they depended on streptomycin and tetracycline to fight the fearful onslaught of fire blight. Removing antibiotics from their toolkit, they claimed, might result in having to burn all apple or pear trees in orchards that had belonged to their families for generations. Their attachment to the orchards was palpable, akin to a dairy farmer's concern for cattle.

Members of the National Organic Coalitions and Consumers Union shared the position of Cornucopia's Will Fantle, Pamela Coleman, and Charlotte Vallaeys in maintaining that, after two decades of antibiotics extensions in orchards, they had to end (NOC 2014; Consumers Union 2014). Antibiotics for tree fruit were on the national list of exemptions to the rule that synthetic substances are subject to delisting in a sunset clause, unless they are relisted by a supermajority. This evoked grumbles about a so-called tyranny of the minority, but some pioneers claimed the 1990 Organic Food Production Act was predisposed against synthetics in all but temporary scenarios. For years synthetics were banned from the processing of items labeled "organic" and only permitted in the "made with organic" products.

Members of the Organic Trade Association and others strengthened consensus to prohibit further use of antibiotics in organic fruit, citing widespread public opinion that overuse of antibiotics increased resistance to bacterial infections in humans, such as methicillin-resistant *Staphylococcus aureus* (MRSA—the infection that has stalked hospital wards). Commercial doom was invoked. The integrity of the USDA organic program could be compromised by the ongoing use of antibiotics. And consumers who understood little difference between conventional and organic fruit could be seduced by meaningless advertising claims of "natural" production.

When the NOSB meeting proper convened in a Hilton Portland conference room on April 9, Tuesday morning, Miles McEvoy showed style as deputy administrator of the National Organic Program (NOP). Laughter is the best medicine for movements potentially riven by schism, so McEvoy began by showing a poignant video titled *Ordering the Chicken* (Portlandia 2011). In it, a politically correct couple relentlessly question a restaurant server on the ethical and humane sourcing of their entrée. They are pleased to hear the heritage breed chicken's name is "Colin," raised to USDA standards just 30 miles south of Portland. But they are nonplussed when the server cannot guarantee that the chicken had plenty of happy friends on the farm or that the farm's owner was not a nonresident on a yacht in Miami. Uncertain they can have a "relationship" with a farm based on so little information, the sensitive couple flees the restaurant.

The hundreds of delegates in all factions erupted in laughter, understanding the frustration of farmers trying to sustain their own livelihoods while satisfying the myriad, sometimes too precious whims of urban consumers. Then McEvoy launched his "NOP Report: Sound and Sensible" (April 9, 2013), praising the first ten years of the NOP for ten years of USDA organics, with 85 accredited certifying agents, and 25,000 certified operations across the nation amounting to $31 billion in U.S. organic sales (up from a third of that in 2003). The NOP had performed tens of thousands of inspections, reviews, and certification decisions. This was remarkable, considering the program employed just one staffer for each billion dollars of sales, but there were serious risks of staff burnout. Fiscal year 2013 saw a 5.1 percent budget reduction from fiscal year 2012.

McEvoy added that there was still no funding for National Organic Certification Cost Share. Small farmers using organic methods occasionally balk at the high cost of USDA NOP certification, complaining that they should be subsidized by the deep pockets of organic-industrial corporations, which they claim skew markets against them. Such smallholders explain they go "beyond organic" by working in local alternative food networks to sustain their incomes. McEvoy's PowerPoint (NOP April 9, 2013: PPt slide 29) acknowledged this: "Some farms that comply with organic standards avoid certification." The sound and sensible way forward was to reduce burdensome paperwork, make certification more consistent and impartial (they might provide broad technical information, but not consult for individual farms), and as affordable, accessible, and attainable as possible for all operations. In closing, McEvoy said NOSB recommendations should be sound (i.e., maintain and uphold organic principles such as biodiversity, continuous improvement, biological pest management, and soil building) and that NOSB recommendations should be sensible (i.e., reasonable for producers and handlers to comply with and significantly pose no undue burden on small businesses). The aim that these all be "implementable and enforceable" was certainly laudable. It was also more credible, on February 7, 2014, when President Obama signed a farm bill that reimburses farmers for up to 75 percent of organic certification fees.

When it comes to research, it is an understatement to say organicists have been chagrined by the low funding available for the scientific research and development of organic crops and inputs, in the context of the many billions of USDA subsidies for conventional commodities, such as canola (rapeseed), corn (maize), cotton, and soy of which crops 70–80 percent are GMO varieties. In his Portland address, McEvoy mentioned aquaculture, nanotechnology, bees, origin of livestock issues, and recusals for NOSB members in potential conflicts of interest.

What McEvoy did not say was that the influence of the Tea Party in the Republican Party had brought political gridlock in Washington, DC. The previous farm bill, called the Food, Conservation, and Energy Act of 2008, expired in 2012. Followed by two years of squabbling over issues such as food stamps, the House finally voted for the 2014 farm bill in January 2014, followed by the Senate in February.

The Tea Party is strongest in the House of Representatives, with its two-year election cycle, but also opposes Democrats and moderate Republicans in the Senate, with its six-year cycle, through well-financed negative media campaigns. The Democratic Senatorial Campaign Committee (2014) blames the two-year gap in farm bill approval on substantial political contributions by Charles and David Koch, billionaire scions to their father Fred's Koch Industries, based on coal and other carbon industries. They mentor an array of conservative and libertarian think tanks, such as the Cato Institute, Federalist Society, and American Legislative Exchange Council (ALEC; see *The Nation* 2011). The Koch brothers advocate lower taxes, privatizing public schools, and trimming social safety nets, such as Obamacare. Greenpeace and other environmental organizations carp that the Koch brothers' initiative Americans for Prosperity have pressured over four hundred members of Congress to oppose climate mitigation unless it is offset by tax cuts.

In such a political climate, the National Organic Program faces strong headwinds. Discussion at the 2013 Portland NOSB meeting about the International Federation of Agricultural Movements (IFOAM) evoked the remark that although the European Union is not an organic paradise, organics benefit from a better policy setting and more funding for research, which is disseminated well. In the United States, a Senate farm bill extension passed June 21, 2012, included $16 million for the flagship Organic Agriculture Research and Extension Initiative (OREI), 20 percent less than the $20 million allotted under the 2008 farm bill. When a farm bill extension was finally passed on January 1, 2013, no mandatory funding was given for OREI, and it would need to reapply for funding the following year. The dearth of research funding threatened to marginalize the National Organic Program. Corporations were writing international definitions on sustainability in Europe and in the United Nations, but U.S. organicists did not have data ready to support the benefits that organic proponents intuitively knew organic products had. It was ironic that Pepsi-Cola could gain points in public relations by reducing the amount of chemical fertilizers in its environment footprint—while organic dairy farmers could not because they did not use such chemical inputs.

The above concerns could not be resolved at one NOSB board meeting. But some long-delayed housekeeping would help safeguard the integrity of the USDA organic logo. The writing was on the wall for antibiotics in organic orchards at the 2013 Portland meeting. Five months later, the National Organic Coalition's Liana Hoodes (NOC 2013: 3) sent a letter to the NOSB regarding public comments at the April meeting, indicating that the phaseout of antibiotics against fire blight "is expected to have significant impacts on multiple stakeholder sectors." Alternatives were not ready. Unfortunately "a promising new copper material has been delayed in its EPA registration and use of a yeast-based material was not as effective as hoped during the 2013 growing season." Hoodes wrote that the NOC suggested adding the following research priorities related to alternatives to antibiotics: "1. Methods for increasing the accuracy of monitoring the presence of fire blight; 2. Systems-based approaches to prevent and control fire blight; 3. Field trials to determine efficacy of new materials when actually used by farmers." The outlook for improved funding was dim and awaited denouement in the 2014 farm bill, to be discussed in a following chapter of this book.

The long-running Family Economics Study at the University of Michigan Institute for Social Research (FES 2014: 4–5) notes a massive decrease in U.S. wealth in the Wall Street crash, the housing debacle, and the unemployment of the Great Recession: "For the top five percent of households, average net worth was higher in 2011 than it was in 2003, but for the remaining 95 percent of households a very different picture emerged." The FES found that "during the main recession years, between 2007 and 2011, over 12 percent of households [lost] $250,000 or more, while over 33 percent lost at least $50,000."

Nationally, arguments raged over food stamps and other safety-net programs. Even though the recession begun in 2007 had the earmarks of depression, the right embraced austerity while the left urged Keyenesian economics in the form of President Obama's quantitative easing, a fiscal stimulus begun under President Goerge W. Bush. The official jobless rate hovered above 7 percent, masking the many people who had given up looking for work. President Obama vowed to fight for the middle class as the bulwark of democracy. But the ranks of employed people—including bankrupted farmers—did not include millions of people who were *under*employed or working part-time and likely never to regain the skilled, middle income positions they once held. Democrats and progressives pointed out that many full-time employees of retail behemoth Wal-Mart were paid so little that they qualified for USDA food stamps. In essence, the U.S. government subsidized the labor costs of Wal-Mart, something that Robert Reich, former secretary of labor under President Bill Clinton frequently pointed out. This was embarrassing to Republicans and others on the political right. At least it helped explain why "Teapublicans" wanted to raise eligibility standards for food stamps.

Smouldering Pasture War

Back at the spring 2013 National Organic Standards Board meeting in Portland, the grazing issues that preoccupied organic activists in the 2000s appeared to be finally resolved in a firm rule by the USDA (*Hoard's Dairyman* 2010). Waiting for the elevator, this author asked Cornucopia Institute cofounder Will Fantle if he was happy with the 2010 USDA NOP Pasture Rule (USDA 2010). Fantle replied, "Well . . . yes."

What about the Cornucopia press releases in mid-2012 that questioned the paltry $3- to $5-million increase in the National Organic Program budget, which reportedly funded a slight increase in Washington, DC, staffing but left little money for the inspection of organic-industrial megadairies and enforcement of the more explicit 2010 Pasture Rule, which mandated a minimum 120 days grazing per annum (more if weather allows), equating a minimum 30 percent of dry matter intake (DMI)? In response to that lengthy question, Fantle elaborated: "Miles McEvoy went to see Horizon's new dairies in New Mexico . . . but the NOP wouldn't tell us what they found."

To our knowledge, no photographs have emerged from the largest corporate organic livestock farms, aside from stock images on their websites (please see chapter 6). We are left to take McEvoy's quiet assurance that organic operations with multiple thousands of cows are providing pasture grazing according to the same standards that family farms routinely provide. Is it possible that the organic milk boycott officially launched in 2007 by the Organic Consumers Association, and unofficially conducted years before that by Seattle members of Puget Consumers Cooperative, along with litigation set in motion by The Cornucopia Institute, tell a happy story in which consumer and activist politics were able to change government policy and agribusiness practices?

Yes, said Fantle, but it seemed provisional agreement. He acknowledged that more transparency on grazing by the organic-industrial companies and the NOP was warranted, and problems of longevity of contemporary dairy cows needed closer scrutiny. It is worth remembering that confined-animal feeding operations (i.e., CAFOs) in the organic sector were not outed publicly until Michael Pollan (2001) wrote about them, and The Cornucopia Institute publicized aerial and ground-level photographs showing how far the industrial reality of the feedlots was from the verdant meadows in which many consumers imagine individually named cows chewing their cud.

Testifying at the full National Organic Standards Board meeting in the Hilton on Tuesday, April 9, 2013 (following the NOC premeeting the preceding day), Will Fantle decried a net loss of organic farmers in the recession and urged better U.S. government inspections of the large influx of food imports from Russia and China, a major concern since the melamine milk powder fraud in 2008 that sickened and killed babies. The National Organic Program (NOP) must return to grassroots farmer and citizen control of the NOSB, said Fantle, which—since the Organic Foods Production Act (OFPA) was passed in 1990—consisted of a diverse mix of public-interest group representatives, an environmentalist, a scientist, and a handler. In late 2006, President George W. Bush's administration appointed employees of Campbell's Soup, General Mills, and Stahlbusch Island Farms to the NOSB. Stahlbusch's Tracy Miedema (formerly of General Mills–Small Planet Foods, who ultimately went to work with Earthbound Farm) became board chair although that company was primarily a conventional agribusiness using nonorganic inputs, with one-third of its acreage, 1,500 acres, as a certified-organic offshoot. This is not to say that the combined organic *and* conventional nous of Miedema, who (in 2001, when this author interviewed her at Small Planet in Sedro Woolley, Washington) was previously an adjunct professor in marketing at Western Washington University in Bellingham, Washington, did not give Stahlbush's sustainability efforts on its farms based near Corvallis, Oregon, a refreshing whiff of innovation that rippled through the industry. Indeed, Stahlbusch's $10-million anaerobic digester for fruit and vegetable waste produced enough methane biogas to generate electrical power for 1,100 homes (half was used to power its frozen food operations). But Fantle was on firm ground when arguing that it was wrong when another NOSB seat, which federal law allotted for a farmer, was taken by a corporate employee of Campbell's Soup. This rising preponderance of corporate agribusiness influence in the NOSB was most likely responsible for the organic certification of massive chicken sheds in California, where 100,000 hens had little or no outdoor access. One-tenth of a million hens in an ammonia-ridden shed was probably not part of what Michael Pollan (2003) called the "original organic dream."

Following the 2010 Pasture Rule, the focus of organic debates shifted from grazing to GMO food labeling, USDA secretary Tom Vilsack's fast-tracking of GMO crop certifications, introduction of new king-size GMO salmon, inclusion of synthetics (such as carrageenan and DHA) with doubtful sunset clauses for their phaseout, and the weakening of consumer and farmer participation in the NOSB. But the NOP's lack of transparency on the inspection of megadairies belonging to Aurora, Horizon, and other companies, regarding proof of compliance with the final Pasture Rule, is surprising. In chapter 8, we will turn to surprising new developments in this lingering Pasture War.

*   *   *

Movements need idealistic firebrands to ignite action whenever campaigns hit the doldrums. Kastel may be considered such. In fact he was honored by the staff of the respected populist magazine *Utne Reader* (2009) in an article titled "50 Visionaries Who Are Changing Your World." Staff writers had this praise: "When you buy organic, you want to trust the label. Kastel and his small but dogged Cornucopia crew make sure that organic food producers are walking their talk by snooping around their barnyards and their balance sheets."

Kastel has influential colleagues in the ranks of organic pioneers who fear organic integrity could be lost without his fire. Goldie Caughlan, former member of the NOSB, decided to join the board of The Cornucopia Institute after retirement from her decades as a nutritionist and public spokesperson at Puget Consumer Co-op (PCC) in Seattle. Caughlan knows Kastel sometimes has a polarizing effect but commented (personal phone communication, April 2013), "After all these years, I still believe Mark is the smartest, best person to carry the fight forward for organic farmers."

To understand The Cornucopia Institute, it helps to rewind to the 1980s when Mark Kastel moved from agribusiness to the 1990s, when Kastel worked with the Wisconsin Farmers Union. This requires recapitulating the story of bovine growth hormone told in chapter 2. In the 1980s Monsanto renamed its synthetic dairy drug (which stimulates milk production an average 15 percent in cattle, but stresses cows) from *recombinant bovine growth hormone* (rBGH) to *recombinant bovine somatotropin* (rBST). The GMO drug, which is here called *rBGH/rBST*, was developed with Monsanto support at Cornell University in New York, but always had public relations problems, which it sought to rectify with name and brand changes. A new acronym and name were less likely to trigger associations with the horrors of growth hormones in science fiction movies involving giant reptiles seen on late-night TV, and thus Monsanto fashioned the brand name Posilac (positive + lactose) for the drug. Under this name, Posilac was marketed inside the United States by Monsanto and in other countries by Elanco. Because Posilac is little known outside advertising, this book uses the combined acronyms rBGH/rBST for the synthetic dairy hormone, which is widely understood.

Due to widespread apprehension about the historic certification of transgenic technology, not just in the dairy sector but also among politicians and the public, the commercialization of rBGH/rBST was delayed until 1994. To monitor its introduction, the Wisconsin Farmers Union installed a telephone hotline with support from the National Farmers Union. In her book *The World According to Monsanto* (2008: 116) Marie-Monique Robin, winner of the 2009 Rachel Carson Award, writes that a litany of cattle ills poured in, and "worse, although Monsanto was legally obligated to report the secondary effects that its product caused in the field, it had, according to Kastel, improperly delayed transmitting some of the reports to the FDA."

On September 27, 1994, the CBS TV network broadcast an "Eye to Eye" segment on problems with what they referred to as *BGH* (bovine growth hormone), eliciting national attention on the dairy hormone controversy—and on Mark Kastel at the Wisconsin Farmers Union. It was a step en route to the founding of The Cornucopia Institute in 2004. In 1995 Kastel won acclaim for a piece published in July by *Rural Vermont* titled "Down on the Farm: The Real BGH Story: Animal Health Problems, Financial Troubles." It detailed symptoms including massive udders linked to excessive milk production, which weakened cows' reserves, brought lameness, and reproductive difficulties leading in some cases to early slaughter. Also in July 1995, *Rural Vermont* published a companion piece by Andrew Christiansen, "Recombinant Bovine Growth Hormone: Alarming Tests, Unfounded Approval." The article alluded to discrepancies found by Canadian authorities in Monsanto's pursuit of commercial profitability (see also Health Canada 1998 in Smith 2003: 95). Mainstream U.S. media had been supplied with reassuringly optimistic reports on rBGH funded by Monsanto. The *Rural Vermont* articles were early bellwethers to what critics saw as a wonder dairy drug's fall from grace and Monsanto's sale in 2008 to Elanco, which had distributed it abroad.

The Food and Drug Administration issued online the "Review of the Safety of Recombinant Bovine Somatotropin" on April 23, 2009, early in the Obama era (see also FDA 1999). *Rural Vermont* and Vermont Public Interest Research Group had challenged the FDA's early 1990s approval of rBGH/rBST (Posilac), citing reviews by its Canadian counterpart agency, Health Canada, which noted that long-term toxicology studies regarding human health safety were not required by the FDA or performed by Monsanto. Health Canada (1998) questioned one test that had been done, a 90-day rat study performed by Monsanto's Searle laboratory in 1989. Cysts were found in thyroid glands that had infiltrated the prostate glands of rats, but the FDA (2009) reported these were not cancerous and did not threaten human health; it also found that increased levels of insulin-like growth factor 1 (IGF-1) in GMO-enhanced milk ranged within the natural variation of cow milk, and claimed that drinking this milk was not a safety concern:

> FDA believes that the Canadian reviewers did not interpret the study results correctly and that there are no new scientific concerns regarding the safety of milk from cows treated with rBGH. The determination that long term studies were not necessary for assessing the safety of rBGH was based on studies which show that: BGH is biologically inactive in humans even if injected, rBGH is orally inactive, and BGH and rBGH are biologically indistinguishable.

As to animal welfare, Health Canada rejected certification of the synthetic hormone in January 1999, after the Canadian Veterinary Medical Association Expert Panel (1998) reported it increased mastitis by 25 percent, infertility by 18 percent, lameness by 50 percent, and culling by 20–25 percent. In response, the FDA (2009) updated online report downplayed the U.S. General Accounting Office (GAO 1992) report finding that cows treated with the

GMO dairy hormone had "a small but significantly greater incidence of mastitis" and "GAO recommended that the degree to which antibiotics must be used to treat mastitis should be evaluated in rBGH-treated cows with respect to human food safety."

Few shoppers like to think about milk cows suffering from mastitis or being given antibiotics to treat it. The result of additional bad publicity about rBGH/rBST was that more processors and retailers asked conventional dairy farmers to desist from using the GMO. The fact that organic farmers had vocally and always abjured it was one factor why, even though general organic sales weakened in the Great Recession after 2008, organic milk sales stayed strong.

Vermont has been a strong dairy state, and its senator Patrick Leahy was a Washington heavyweight and key ally to small farmers in their resistance to agribusiness's appropriation of their livelihoods. On Leahy's staff was Kathleen Merrigan, a staunch but realistic supporter of organics, with a PhD from Massachusetts Institute of Technology and a master's in public affairs from the University of Texas. As a staff member for the U.S. Senate Committee on Agriculture, Nutrition and Forestry for six years, Merrigan wrote the law establishing national standards for organic food (NODPA 2014b). In 2009 Merrigan was unanimously approved by the Senate to serve as deputy to USDA secretary Tom Vilsack in President Obama's first term. *Time* (2010) magazine hailed her as one of the "100 Most Influential People in the World." At the USDA Merrigan presided over daily operations, its $149-billion budget process, established priorities, and monitored progress and—crucially during the Pasture Wars—drove its rulemaking process. Merrigan had several achievements in the first Obama term, which included creating and leading the Know Your Farmer and Know Your Food Initiatives to support local food systems. She also helped design First Lady Michelle Obama's "Let's Move!" campaign against obesity and served as U.S. representative to the United Nations Commission on Sustainable Development. In terms of organic farmscapes, Merrigan's principal contribution was ushering the Pasture Rule to an acceptable conclusion in 2010. After her departure from the USDA, Merrigan worked with First Lady Michelle Obama on healthy food and exercise campaigns before returning to academia at George Washington University.

*   *   *

On the other side of the Atlantic three decades earlier, in the 1980s, it was already apparent that Europe would reject GMO dairy hormones when this author interviewed farm leaders and social scientists who were studying the hormone for the European Economic Community. With laughter, Willi Kampmann, head of the German Farmers' Union (Deutsche Bauernverband) in Bonn, told me (Scholten 1989b), "Europe already has enough milk!" The budget of the Common Agricultural Policy was stretched by existing dairy subsidies to farmers and burdened by storage and transport costs and export subsidies for its surplus Butter Mountain, before milk quotas were established on April 1, 1984 (Scholten 1989a). Europe's dairy stocks posed a chronic danger to developing countries, such as Bangladesh, whose dairy system was disrupted by dumping (Scholten and Dugdill 2012).

On the other hand, as the previous chapter details, India's dairy farmers' cooperatives, led by the Amul brand of the Gujarat Milk Marketing Federation (GCMMF) under the leadership of Dr. Verghese Kurien, preempted a tsunami of European dumping by negotiating the world's biggest ever dairy program called Operation Flood 1970–96. In this clever way, India's White Revolution, partly funded by the World Bank, sold European dairy commodities to invest in India's own infrastructure and reformed price structures to incentivize domestic production, to the point that India had surpassed the United States as top world milk producer by 1998 (Scholten 2010c: 10).

Relative to industry, U.S. farmers' organizations were in some ways less powerful than those in Europe or India (though not the United Kingdom). The imposition of European milk quotas on April 1, 1984, encouraged dairy farmers who had been lobbying the Republican Reagan and Bush administrations for production management to boost their incomes, which were suffering from surpluses, but they were to be disappointed. "Production management" was a quiet synonym for Canadian or European milk quotas (Scholten 1989c), and it was anathema to the neoliberal market economics dominating Washington, DC, which disdained quotas. Some farmers admired the avowedly conservative moral values of the Republican Party. Many appreciated Reagan's reversal of the grain embargo imposed on the Soviet Union by President Jimmy Carter, after the communist country's invasion of Afghanistan in 1979. However, the Reagan-Bush administrations rejected pleas for U.S. quotas, on the basis that they violated principals of free trade.

Dairy farmers were aggrieved. What could they do? They envied nonfarming industry's ability to substitute synthetics for traditional inputs. Many were angry that what they saw as misguided, unscientific consumers blocked their cost-cutting innovations but were reluctant to pay a penny more for a gallon of milk in supermarkets, while the costs of farm inputs soared. Every time farmers devised a modern method to boost net profits—be it prophylactic antibiotics to accelerate heifer growth or estrous synchronization to save labor costs on detecting heat and performing artificial insemination (AI)—it seemed that the influence of greens' concerns disconcerted more consumers at the dairy case.

Some, but not all, farmers demanded that Monsanto's GMO dairy hormone be certified to give the drug a chance. (Upjohn, American Cyanamid, and Eli Lilly developed similar drugs, but Monsanto was first to commercialize it.) Mark Kastel of The Cornucopia Institute recalls (personal communication, March 31, 2014) university research at the time that reported: "Over 60 percent of farmers objected to it . . . There was a scale difference. The large lobbies, like Farm Bureau and the International Dairy Foods Association, all supported it while family farm organizations generally opposed."

Some questioned the ultimate efficacy of the drug, but conversations with farmers on or off farms reflected a consensus that government and consumers give the innovation a chance to prove itself. It was a policy that the USDA was happy to support, as the dairy hormone was—after the introduction of recombinant insulin for human diabetics—a spearhead for biotechnology as a national export champion.

GMO success with dairy hormones in the U.S. is debatable. Its zenith was around 2009 when perhaps one-third of U.S. dairy cattle were administered rBGH/rBST, but its use ultimately declined. Elanco, which had sold Monsanto's Posilac overseas, bought Monsanto's remaining stake in it in 2008 (see above), as the latter firm refocused its energies on marketing GMO crops.

These crops included alfalfa, which is the fourth largest U.S. crop after corn, wheat, and soybeans. The dairy industry is dependent on alfalfa, and contamination threatens conventional and organic users of non-GMO crops. Against the hopes of greens, President Obama's USDA secretary Tom Vilsack gave the go-ahead for GMO alfalfa in early 2011 (*Grist* 2011). Its pollen threatens the purity not just of nearby organic stands of alfalfa. It also polluted at least one shipment of *conventional* non-GMO alfalfa slated for Japan, which imports about $4 billion of it annually. That such a lucrative market could be ignored by the USDA suggested so much political support in Washington, DC, that GMOs could not be stopped. Critics noted that alfalfa, a perennial plant that returns every season for half a decade and is normally pollinated by free-ranging honeybees, can easily crossbreed with organic or conventional alfalfa. The geographical separation of GMO alfalfa once promised by the USDA was not in place or, at any rate, was ineffective at containing its pollen. "It's telling that these things keep happening repeatedly," said George Kimbrell, senior attorney at the Center for Food Safety in Portland, Oregon. "It's a systemic problem. We have a failed regulatory system for these crops" (*Guardian* 2013b).

The USDA's decision to permit GMO alfalfa has increased the paranoia of critics who fear federal political collusion with an unspoken biotech-industry strategy. At some point so much of the biosphere could be contaminated that resistance is futile, and GMO hegemony is a fait accompli.

Organic Milk Boycott and Pasture Politics

Like political entities, nongovernment organizations (NGOs) can accumulate power, prestige, constituencies, and funding by exerting leadership in crises. Matt Reed (2001; 2006) found this was the case of the Soil Association, the leading organic organization in the United Kingdom, in widespread protests of GMO crop experiments in the 1990s. In the first decade of the twenty-first century, The Cornucopia Institute became a prime actor against conventional dairying, which allowed the use of GMO fodder and hormones in cows, unlike the USDA National Organic Program (NOP), which did not.

The Cornucopia Institute often works in concert with the Organic Consumers Association. In 2013 the OCA numbered more than 850,000 members, subscribers, and volunteers, as well as thousands of natural food and organic businesses. The OCA and Cornucopia are a good fit when opposing disempowerment of family-scale farmers and consumers. Cornucopia's (2012c) membership of about 10,000 is smaller than OCA's, but about 70 percent of Cornucopia members are farmers, more than any other comparable U.S. group. In an interview with health guru Dr. Joseph Mercola (2011), Kastel recalled helping the launch of the Organic Valley cooperative in the 1980s and 1990s, before cofounding The Cornucopia Institute with Will Fantle around 2004, "when the giant corporate agri-businesses that have squeezed family farmers out of conventional farming, and that were responsible for the deterioration in the nutrient level and the safety of our food, were buying out, on a wholesale basis, all the brands that had launched the organic commercial movement."

A decade ago deep greens were already complaining of *green washing* in the USDA National Organic Program (USDA-NOP 2002). They deplored organizations happy to take green price premiums, but paying only lip service to deep organic methods. The Cornucopia Institute's first campaign was directed against the 10,000-cow Vander Eyck Dairy that supplied certified-organic milk to Horizon Organic Dairy. Horizon had been bought a few years before by Dean Foods, a $12-billion corporation as big as Monsanto, which Cornucopia identified as a *green washer*. Dean Foods repeatedly bought brands successful in the organic and natural foods sector, including Alta Dena dairy and Organic Cow of Vermont, before altering them for higher profits in a more mainstream demographic. In his book *Organic, INC*. Samuel Fromartz (2006: 157–59) details Dean's acquisition of White Wave Silk brand organic milk and tofu in 2005. Steve Demos had developed the organic company in Boulder, Colorado, 1977–2005. Demos, a Buddhist fascinated with the concept of *right craftsmanship*, was saddened when Dean-Horizon began sourcing conventional soybeans from China, instead of the organic American farmers he'd formerly depended on. Dean-Horizon repeated this pattern in acquiring Rachel's Organic Yogurt (2006; see *Dancing Cows* video) in Wales, before eventually turning it into a non-organic-certified product in the U.S. marketed as natural.

While The Cornucopia Institute (2007a; b) and the Organic Consumers Association (2007a; b) cooperated on the consumer boycott of organic-industrial scofflaws, Cornucopia filed formal legal complaints with the USDA in 2005 and 2006, targeting plaintiffs Aurora, Dean-Horizon, and Vander Eyck organic dairies for noncompliance with pasture rules, among other alleged violations of the law. The consumer boycott joined by Cornucopia was led by OCA national director Ronnie Cummins. The OCA had been formed partly to shape organic rules. Its membership surged after 1997 when the USDA, under Secretary Dan Glickman, mooted for public comment tentative organic standards allowing the Big 3—GMOs, sewage sludge replete with heavy metals, and irradiation. Some academic and corporate observers were startled that hundreds of thousands of consumers were acting as citizens (Bonanno 2000) in a teat-to-table struggle over the rules concerning what the government admitted was the fastest-growing economic sector in agriculture and food retailing. Fighting what it called a corporate agribusiness attempt to highjack organics, the OCA waged a public relations war against Monsanto. Related to this was the OCA's support of local farmers in its Breaking the Chains campaign, calls for GE-free zones (i.e., free of genetic engineering or modification), and attacks on loopholes in prohibitions on feeding animal blood products to ruminants in its Mad Cow USA campaign (*USA Today* 2003).

The strident, progressive voice of OCA's website would not have been out of synch with the socialist-allied United Farmers Movement of the Canadian prairies after the First World War—or even French anarcho-syndicalism in the time of Pierre-Joseph Proudhon in the late 1800s. But the OCA's tactics are resolutely peaceful, akin to the evolutionary socialism of German Eduard Bernstein rather than the violent revolutionary communism of V. I. Lenin, which, according to Sidney Hook, Bernstein thought unfaithful to the original principles of Marxism (Bernstein 1961: xix, 58, 70,157). Marx (1879; Tucker 1978) was less admiring of Bernstein, whom he portrayed as a petit bourgeois chatterer expecting the movement to be led by philanthropists rather than workers. The Social Democratic Workers' Party of Germany was formed in 1875 and in 1890 changed its name to the Social Democratic Party, the name it carries today. Although the SPD rose on a tide of industrial discontent, its support of farmers seems comparable to kinder aspects of U.S. Democratic and Republican administrations' farm policies, exemplified by its support of green rural policies, such as multifunctionality (see Wilson 2007). To its credit, the SPD eschewed the tragic and politically stupid persecution of religion that marked communism in the USSR.

The Organic Consumers Association's Millions against Monsanto campaign encouraged the public to voice opinions on the company's actions regarding "sustainable agriculture and farmers' rights." Presumably naming and shaming what it called a "biotech bully" would alter its policies. The OCA claimed two million people worldwide participated in a March against Monsanto on May 25, 2013, and threatened to out 71 U.S. Senators who voted against guaranteeing states the right to label genetically modified organisms in food. The outcome of Washington State's 2013 Initiative 522 to label GMO foods in November 2013 was a repetition of California's Propoposition 37. The Grocery Manufacturers Association (GMA) dispensed about $11 million, resulting in voters' rejection of I-522 by a narrow margin. On a per capita basis, the GMA spent more in Washington State than California.

The OCA rallied consumers and allied itself with NGOs in issues such as the protection of small family farmers, bovine growth hormone (rBGH/rBST), PCBs, Agent Orange and Roundup (in Third World military or drug war operations), the danger to biodiversity posed by Monsanto's GMO crops resistant to Roundup (its glyphosate-based pesticide), water privatization, and new rounds of farm bankruptcies.

Most germane to this chapter is the OCA's cooperation with The Cornucopia Institute in organizing a consumer boycott in the 2000s of products from Dean Foods/Horizon Organic and Aurora Organic Dairy until the USDA issued a final access to pasture rule in 2010. Some activists feared these organic-industrial giants could weaken organic rules to the point that the organic label masked products unworthy of the organic appellation. As mentioned, Dean Foods previously acquired Silk and White Wave Foods, which had used organic U.S. soybeans before substituting cheaper, conventional soy from exporters such as China and then marketed the products as "natural," trading on Silk's reputation as an organic pioneer.

Activists identified the chink in the armor of the organic-industrial firms. Consumers expected cows to graze on grass in pasture. Thus the profits of companies that confined their cows in feedlots were at risk.

Seattle Joins the Pasture War

This author's decades of living and studying in Washington State and the Pacific Northwest inform debates between pastoralists and proponents of zero-grazing in confined feedlots. From the 1950s, debate grew more pointed as dairy science slowly advanced. Into the 1960s the notion persisted that cows obtained the most nutrition from gloriously tall grass, three or four feet high, like a hayfield ready for cutting. Others argued it was a waste to have cows tramp through—and soil—a field that could be hayed. Some plant nutritionists presented evidence that shorter grass, perhaps a foot high, was better for pasture. This view led some farmers to green chop and fill wagons daily and tractor them to cows in feedlots near the milking barn. Relatively low real prices for gas or diesel fuel in the 1960s and 1970s encouraged mechanization and petroleum use for doing so.

Most of the background noise in dairying, emanating from government extension agents, equipment dealers, and farm magazines, pushed farmers to scale up. After all, President Nixon's USDA secretary Earl Butz had told them bluntly to scale up or leave farming. Most farmer members of the regional conventional cooperative Darigold got the message. They gradually shifted from the spring-to-autumn pasturing of cows to year-round confinement of cows. Not that farmers felt they had much choice. When they couldn't pay their bills, it seemed the only response for farmers was to add buildings, add equipment, add cows, boost yield, and pray for enough margin to stay solvent.

Across the United States the dairy outlook was grim. Given low farm gate milk prices, farmers tried to buy or lease enough nearby land to feed more cows to keep their family farms economically sustainable. Their places required constant investment, needed just to maintain their million-dollar operations—for that is what a family farm now was—much less expand them. Soon many of the farms that increased stock no longer had enough room to graze them all. So cows were confined in feedlots while farmers bought hay, alfalfa, and commodities for totally mixed rations (TMR) from outside sources and heaped the fodder they still grew on the farm in bunkers. Farmhands carefully mixed the rations for maximum milk yield and dispersed them in feedlots—sometimes with vehicles as massive as 16-wheel trucks. A family dairy farm became a crowded factory.

Finally, like a bolt of lightning in the upper Midwest, a new economic model appeared. It suggested a business model based on the quality of organic milk and strict standards for pasture grazing, rather than the "stack 'em high, sell 'em cheap" logic of intensive quantity-obsessed conventional dairying. In 1988, the Organic Valley brand was established in La Farge, Wisconsin, with just seven dairy farms operating as the Coulee Region Organic Produce Pool (CROPP). The brand gathered momentum and spread from the Midwest, numbering 35 states and three Canadian provinces by 2011, with $715 million in sales. By January 2013, OV counted 650 employees and about 1,800 member family farms. The OV story appealed to consumers ready to pay a price premium for milk produced the traditional pastoralist way, without herbivore cows ingesting the carnivorous feeds that led to mad cow disease in the United Kingdom and that still permeate conventional U.S. feed supplements in the form of avian blood from poultry (Scholten 2007). Once winter ends Organic Valley cows enjoy "first grass" (OV 2013).

Many Pacific Northwest and Seattle foodies were uptight about conditions in industrial dairying. Horizon Organic Milk was a good seller, partly due to the location of its principal company farm in Idaho. But its popularity waned when Michael Pollan (2001) exposed massive confinement at its Idaho operation, converted from a previous conventional megadairy of about eight thousand cows. When Horizon admitted to milking its cows three times a day, astute consumers doubted the thousands of cows on its organic megadairy actually walked three daily roundtrips from milking barn to pasture. That bothered shoppers in organic hotbeds like Seattle, populated by people with perhaps the highest average level of academic qualifications in the country.

Many Seattleites uptight about the globalization of food systems would pass tests on civic participation posed by Alessandro Bonanno (2000), a social scientist favored by the political left. Seattle greens were less inclined to be passive consumers than citizen soldiers or ecowarriors in the dairy case. By paying attention to the provenance of their food, many of them also followed the call of the political right's favorite monetarist, Milton Friedman, for people to "vote with their dollars." (This book is not the first to observe that the political left and right often shake hands over organic food.)

Seattle consumption is more akin to that of San Francisco than Los Angeles, says geographer John Agnew of UCLA (Scholten 2011). In 2002 Agnew gave me his impressions of their very different organic markets. In LA, said Agnew, organic buyers are focused on personal health, while San Francisco organicists are more altruistic, concerned with the ecology of animals, the environment, and social justice in healthy food systems (Scholten 2011: 187–88).

Aurora Organic Dairy (AOD) had a lower profile than Horizon in the Pacific Northwest, though it supplied organic milk to prominent supermarkets, such as Safeway, and hypermarkets, including Walmart, Target, and Costco. Seattle and Washington State critics suspected AOD relied on confined megadairy feedlots instead of family-scale farms that pastured their cows traditionally (Scholten 2002; 2006a; b; c; d; 2007a; b). Their suspicions were later confirmed when it emerged that Aurora interpreted NOP rules in such a way that it did *not* graze cows during lactation (10 to 11 months of the year) and put them on grass only during their "dry" period before giving birth. Even if lactating cows had access to exercise areas, that is not the same as pasturing cows. Many consumers were horrified that some USDA certified-organic milk came from cows that were in practice confined.

Puget Consumer Co-op heard customer complaints and joined the milk boycott with scores of other groups around the nation. After Goldie Caughlan retired from PCC in 2011, she recalled that Horizon milk was banned from its eight Seattle natural food supermarkets because consumers complained that Horizon's practices belied its familiar happy cow logo (Scholten 2010b: 145; 2011: 130). Nationwide, the sides in the Pasture War were forming up, with consumer boycotts of brands such as Aurora and Horizon in important markets. PCC banned Aurora and Horizon brands from its dairy cases. At the same time, it featured Organic Valley milk, cream, butter, and half-and-half prominently. PCC also carried nonorganic milk from local Wilcox Family Farms even though they did not at that time pasture their cows but kept them in feedlots in eastern Washington. Was this logical? Caughlan explained it made perfect sense "because Wilcox was not claiming anything it did not do on the label" (Scholten 2010b: 145).

Victory in the Pasture War depended on the USDA's clarification of "access to pasture," a phrase in National Organic Program (NOP) rules that was interpreted with drastically different outcomes (2002; USDA 2001a; b). Most family-scale farmers claim grazing as basic ruminant behavior. Large-scale producers note that some pre-NOP organic certification programs did *not* mandate grazing. These programs generally represent the post-1970s rise of feedlot dairies in arid California, Colorado, and Idaho, which were dependent on irrigation and inputs sourced nationwide. This followed the petroleum dependency of conventional dairying in a westward shift from rainier northern states to the burgeoning population centers in the dry Southwest (Scholten 1997; 2011).

Definitions and rules on organics differ slightly worldwide, but one reason that governance of the huge U.S. market is so important is because it influences organic policies in its trading partners. That includes the European Union, where organicists worry that new rules could shift the habitual focus from process to technical rules that hamper small organic farmers in less developed regions of the EU (*Sustainable Food News* 2014). Since "access to pasture" implies management of landscape, U.S. rules affect what geographers call the production of space in ruralities worldwide (Lefebvre 1974; Elden 2004).

Now we turn to some confusing, even counterintuitive developments by key actors in the "access to pasture" wars, in the context of consumer efforts to participate in governance of the food chain. This chapter explains these, relying on various sources, including Internet searches and also interviews, conversations, and email exchanges. Some of these date back to the late 1990s when a wave of previously successful dairy farms exited from conventional pasture farms and even from (theoretically more lucrative) confinement operations. Some converted to conventional horticulture, such as raspberry production, and a few converted to organic horticulture and dairying.

It should be noted that USDA terms any farm that houses cows in barns for the winter a "CAFO," even if cows graze the other 8–10 months. But for many actors in the pasture debate, CAFOs connote year-round confinement or zero-grazing feedlots, and that popularly understood meaning is used here. Confinement is less common in Europe. German consumers often refer derisively to such a confinement farm as *eine Massentierhaltung*, loosely translated as "factory farm." The Cornucopia Institute's Mark Kastel observes that, because of breeding problems, heifers cannot be raised fast enough to replace stressed, burned-out cows in such "organic" CAFOs. Organic consumers aware of these conditions were dismayed.

Background: Awaiting the Pasture Ruling

When my presentations on the USDA Pasture War were prepared for academic conferences in 2006, a USDA ruling was expected at any time to clarify "access to pasture" (Scholten 2006a; d; 2008). Fortunately no one held their breath. A year later, at an NOSB rules meeting in March 2007, an agriculture department official said, "We hope that the proposed pasture regulation is out before the end of the year." Hope is eternal, but governance takes its own time. *Sustainable Earth News* (2007) was savvier, predicting it could be 2009 before a pasture rule was enforced. Even that was optimistic. At this writing in 2014, not everyone is convinced that a proper pasture rule is being enforced on megadairies. National Organic Program rules are supposed to be scale neutral. Even the gadfly NGO The Cornucopia Institute claims to be scale neutral when assessing the effects of megadairies on animal welfare, the environment, and social justice. But the stocking rates of megadairies run by Aurora and Horizon remain under scrutiny. As mentioned, Horizon's Paul, Idaho, megadairy was converted from half of a six- to - eight-thousand-cow conventional farm in 1994. According to a company history (*Funding Universe* c. 2001), the approximately three thousand organic cows were milked three times daily, grazed on pasture during the day, and spent nights in barn stalls when organic farming began. Its pasturing plans have been the subject of speculation ever since.

What mitigates against what is arguably efficient use of space on megadairies? Studies by M. J. Haskell and his coauthors (2006, 2007) in the *Journal of Dairy Science* note that large animal populations suffer if they spend too much time inside, on concrete-floored barns. High milk production resulting from high-protein, high-energy feed, such as corn, and the use of growth hormones do not always bring premature bovine maladies. But many confined, zero-grazed cows suffer stress, which weakens their immune systems and makes them more vulnerable to lameness, mastitis, breeding problems, and truncated longevity than pastured cows.

Early lameness is difficult to detect (*Hoard's Dairyman* 2013d: 569). As prey animals, cows have evolved to betray few signs of weakness. Keen cowhands look for signs of lameness, such as cows quietly shifting weight from one hoof to others to relieve strain. Cows' hooves are not only vulnerable on concrete inside the barn. Cattle in crowded outdoor stockyards with a high manure buildup are more susceptible to hoof infections than those on grass.

A step toward transparently stricter pasture rules came in early 2007. The USDA rescinded the organic certification of the Vander Eyk megadairy, which supplied Dean/Horizon, for pasture noncompliance when its ten thousand cows were found to exist in feedlots. The Odairy email list and other media buzzed with speculation on how the organic giants would react.

Shortly afterward, Horizon Organic vice president Kelly Shea announced support of the proposed 120 day/30 percent dry matter intake (DMI) rules. *Business Wire* (2006; Horizon 2007) reported that Shea and Horizon included lactating cows in their support of the clarified rules and urged the industry to adopt stricter standards. This was a significant endorsement of pasture grazing by an industry leader, a benchmark use of what could be Horizon's bully pulpit if and when it wished. But some questions were far from resolved.

In response to Horizon's claim that 80 percent of its milk was supplied by 350 family partners, Cornucopia's Mark Kastel remarked that consumers "should know that 20 percent is milk shipped from Horizon's two corporate-owned facilities . . . [and] they count CAFOs, milking thousands of cows, as 'family farmers.'" Kastel also claimed that "before Horizon quit buying from the 10,000-cow Vander Eyk organic certified dairy in California [in 2007], this setup was part of their 80 percent of 'family farmers'! They are still purchasing milk from a growing number of megadairies" (personal communications 2013).

Scale and Pasturing

For over a decade, Ed Maltby has been executive director of the Northeast Organic Dairy Producers Association and list manager of NODPA's Odairy online email discussion group. Based in Deerfield, Massachusetts, Maltby draws on his personal experience of managed intensive rotational grazing (MIRG) to say that it is indeed possible to graze large herds on pasture. But it is a tricky operation requiring great skill balancing cattle and forage—and the practical limit is about one thousand cows, depending on conditions. Awaiting the USDA final pasture rule, Maltby wrote this author (Scholten 2010b: 142–43, 147; also personal communications 2006): "The strict enforcement of pasture standards is where smaller farms (80–100 cows in the East, 100–300 in the Midwest and 500 cows in the West) see the ability to maintain the integrity of the organic standards. Unfortunately none of the processors do that right now, although they all pay varying amounts of lip service."

Maltby continued with insights on the cutthroat tactics that characterized industry battles for market share. Strong pasture standards were necessary not just for animal welfare, but also to protect the economic sustainability of small farmers:

> There are no good guys. All the processors have some farms that do not meet the pasture standards and have purchased milk from [noncomplying] mega-dairies; there are no good guys, just a varying amount of grey. The varying levels of interpretation of organic standards need to stop being a marketing tactic and become the base standard that allows entry into the market . . . Industrial agriculture will not disappear; we just have to fight realistically . . . to maintain a sustainable way of life for farmers.

Officially, The Cornucopia Institute is not anticorporate or antimegadairy and merely insists that enforcement of USDA organic rules be scale neutral across family and corporate farms. As a metric understandable by producers and consumers alike, Cornucopia devised its Dairy Scorecard (see following chapters for details) from a19-question survey (81 percent return rate) of 68 name-brand marketers on life span, stocking density, milking frequency, and so forth. In 2006 Kastel said 90 percent of organic name-brand dairy products met the letter and spirit of the law, but, unfortunately, "large corporate farms are gaming the system at the expense of ethical family producers" (Cornucopia 2006).

Corporations complained that USDA delays in making a Pasture Rule were costing them money. Farm plans for multimillion megadairies were put on hold, pending a final Pasture Rule. Organic pioneers replied that industrial free riders were free riding on their market niche, one the public expects to include traditional pasture grazing (Fromartz 2006).

Delay could be partly explained by government reluctance to regulate without industry consensus. Under the Federation of Organic Dairy (FOOD 2007) farmers umbrella, Northeast, Midwest, and Western Organic Dairy Producers Alliances (ODPAs), processors, and others lobbied the USDA to accept the recommendation, made years before, of the National Organic Standards Board (NOSB) for these rules:

1. Organic dairy livestock over 6 months of age must graze on pasture during the months of the year when pasture can provide edible forage;
2. The grazed feed must provide significant intake for all milking age organic dairy cows. At a minimum, an average of 30 percent of the dry matter intake (DMI) must come from grazed pasture during the region's growing season, which will be no less than 120 days per year. (NODPA June 9, 2007 ANPR)

Organic experts, including Ed Maltby and Mark Kastel, saw these standards as bare minimums that could be easily met. Farmers interviewed in the Pacific Northwest and New England agreed, saying their cows were often out more than 270 days a year, depending on that year's weather.

The threat of peak oil (Kunstler 2007), U.S. and Brazilian biofuel programs, drought in Russia, and growing meat demand in China and India increase pressure on land. Conventional corn and soybean prices doubled after 2005 when the U.S. biofuel program began inflating prices for human food and cattle fodder. There was even more pressure on organic grain prices, which soared, decimating the net profits of organic dairy farmers dependent on out-sourced fodder (Kunstler 2007; *Hoard's Dairyman* 2007a; *Seattle P-I* 2007; *Guardian* 2007; Scholten 2010a).

A tough propasture USDA decision would indirectly, but significantly, affect energy consumption and food miles in organic dairying. If megadair-ies legally had to pasture their cows, they might have to buy more land for grazing, or even move location if land were not available. A soft ruling could legitimize organic CAFOs, discourage campaigns on the welfare of sentient animals (IFOAM 2006; see below), disillusion consumers, and ruin the lucra-tive USDA certified-organic program, estimated to be $15 billion in total sales in 2007 (over $30 billion in 2012); although organic milk represented only about 2 percent of national milk volume, it led the livestock segment with 25 percent annual growth.

Margaret Wittenburg, of Whole Foods supermarkets, explained that organic milk is an "entry point" for consumers who expect cows producing it to graze on pasture (NODPA 2006; Hartman 2004; 2006; Hall et al. 2004). When processors' group NMI claimed that consumers prioritized the absence of GMOs or pesticides higher than pasture, Wittenburg pointedly asked why so many of them used pasture images on milk labels.

The meaning of the 2000s Pasture War in the U.S. organic dairy sector is found not only in the context of the U.S. pasture wars of the 1800s, in essence competition between beef and sheep ranchers for control of the open range, but also in other issues pitting greens against agribusiness. Fred Buttel (2000) described the contested process of the USDA's approval of genetically modi-fied (GM) recombinant bovine growth hormone (rBGH) in 1994 over green opposition. David Goodman and Melanie DuPuis (2002; DuPuis 2000) identified growing consumer "not-in-my-body" resistance to the drug and linked it to the 275,000 protests to the USDA after 1997 when Secretary Dan Glickman proposed that GMOs, sewage sludge, and irradiation be allowed in organic certification (see chapter 1).

Resistance to the corporate appropriation of organics rose. In the height of the Pasture War (2005–07), the USDA received about 80,000 comments, from farmers to consumers, most in favor of pasture and against confined zero-grazing.

Boycott Maneuvers

Ironically, the organic boom boosted incentives to weaken organic process standards and replace them with quantifiable rules (Schlosser 2001; Pollan 2001; Guthman 2004; Morgan, Marsden, and Murdoch 2006). Central to arguments between pastoralists and agribusiness investors is the fact that rising demand for certified organic food encourages producers to increase output and seek economies of scale, such as increasing cattle density, to the point that pioneers claim their rivals are stretching the meaning of "organic" (Krawczel et al. 2008).

In his 2006 book *Organic, Inc.*, Samuel Fromartz suggests Wal-Mart's plan to price organics within 10 percent of conventional fare pressured Horizon and Aurora to adopt three-times daily milking typical of intensive conventional dairying. An extra trip a day between barn and pasture would have reduced the cows' time for grazing and committed more of their energy to walking instead of producing milk. The economic decision would have been to reduce grazing on their megadairies in Idaho, Colorado, Maryland, and Texas.

According to the Organic Consumers Association (OCA) and The Cornucopia Institute, conditions on multithousand cow dairy farms deny cows' instinctive, natural behaviors on pasture and are associated with sore hooves, breeding problems, curtailed lives, and mastitis (Vaarst 2001).

A veterinarian who left Horizon's Maryland farm after eight years claimed, "They portray to their customers they've got this happy cow out on grass, this pastoral idyllic scene, but that's not the case" (*Baltimore Sun* 2006; OCA 2006; *Hoard's Dairyman* 2006: 736). The farm was home to about eight hundred head, with about five hundred milk cows and replacements. On my visit in November 2007, guided by the farm manager in a pickup truck, conditions for calves and heifers in exposed pens seemed good, as some dry cows grazed in fields. The milk herd was not visible because they were inside for the winter, according to the manager.

Horizon spokespersons said its large-scale farms supplied just 20 percent of its milk, and 80 percent came from small family farms.

In May 2014, Organic Valley (OV) was the largest U.S. farmer co-op, with 1,834 farm families supplying supermarkets, including Whole Foods. (Horizon Organic Dairy counted about 600 farm suppliers at this time, including the megadairies it owned.) Sometimes farms change their production practices in order to honor commercial commitments. In 2004, when OV was struggling to meet consumer demand, it relinquished a contract with Wal-Mart (*Economist* 2006; *Inc. Magazine* 2007). As mentioned above, the Wal-Mart contract fell to Dean-Horizon. Signs of struggle for market share soon appeared. *Inc. Magazine* (2007) reported that Horizon remained a market leader, outselling Organic Valley by $339 million to OV's $232 million. But OV led in natural food stores, such as Puget Sound Co-op, with $124 million in sales to such outlets and 28 percent growth per annum, compared with Horizon's $91 million in sales to similar outlets and a 9.5 percent decline per annum. In 1999 Horizon bought Welsh family firm Rachel's Organic Yogurt (mentioned earlier). In 2003–04 Horizon-Rachel's was bought by multinational Dean Foods (2006 profits $822 million on $10 billion sales). These Dean acquisitions troubled organicists, who believed Dean switched to nonorganic beans costing "about two-thirds less than organic" after acquiring White Wave Silk soymilk, according to Fromartz (2006: 186). They were vexed again in 2007 when Dean-Horizon launched Rachel's yogurt in America as a conventional, so-called natural product—not organic as it still is in Britain. Organic farmers boasted that their products were certified by the USDA, while the term "natural" was meaningless. Consumers were confused. Some revealed they trusted the term "natural" more than "organic." They did not realize that under USDA rules, products advertised as "natural" could be produced with antibiotics and GMOs, irradiation, and sewage sludge, unlike certified-organic products.

In 2006, when The Cornucopia Institute, led by family farm advocates Kastel and Fantle, filed formal legal complaints targeting Horizon and Aurora for noncompliance with NOP pasture regulations, Kastel said, "What we're trying to counter right now is a corporate hostile takeover of organics." Cornucopia worked with the OCA, which soon enlisted support from consumer groups, such as Puget Consumer Co-op. Headquartered in Seattle, with 43,000 co-op members and annual sales of $93 million, PCC claimed to be even larger than New York City's respected Park Slope Co-op. In the Pacific Northwest region, the Horizon boycott allowed Wilcox Farms (a conventional family farm at rural Eatonville, Washington, which met consumer preference for rBGH/rBST hormone-free milk before rival Darigold) to replace Horizon in co-op dairy cases due to its truth in labelling, as noted above. Nationally, the suit lost Horizon sales in natural food stores and jeopardized its image, leading to advertisements in *Utne Reader* (2006) in which Horizon promised to buy 2,500 more acres of pasture and reiterated their commitment to animal welfare.

While Horizon appeared sensitive to the public's understanding that cows are healthier on pasture, Aurora Organic Dairy took a more technical approach. Aurora's origins are mingled with Horizon's: president Mark Retzloff was a cofounder of Horizon in 1991 and left in 2001 to cofound Aurora with Marc Peperzak, former chair of the Horizon board. Peperzak raised $18.5 million in capital from Charlesbank Boston (which invests for Harvard University) to convert a five-thousand-cow, five-hundred-acre conventional dairy in Platteville, Colorado, to organic, selling most of its milk to institutions, private labels, and megastores, such as Wal-Mart. (Platteville's average annual rainfall was under 15 inches, less than the Colorado state average and less than half of traditional dairy states, like New York, Wisconsin, or Washington. In such arid conditions, farm managers often kept herds in feedlots since pasturing was difficult.) Aurora kept its cows off pasture during their ten-month lactations, while its veterinarian and executive vice president, Dr. Juan Velez, said, "Pasture can have a positive impact on animal welfare, if managed properly," and stated that USDA rules must accommodate "variability between farms, climates, geographies, facilities, etc." (Aurora 2006b: 11). Grazing dairy farmers interpreted these comments as damning pasture with faint praise, a tactic to weaken their advantage in grazing.

Across the Atlantic were hints that UK actors in the Dean-Horizon-Rachel's network were wary of public relations fallout. About a year after Horizon-USA bought Rachel's, the labels of Rachel's Organic UK milk cartons reverted from Horizon's cartoon cow to Rachel's former distinctively matte-black logo and images of real cows touting donations of cattle to Africa. Another hint that Rachel's Organic UK was distancing itself from its American partner's role in the USDA pasture wars was that the UK website featured the video *Dancing Cows—Born to Graze* (2006). Still found on YouTube, it is a cheeky "two-hooves-up" to confinement.

Consumer wrath demanded a sacrifice. In spring 2007, when the ten-thousand-cow Vander Eyck Dairy was decertified from organic production by USDA for not grazing, The Cornucopia Institute was praised by Ed Maltby of NODPA for its watchdog function. But some observers deemed it a token victory to placate sensitive shoppers, far from triumph in a war that could be won by agribusiness.

Counterintuitive moves were afoot. Most curious was Organic Valley (OV) leader George Siemon's cooperation with rival processors, including Aurora and Horizon, when he signed a "final alliance letter" to USDA secretary Mike Johanns asking for a quick ruling on access to pasture. The letter was supplied anonymously to and released by Cornucopia (2006b). When the letter became public, it spawned a flurry of emails. Hans and Colleen Wolfisberg, organic dairy farmers in Washington State, told this author that in regional co-op meetings, members asked Siemon why, when he joined corporations in asking for an easily-met 120-day grazing rule, he did not simultaneously demand strict dry matter intake (DMI) standards. This vagary might allow organic-industrial farms to fake pasturing as they built market share.

Ronnie Cummins of the Organic Consumers Association used the Internet to alert consumers to Organic Valley's unusual action (OCA 2007a; FOOD 2007):

> The USDA will soon propose new federal organic dairy standards that allow so-called organic factory farms to create the impression that their milk cows are being grazed on pasture, while in fact unscrupulous certifiers and bureaucrats in the USDA National Organic Program (NOP) will allow them to get away with "symbolic access to pasture" i.e. intensively confined, stressed-out dairy cows briefly chewing their cuds outside giant milking parlors in between their 3-x-a-day milkings. What is surprising to learn is that three highly respected organic dairy brands have joined with Aurora & Horizon to lobby the USDA for this "Big Fix" . . . We have no evidence that Stonyfield Farm, Organic Valley, and Humboldt Creamery are deceiving the public—as Horizon and Aurora are—by not requiring their farmers to pasture their animals and provide them with at least 30 percent of their diet with pasture grass, but we certainly do have the evidence that they are jointly lobbying the USDA for the continuation of *vague and non-enforceable standards* [italics added; see Processors Final Letter to USDA] . . . Otherwise consumers will continue to lose faith in the already tarnished "USDA Organic" label on dairy products.

Michael Funk, of United Natural Foods, Inc. (UNFI), urged industry to pressure the OCA to stop the boycott because it damaged the market (*Sustainable Food News* 2007). Cornucopia countered that the reason for Funk's position was commercial since UNFI's private-label brand of milk, Woodstock Farms, was produced by Aurora. Cornucopia added that UNFI was also the largest distributor of Horizon products.

It seems logical that the OCA and even OV farmers question OV actions. But it is also helpful to view actors through the lens of Thomas Rochon's (1998) social movement theory. Jeanne Merrill (2005), formerly of the Michael Fields Agricultural Institute, founded in 1984, says Rochon shows how actors, starting with similar ideologies, follow varying trajectories depending on their unique resources and political prospects. Consider Horizon, a firm begun with concerns for environmental and social *stakeholders* but also with concerns for the profits of *stockholders* in its university endowment fund (personal communication with Chuck Marcy, Horizon CEO, at National Dairy Leaders Conference, Sun Valley, Idaho, September 9–11, 2001).

Social movement theory would expect Aurora and Horizon to seek economies of scale on their farms, in order to lower organic sticker shock for consumers and raise Horizon profitability for their stockholders—just as it expects NGOs Cornucopia and the OCA to embrace family farmers, greens, and animal-welfare advocates as their stakeholders. Similarly, the pro-organic membership base of the UK Soil Association sharpened opposition to GMO field trials in the United Kingdom, which, as mentioned above, increased public participation in GMO protests that made the SA the primus inter pares of UK environmental politics (Reed 2006).

On consideration, it is unsurprising that Organic Valley cooperative leader Siemon quietly worked with corporate rivals in asking the USDA for a pasture rule *to reassure consumers* on organic dairy product integrity. It bought time to secure market share from OV rivals that cut costs by not grazing their cows. Once the 120-day grazing rule was won, the logic might run, cudgels could be taken up against free riders on a legitimate mantle of sustainability, in the fight for a meaningful DMI rule.

A slightly less combative view of Organic Valley strategy derives from Sun Tzu's ancient essays in *The Art of War*: it was wise for OV to avoid conflict and maintain diplomatic relations with its business rivals, while it gained strength and geopolitical reach by absorbing members from weaker organizations. After all, in 2007 Organic Valley co-op farms nationwide numbered barely half the 1,834 plus of 2014. Biding its time, OV was doing fine, building market share and consumer trust, promoting its farms as offering the best value in terms of animal welfare, environmental care, and income for family members and communities.

Outlook on Grazing and Future Welfare Issues

From their nadir in 1994 in protests against USDA approval of the recombinant dairy hormone rBGH/rBST, greens have rallied and felt vindicated by a public backlash against the drug. The bell tolled against the dairy hormone in 2007 when nationwide supermarket chain Safeway asked processors to supply conventional milk without GMOs (*Hoard's Dairyman* 2007b). Conventional cooperative Darigold eventually caught up with the change in public mood and growing disenchantment with the drug among megadairy managers and asked its farmer members not to use it.

With price premiums to be made in organics, the Pasture War dragged on. The Cornucopia legal complaints and the OCA milk boycott continued despite calls by UNFI head Michael Funk not to rock the boat. The OCA claimed successes when many retailers and consumers dropped Horizon and Aurora, "as well as the private label milk brands supplied by Aurora and sold by Walmart, Costco, Wild Oats, Safeway, Giant, UNFI, and others" (OCA 2007). But NODPA director Ed Maltby agreed with Funk in summer 2007 that milk boycotts had been less harmful to Horizon and Aurora profits than to the consumer perception of organics. That may have reassured stockholders in the short term, but consumers' trust seemed damaged. It was more likely that the USDA would invoke strong land-animal rules if consumers held errant processors accountable—and didn't buy their products.

Grazing and CAFOs

As mentioned above, prograzing actors in the Pasture War marked a major victory on May 25, 2007, when Kelly Shea, vice president for organic stewardship at Dean-Horizon, announced support for USDA pasture rules of 120 days and 30 percent dry matter intake (DMI) and urged the industry to exceed those standards. This act of leadership by the dominant player in U.S. organics was welcomed by pastoralists. It was clear that the 120-day rule was desired by all organic macroactors, including The Cornucopia Institute, the OCA, Aurora, Horizon, the Organic Valley co-op, the ODPAS, FOOD Farmers, Center for Food Safety, PCC, Whole Foods, and the USDA itself. (Meeting the DMI minimum was also a necessary challenge.)

But waiting continued for the USDA to actually publish a Pasture Rule. Meanwhile, Ed Maltby of NODPA said the 120-day grazing minimum was easily met across the country and already part of Aurora and Horizon farm plans—not to mention grassroots co-ops, such as Organic Valley, whose cows often graze 300 days a year.

A strong pasture ruling would encourage retention of pasture around rural towns, such as Lynden (pop. 9,000) in Whatcom County, introduced earlier in this book. In this farming community, links to Dutch family dairying are a strong and obvious part of local identity. With its windmill themes (see photos), Lynden's Front Street resembles Volendam in the Netherlands, and attracts nonfarmer home buyers from the United States and Vancouver, BC, Canada, as well as immigrants from Hong Kong and India. Activists fighting to retain agricultural zoning east of the Northwood Road in Lynden warn that it is wiser to encourage more housing in the nearby university city of Bellingham (pop. 75,000) than to build homes on rich pasture land in the Nooksack River valley. Though most local cow herds now inhabit CAFOs year-round, rising energy costs could induce farmers to forsake trucked-in fodder and return to extensive pasturing of their cows. Farmers won't have that option if houses have been built on pasture land.

Whatcom County is also a magnet for tourists seeking bucolic vistas of animals, barns, and landscapes recalling previous generations—this was a pattern found in England's Lake District by Ken Willis and Guy Garrod (1992; Scholten 1997). Joyce LeCompte-Mastenbrook (2004; also see next chapter), an anthropology student at the University of Washington in Seattle, conducted interviews among Dutch Americans around Lynden on the topic of the "stewardship" of natural resources. Farmers' care for land and animals imbued in Lynden's culture is attractive to vacationers. Thus agritourism remains a viable economic base in Lynden and strengthens the need for Lynden and Whatcom County to retain its agricultural heritage in an economically sustainable manner.

MIRG Weakens Need for CAFOs

But what is economically sustainable? Can dairy farmers survive if they do not industrialize? In the mid-2000s, managed intensive rotational grazing (MIRG) had already been touted in venerable farm magazines, such as *Hoard's Dairyman*, as a way for farmers to eschew the idol of spectacular milk yields in favor of net, bottom-line profits. The reintroduction of intelligent grazing knowledge into mainstream dairy discussions also increased chances that the USDA would rule strongly on the issue of dry matter intake, which the 2007 rulings had not addressed (on DMI, see *Hoard's Dairyman* 2000; CIAS 2005). In support of MIRG, NODPA's Ed Maltby (personal communication, 2007) recalled working on a Massachusetts organic farm where one thousand cows grazed intensively: "So I know it can be done if you have a commitment to pasture-based systems."

That might sound similar, albeit on a smaller scale, to the "green field" system of Aurora Organic Dairy's High Plains farm in Colorado, with 3,200 cows on eight hundred acres, and barns, milking parlors, and pastures arranged so "all animals have year-round, daily access to organic pasture and outdoor exercise" (Aurora 2005b; 2007). But conversation with Maltby suggested 1,000-cow herds were the upper limit to the effectiveness of rotational grazing.

The Cornucopia Institute deemed the 4-to-1 animal per acre ratio on Aurora's High Plains farm unsustainable because conditions were so arid— far dryer than in verdant Massachusetts. Cornucopia (2007a; Scholten 2010b: 147–48) even claimed that Aurora's original Platteville, Colorado, farm should be decertified by USDA because aerial pictures and farm visits by Mark Kastel and others showed only "1-to-2 percent of their cattle were actually grazing." The dominant color of the photos was brown, not green. The visitors found exercise areas near feed bunks, but insufficient extensive pasture that pastoralists deem necessary for traditional grazing.

The Battle Turns

Animal welfare issues are complex. In organics, as in other sectors, today's hero is tomorrow's zero, and vice versa. Some animal welfarists critical of Aurora Organic Dairy's pasture policies welcomed its claim that it is "among the only dairies in America, organic or conventional, which rely completely on natural breeding rather than artificial insemination" (Aurora 2005a; 2006a). Others remained critical, claiming that, as on many large CAFOs, they have so much difficulty breeding cattle they resort to on-farm bulls rather repeat more expensive artificial insemination, which might again fail to prompt conception.

Nevertheless, Aurora seemed to be expanding inexorably when it opened another megadairy in Stratford, Texas, on July 10, 2007. The firm claimed this was good news for low-income consumers. But small organic farmers riposted that the scale of this new operation threatened their price premiums and the integrity of the organic label in the eyes of many consumers. Less than two months later, the USDA concurred with Cornucopia's complaint that Aurora's Platteville, Colorado, farm was indeed in noncompliance. On August 30, 2007, *The New York Times* reported that Aurora "agreed yesterday to stop applying the organic label to some of its milk and make major changes in its operation after the USDA threatened to revoke its organic certification for, among other problems, failing to provide enough pasture to its cows."

When Aurora's public relations department attempted to claim that the pasture suit brought by Cornucopia and the OCA had been dismissed, it evoked reactions from chortling to consternation. Soon the USDA (Aug. 30, 2007; Scholten 2010b: 148) made it clear in a press release that the firm had a "one-year probationary review period" to improve grazing and the replacement of organic stock, or the USDA could "withdraw from the agreement and reinstate the Notice of Proposed Revocation." The USDA forced Aurora's Platteville farm (not to be confused with its High Plains farm in Colorado) to about 1,075 milking cows. Aurora then had to bulldoze most of the farm's feedlots to increase pasture to about four hundred acres. The firm was on the defensive when the Federation of Organic Farmers (FOOD 2007) sent a letter to the USDA complaining that the consent decree was insufficient to bring closure to Aurora's "willful violations" of organic rules found by USDA investigators, including neglect of grazing for lactating cows. In the end, Aurora was allowed to continue its $100-million enterprise and was not fined for organic improprieties.

In the aftermath of the USDA's August 2007 decision, Aurora president Mark Retzloff defended his company, which proclaims its mission to make organic milk more affordable to mainstream America. Retzloff told *Fortune* (2007) senior writer Marc Gunther: "We don't think we did anything wrong." Noting that the USDA did not fine Aurora, he said: "People are saying you're not putting your cows out to pasture. Well, we are. Just not the way you'd like us to." Retzloff explained the organic rule violations as trivial or committed by a former supplier and threatened to sue critics who accuse Aurora of fraud.

Final Pasture Rule 2010

The USDA organic Pasture War of the 2000s was waged in competition for market share in supermarket dairy cases. (Not that this problematic competition is entirely settled. Please see later chapters.) What swung processors' focus back to sustainability, represented by strict pasture rules, were (1) The Cornucopia Institute's formal legal complaints (such legal complaints are first adjudicated by the NOP and then an administrative law judge for the USDA's Agricultural Marketing Service [AMS] if they are appealed) and (2) the boycott launched by the Organic Consumers Association joined by organizations such as Puget Consumers Cooperative in the Seattle area. It was not foreordained that shoppers could materially affect industry or governance, but consumers' boycott of milk from confined organic cows improved power relations in the countryside for small farmers, who had a competitive advantage in grazing.

The penny dropped in 2007 when USDA decertified Dean-Horizon supplier Vander Eyck Dairy from organic production. Later that year the department came close to decertifying the much larger, more powerful entity of Aurora Organic dairy, which was found violating 14 provisions of the Organic Foods Production Act. Enforcement actions by the NOP targeting producers, processors, and certifiers increased, if only slightly, after the 2008 farm bill doubled staff levels from a very meager base of about a dozen.

Following up their adjudication victories at the USDA, on October 17, 2007, The Cornucopia Institute announced new class-action lawsuits against Aurora and its distributors on behalf of consumers in 27 states, asking damages from Aurora for "consumer fraud, negligence, and unjust enrichment concerning the sale of organic milk by the company." The lawsuit was applauded by Joe Mendelson of the Center for Food Safety (2007). When the lawsuit was eventually quashed by the courts, organic pastoralists may have felt their pockets had been picked, but Aurora eventually paid out millions to settle the suits.

On June 7, 2008, NODPA and the Coalition of Organic Groups, including FOOD, White Wave Foods (Horizon Organic), the Organic Valley cooperative, Stonyfield Farm, Humboldt Dairy, Organic Choice, Pastureland Cooperative, and Organic Dairy Farmers Coop, urged the USDA to publish the access to pasture and origin of dairy livestock standards immediately (NODPA 2008). They noted that delay harmed the organic market, and farms would need a lengthy transition period to comply with any pasture rule with teeth.

In retrospect what had already occurred might prove to be as important as the eventual pasture rule. That was the key development, in May 2007, of the decision by industry leader Dean-Horizon to support the National Organic Standards Board (NOSB) proposal that the USDA-NOP adopt rules of at least 120 days grazing and 30 percent dry matter intake (DMI) from pasture (Horizon 2007). Horizon's decision—perhaps prompted by the bad publicity that surrounded the boycott by Cornucopia and the OCA—could help Horizon regain sales among reflective consumers. More significant to the overall sector was that Horizon's leadership probably induced other firms, including Aurora, to take the grazing of lactating cows seriously. It is questionable whether the 120 day/30 percent DMI minimum rules in time and forage represent dairy farming sustainable enough to satisfy critical think-ers like Wendell Berry (1970/1972). Berry might stipulate closed systems in which cropping and grazing patterns replenished each other, cow longevity was closer to ten years (than the three to five on many U.S. farms), and that milk cows be replaced only by heifers raised within the herd. It remained to be seen whether Horizon would lobby for even more grazing, as Kelly Shea hinted when she urged competitors to exceed the minimums.

In the long run, macroeconomic forces could encourage grazing. These forces are chiefly the rising cost of energy. In June 2008, record prices of $138 per barrel oil, and U.S. corn at $7 a bushel in the biofuel boom, forced many small and large actors to adjust their business plans (*Hoard's Dairyman* 2007a: 129). A Washington State University agent recalled how her father deplored petroleum wasted in intensive dairy operations (personal communication, 2007): "Dad thought it was crazy to take cows off pasture, bring fodder to the barn with a tractor, and haul waste back to pasture when cows do that better themselves."

In 2010 the USDA finally issued its final Pasture Rule. It was a feather in the caps of NOP administrator Kathleen Merrigan and deputy administrator Miles McEvoy, but the rule had a long provenance. Mark Keating, who had been the NOP livestock standards specialist at the time the first organic rule was passed in 2000, said Richard Matthews, as senior agricultural marketing specialist, had been the single most important official within the USDA for advancing the access to pasture rule (personal communication, May 11, 2014). The final 2010 Pasture Rule included the 120-day grazing minimum and, also very important, the 30 percent dry matter intake minimum demanded by pas-toralists (Scholten 2010b). However, the NOP's budgetary ability to inspect farms and enforce existing rules remained weak. Deputy administrator Miles McEvoy reportedly journeyed to the American Southwest to see new mega-dairies built by firms such as Horizon (then owned by $12-billion corporation Dean Foods before it was spun off as White Wave/Horizon in 2013). But few if any photos of the new operations were widely available, and McEvoy seemed closemouthed at public meetings. Statistics on land stocking rates, milking times per day, pasture usage, and cow cull rates were hard to obtain.

Some USDA documents obfuscated details by refusing to disaggregate some statewide data. For one reason or another, this seemed designed to protect proprietary information of those corporations owning operations so large that one or two megadairies can skew statewide statistics.

At this writing (May 2014) in the aftermath of the 2010 USDA "final pas-ture rule," not everyone has seen enough evidence to convince them that the biggest players are honoring the rule. Are megadairy organic cows obtaining at least 30 percent of their diets on pasture? Horizon claimed their large Idaho farm achieved that a couple years before the rule went into effect. (Recall that this was after they were forced to lower stocking rates. A great deal of infor-mation can be found online.)

An abiding concern of small organic farmers is an origin of livestock rule. The NOP promised one in 2010, but four years later critics suspect some organic certified megadairies were still burning out overstressed cattle and replacing them with heifers raised cheaply in conventional herds, via loopholes in existing rules. The likelihood is that such practices are dwindling, but the extent of their continuation is unknown, partly due to the underfunding of NOP staff (roughly one staffer for each billion dollars of the U.S. organic market). With so little funding for monitoring and rules enforcement, transparency suffers. At NOSB meetings and on the Odairy email discussion list, sage advice often comes from senior women farmers. They remind dairymen that despite organic milk's high profile in dairy cases, it remains a relative drop in the bucket compared to the overall conventional sector, and that organic dairy is a poor shirttail cousin to the USDA's obsession with biotechnology. Mark Kastel is unsurprised by USDA secretary Tom Vilsack's embrace of GMOs and lax enforcement of the Pasture Rule (personal communication, July 31, 2013): "First, under the Bush administration, they dragged their feet on pasture enforcement. The number of CAFOs, and the percentage of milk they produced, exploded. Now, they are continuing to drag their feet on prohibiting conventional replacement cattle. We are still years away from enforcement. It doesn't matter who's in charge in Washington, there is a pro-agribusiness bias."

Cornucopia's Dairy Scorecard

The Wisconsin-based Cornucopia Institute calls itself a watchdog for family farmers. It insists that it has consistently emphasized that "the *vast majority* of *dairy* brands in the marketplace are from highly *ethical* companies," and it is on their behalf that it pursues scofflaws. In the mid-2000s, The Cornucopia Institute (2006c) developed its well-known Dairy Scorecard, rating organic dairy corporations, cooperatives, and so forth using cow icons: from 5 cows (Outstanding) to 0 cows (Ethically Deficient). The scorecards are in the public domain, published online with permission. Note: some dairies refused participation, and ratings are based on data before clarification of the USDA "access to pasture" rules in 2010. See below for how further points are awarded. Here is a sample from the web (accessed September 16, 2013).

- Aurora Organic Dairy in Boulder, Colorado, was rated by Cornucopia's scorecard at 0 cows, with no points awarded (described as "Largest conventional/organic factory-farm operator. Largest Private label manufacturer.").
- Horizon Organic Dairy of Dallas, Texas, was rated 0 cows, with no points awarded (described as "$11 billion—the nation's largest conventional & organic milk marketer. Owns 4,000 cow 'farm.'").
- Organic Valley cooperative in La Farge, WI, was rated by Cornucopia as 4 cows, with 1,115 points awarded out of 1,200 possible points (described as a "full line dairy").

- Organic Pastures Dairy Company of Fresno, CA, was rated by Cornucopia as 5 cows, with 1,200 points awarded (described as "fluid milk products (raw) butter, colostrum, kefir [in] California or sold as pet food"). This author's visit in 2007 revealed contented cows, able to graze 24/7 because mobile vacuum milkers reached them in fields (a common practice in Europe). Owner Mark McAfee is a leader of the raw milk movement, adamant that agribusiness not usurp family farmers via spurious hygiene rules. Especially galling to McAfee is what he perceives as the persecution of producers of raw milk products when studies show cheese from grass-fed cows contains high amounts of conjugated linoleic acid (CLA), which he extols as a cancer fighter.
- Fresh Breeze Organic Dairy near Lynden, WA, was rated as 5 cows by Cornucopia, with 1,195 points awarded (described as "fluid milk products, butter"). Perhaps because of annual "Northeaster" snowstorms via Canada, the farm, which pastures its cows in season "April to October, weather permitting," lost 5 points for enclosing cows in winter. The fifth-generation farm is a vertical operation that produces, pasteurizes, processes, and bottles milk for buyers and stores: "We provide our friends, neighbors and the community with local, quality, certified organic milk that is fresh!" Traditional organic values are evident in their brochures, vaunting links between healthy soil, animals, and people.

The Cornucopia Institute rated the farms according to these qualitative criteria for each of which the maximum points was 100. For Fresh Breeze, points were awarded as follows: ownership Structure (100, family farm), milk Supply (100, single farm), disclosure of information for verification (100, full and open disclosure), certifier of farms (100, e.g., state of Washington), certifier of processing (100, state of Washington), cows on pasture time/acreage provided (95, good pasture compliance), health and longevity of cows (100, extremely low cull rate), replacement animals only from organic farms (100, closed herd), antibiotics used on young cattle (100, never), reproductive hormones used (100, never), farm support oversight (100, owner-operator), outside dairy ingredients purchased (100, none). Great Goodness earned a total score of 1,195 out of 1,200, and 5 cows (outstanding). Its good pasture grazing practices are reflected in the extremely low cull rate of its long-lived cows, who are born and grow up in the same herd. These cows are so healthy and fertile that the farmer of this closed herd needs to buy no replacements from other farms. All the cows' hay, grass, and corn silage and grain supplements are grown on farm, putting its nutrient management system in line with the ideals of organic pioneers.

Extracts from Cornucopia's Dairy Scorecard, above, show a range of organically certified farms, and they find more likelihood of better sustainability practices on smaller farms than very large operations. Below, an Organic Timeline suggests dairy wars could drag on indefinitely, shifting focus from one salient to another.

Organic Timeline (Scholten 2006–14)

1988 CROPP/Organic Valley farmers' cooperative formed.

1990 Organic Foods Production Act (OFPA).

1991 Horizon Organic founded.

1994 USDA certifies synthetic recombinant hormone rBGH/rBST.

1997 USDA moots organic rules; 275,000 protest Big 3 of GMOs, heavy metals, irradiation.

1998 Organic Consumers Association (OCA) formed to fight Big 3, etc.

1999 Horizon Organic Dairy buys Welsh firm Rachel's Organic Yogurt.

2002 USDA publishes National Organic Program (NOP) rules sans Big 3.

2002 Dean Foods buys White Wave, maker of Silk organic soymilk.

2003 Aurora Organic Dairy set up to supply store-brands. USDA finds first BSE cow.

2004 Dean Foods buys Horizon and Rachel's.

2004 Dean consolidates Silk, Horizon Organic & other brands as WhiteWave Foods, in Broomfield, CO.

2005–06 USDA files Cornucopia Institute legal complaints vs. Aurora & Horizon on pasture.

2005–07 USDA-NOP gets 80,000 protests on pasture in OCA milk boycott.

2006 OCA boycotts Aurora and Horizon, claiming little grazing.

2007 USDA decertifies Vander Eyck dairy. Horizon supports 120 day/year grazing.

2007 USDA finds Aurora dairy violating OFPA, but milk declared organic.

2010 USDA-NOP final Pasture Rule: ≥120 days pasture grazing/30 percent dry matter intake.

2010 Dean Foods sells Rachel's to French firm Lactalis.

2013 Dean Foods sells WhiteWave Foods.

2014 WhiteWave/Horizon: WW buys Earthbound, and later So Delicious plant-based companies. Horizon touts consumer choice in mixed organic and made-with-organic Mac & Cheese lines. The Cornucopia Institute files complaints alleging violations of Pasture Rules on an Idaho farm supplying milk.

(Sources: Agriculture and Agri-Food Canada, Aurora Organic Dairy, Cornucopia, Dean Foods, Defra, FDA, OCA, Odairy, OV, PCC, USDA (2002c) WW/Horizon.)

4

Animal Welfare

Photo 4.1   Organic Valley cows mob farmers, hoping for a scratch behind their ears.
Photo Credit: Bruce Scholten.

4

Animal Welfare: From Rudolf Steiner to the St. Paul Declaration

The center of the chessboard in U.S. dairy wars is pasture. As regional economies rise and fall, farmers still exploit their factorial endowments of soil, sun, and water to compete for local and distant markets. Trends since the 1960s have turned much of the middle ground of pasture into fields of fodder, to be tractored to cows massed in on-farm feedlots or trucked to other farms hundreds of miles away. This chapter explores the relationships between pasture systems, their opposite confinement, and surrounding aspects of animal welfare, such as longevity, fertility, lameness, mastitis, and antibiotics. These affect cows, but they ultimately affect human health, too. Dairy farming on different scales shows that animal welfare is linked to environmental impacts on air, land, and water. It is also connected to rules on drugs, such as antibiotics, which differ in the United States, Canada, and Europe, in conventional and organic dairying.

Some conventional dairy farmers resist the trend to confinement with the managed intensive rotational grazing (MIRG) mentioned several times in previous chapters. Promoted by Ed Maltby, executive director of the Northeast Organic Dairy Producers Association (NODPA), MIRG is a modern scientific version of grazing that has been practiced by astute dairy farmers for generations, abetted by contemporary understandings of the biology of plants and legumes subject to the iterative compaction of cows' hooves and renderings of manure. MIRG can improve herd health while increasing farmers' net profits by cutting the extra capital investments in machinery and buildings involved in confinement dairying. But the strongest force for grazing cows on pasture has been the organic dairy movement, led by organizations such as the Organic Valley/ CROPP cooperative, which has mandated pasture since it was established in 1988. All livestock farms certified by the USDA National Organic Program (NOP) prohibit the use of antibiotics, and, as the last chapter detailed, all U.S. organic farms are now required to pasture cows a minimum of 120 days per year, to result in at least 30 percent of their dry matter intake.

The use of antibiotics in animal agriculture has become a widespread public concern. In this and the following chapter, discussion draws on groups including (in alphabetical order) The Cornucopia Institute, Dairy Farmers of America, regional Organic Dairy Producer Associations, the Organic Consumers Association, Puget Consumer Cooperatives, and Tilth Societies in the United States, and groups such as the Food Ethics Council and Soil Association in the United Kingdom. Remarks by organic and conventional actors from farm to plate enliven debate. Comments by leading organic veterinarian Dr. Hue Karreman, author of the 2004 manual *Treating Dairy Cows Naturally*, ground the discussion of antibiotics and of alternative natural treatments for mastitis and other maladies.

Confinement and Disease

It is an axiom of agriculture that wherever animals congregate disease vectors multiply. Cows prefer free grazing to confinement in feedlots, but there is a material link between their sense of well-being and overall welfare with herd health and thus human health. In an article titled "Overcrowding Invites Disease," conventional and organic veterinarian Dr. Paul R. Biagiotti writes that cows feel stress in overcrowded facilities: A cow's time budget is skewed when she has to queue to "eat, lie down, socialize, drink and be milked" (*Hoard's Dairyman* Aug. 25, 2013: 540, book forthcoming 2015; see also Grant 2009). This stress has further effects: "Social stress may result in the production of stress hormones such as cortisol, which suppresses the immune system. Nutritional stress resulting from inadequate opportunities to consume feed also can contribute significantly to immune system suppression by causing subclinical ketosis or rumen acidosis" (ibid: 540).

Studies by M. J. Haskell and his coauthors (2006; 2007) note large animal populations generally suffer if they spend too much time on concrete-floored barns. Haskell writes that high milk production by itself does not always bring lameness but agrees with Biagiotti, above, that zero-grazed cows suffer more stress than pastured cows and consequently experience more lameness, as well as breeding problems and truncated longevity. As pointed out in chapter 3, farmers do not always detect early lameness because, as prey animals, cows evolved to hide weakness.

Cows' teats are the prime avenue for infections, such as mastitis, and the more crowded a confinement operation is, the more likely that urine and leaked milk will transfer such an infection from one animal to the next. As herd sizes increased after the Second World War, farmers and veterinarians increasingly treated cows with the wonder drugs developed by Scots biologist Alexander Fleming in 1928 and taken up by the U.S. pharmaceutical industry thereafter. Antibiotics, such as penicillin, became synonymous with concentrations of cattle. Antibiotics suddenly simplified veterinary care, but not for long, sadly, because microorganisms (as if they'd read Charles Darwin on natural selection) evolved defenses against these wonder drugs. Soon new types of antibiotics had to be developed by pharmaceutical companies, as veterinarians found older varieties impotent. Cows are also given pain killers; for example, the FDA (2008: 10) predicted "the development and approval of much-needed analgesics for food-producing animals." But organicists fear that industrial agriculture can succumb to moral hazard in such a scenario. Just as irradiation could be used to sanitize unhygienic food produced in sloppy circumstances, analgesics could mask pain from animal mismanagement.

On top of the clinical (emergency) application of antibiotics against disease grew the prophylactic use of antimicrobials in animal agriculture. Specialist farms boosted profits by mixing small amounts of antibiotics in feed to maximize weight gain and minimize the natural two-year (or so) span from calf to springer heifer ready to bear her own calf (often in spring, hence the name), and begin commercial lactation on the farm that bought her. U.S. livestock now receive 80 percent of the antibiotics sold in the country, four times more than the human population of 320 million people.

This medicalized approach to dairying became de rigueur in state agricultural colleges. Farmers attempting to remain organic were alternately regarded as economic fools or saints bucking the tide. Most consumers remained oblivious to the switch to CAFOs, unless media messages from animal welfare groups, such as People for the Ethical Treatment of Animals (PETA) or Compassion in World Farming, alerted them. The switch to confinement also seems to have escaped the notice of many people reared in the country. They knew some farms were merging into confined megadairies. Dry cows and heifers that still grazed in fields masked the fact that more lactating cows were confined in feedlots than people knew. Then one day they realized that almost no milking cows were grazing on pastures.

Pathogens for mastitis that infect cows' udders and other organisms that attack the joints of cows' hooves accumulate in the bedding of free-stall barns (cows move freely in free-stall barns, unlike barns where they are tethered or in stanchions). Farmers have experimented with wood shavings and sawdust and sand as a replacement for traditional straw, which was in short supply as mixed farming (with cows, pigs, and crops) subsided in the face of intensive dairying and monocultures (*Hoard's Dairyman* 2012a: 372).

Once pathogens take up residence in bedding, they are hard to eradicate. Pathogens can be introduced by animals bought from another farm. If the herd is a closed one, in which all the heifer replacements are born and raised on the same farm, the farmer can more easily monitor and maintain its level of health. Farmers who pass a healthy closed herd between generations boast of their accomplishment.

Biosecurity was the buzzword on everyone's tongue, at a National Dairy Leaders Conference attended by this author in Sun Valley, Idaho, one fateful week in September 2001. Farmers asked if the public relations advantages of inviting city slickers to on-farm visits compensated for the dangers of importing, say, foot-and-mouth-disease germs from Britain (suffering a severe FMD epizootic in 2001), or dreaded BSE (Scholten 2007). When the dairy conference finished on September 11, the definition of biosecurity focused more than before on terrorism and the possibilities for terrorists to assault a nation's livestock. In a sad turn, it became common for farm visitors to don clear plastic "moon boots" and avoid petting friendly cattle. The world turned fearful, as CAFOs that distanced cattle from the public assumed a patina of sensible hygiene.

The USDA had long proclaimed the nation's food supply the purest, most nutritious the world had ever seen, and rising human longevity seemed to support government claims. But rising rates of cardiovascular diseases, diabetes, and obesity suggested that the millennial generation could have shorter lives. The Rachel Carson effect grew in time, long after her book *Silent Spring*, about the effects of pesticides, was published in 1962. Although the organic food movement was marginalized and equated by stand-up comedians with "fruits and nuts on the Left Coast," food scares kept it growing. More shoppers among the baby boomers began to read food labels in supermarkets as more of them resisted chemical inputs in agriculture and sought more natural food. Food scares, allergies, and increased reports of irritable bowel syndrome (IBS) and celiac disease left more people questioning the mainstream food system. Even the humble loaf of bread became suspect as modern strains of wheat were suspected as sources of digestive disorders. As we have seen, consumers were also disgruntled to learn that "nature's perfect food" (DuPuis 2002) was often produced by cows that seldom grazed natural pasture, and switched to organic milk. The last straw for pasture zealots came when they realized some USDA certified-organic milk came from cows in feedlots, not pasture. Such consumers demonstrated their expectation that cows eat grass by participating in The Cornucopia Institute and Organic Consumers Association boycott of organic brands that kept cows in feedlots.

Not everyone is sanguine about grazing. Defenders of zero-grazing operations resent the pejorative stereotype of the term "confinement" brandished by critics. Zero-grazing defenders may claim that cows thrive in CAFOs just as humans enjoy the ambience of dining inside the walls of a fine restaurant. They contrast the lives of their cows walking between the "salad bars" of feed bunks to comfortable, dry free stalls under a roof with traditional pasture-grazed animals, some of them ageing and hobbling in bad weather from the "back 40 acres" to the milking barn through long muddy cow lanes.

Arguments

Those who have discussed organic dairying with adherents or cynics in the United States may recognize some of these exchanges. This is not unique to the United States. Increasing intensification of farming everywhere means that these exchanges echo dialogues between organicists and conventional farmers in countries as far afield as Germany and Argentina.

Pastoralists claim cow welfare and well-being are better on pasture. They also maintain that, all things considered, low-input/low-output dairying has far less of an environmental impact than the highly capitalized high-input/high-output systems that characterize megadairies.

Zero-grazers argue that, because of the high productivity of intensive confinement dairying, their method uses the least amount of resources per unit of milk and is therefore the best route to environmental sustainability.

Pastoralists claim zero-grazers fail to include all resources in a life cycle analysis (LCA) of confinement dairying (see Arsenault, Tyedmers, and Fredeen 2009). Both sides tend to believe proper LCA will eventually validate their positions.

Pastoralist claim zero-grazers overlook cow welfare and effects on longevity.

Zero-grazers answer that high animal welfare standards can be attained in confinement. For example, when a link was proven between cattle lameness and time spent on concrete, farmers deployed rubber sheets on feedlot walkways (Robinson 2010) to cushion the hooves of cows, whose ancestors grazed on turf. Sheets that were too slick caused slips and injuries. More pliant sheets gave more traction for increased safety but wore out more quickly. Eventually, studies showed that cows suffered fewer slips and injuries on sand than sawdust and other materials (DeLaval 2013). However, sand was hard on machinery and could be hard to source.

Pastoralists claim that grazed cows suffer less stress, pain, mastitis, lameness, and infertility than confined cows, who, despite their use of antibiotics, burn out young. All farmers, conventional or organic, manage their herds by voluntary culling to maximize production, and by involuntary culling, or sending to slaughter those cows that produce low amounts of milk or none. Failure to conceive is the primary reason that young, stressed cows are culled.

In considering the causes of cow stress, there are some surprises. Paul Robinson is a UK cattle nutritionist who was raised on a 32-cow family dairy farm in the south Pennine Hills, and he advises on fertility in high-yielding dairy cows. As a Nuffield Foundation–funded scholar, he studied aspects of fertility, cow comfort, and general welfare among herds of various sizes in the dramatically varied climates of Saudi Arabia, Sweden, the United Kingdom, and the United States. Robinson (2010: 2) notes that bovine dairy genetics have accelerated in the past quarter century with many improvements, but the greatest genetic gains are in the amount of milk a cow produces; this comes at a cost to the cow: "Increased milk yields have also been helped by improved management and nutrition. This increase in yield has come at a price. Cow longevity has been reduced, which is another way of saying the culling rate has increased. Possibly the worst impact of higher yields has been significantly poorer fertility."

The prospect for a high-producing cow that might have calved one or more times but then cannot calve again is slaughter. At the annual Animal Health Conference at Newcastle University in northeast England, November 27, 2013, Robinson stressed the impact that the stocking rate in the physical environment of barns and sheds has on animal health, including fertility. One of the CAFOs he observed in the United States had found a way to save money on lumber. Instead of running a wooden two-by-six-inch header board above a manger, a photo showed barbed electric wire above cows' heads that instead discouraged them from getting too rambunctious. Consequent stress lowers cows' immune systems, making them more susceptible to disease.

Conventional Lives Worth Living

Dairy Cow Housing, a 2010 report prepared by The Dairy Group of Somerset, United Kingdom, explains how cow stress is triggered by environmental and social factors that are harder to manage in confinement than free-range pasture, where cows have a chance to avoid rivals. (The report was prepared for Arla, Morrisons and Dairy Co., a multibillion-dollar consortium of conventional Danish, German, and UK cooperatives, processors, and supermarkets, including the Anchor butter brand.) The need for careful management is imperative for Holstein-Friesians, big-boned cows typically weighing over 1,200 pounds that were originally bred by northern Europeans to thrive on grass. Most U.S. milk comes from such cows. These black-and-white cows' yield has long been prized by conventional, production-oriented farmers, but the very size of these bovine giants means that care must be taken to ensure confinement barn stalls and walkways are large enough to accommodate them comfortably. If that sounds simplistic, consider this: The dairy industry is cyclical, with periodic slumps. When prices finally return to profitability, farmers are tempted to add more cows to their CAFOs than they can comfortably contain. Nondominant cows suffer stress-related decreases in production, yet farmers hesitate to cull them because maximum income, from high and low producers, is needed to pay debts.

Cows are hierarchical animals, with heavier, more mature animals generally dominating new entrants to a herd. Negotiating dominance in a new herd, generally in face-offs or actual head butting and horn wrangling among cattle that are not dehorned, may take three days to a week. Both dominant and subordinate animals, especially the weaker ones, suffer stress till the outcome is determined. Barn architecture that fails to allow sufficient space for these natural activities can severely affect herd health and in turn farm profit. If, for instance, a barn is poorly designed and a heifer who has just had her first calf cannot avoid dominant females, her transition period from pregnancy through calving to full lactation will lengthen, with "adverse effects on animal health, welfare and production" (The Dairy Group 2012: 7). Cows need more shoulder room at feed bunks than mathematically needed, at least 5 percent more manger length than the sum of the cows standing shoulder-to-shoulder at a theoretical manger would suggest. The same rule applies to free stall space. Cows tend to "synchronise their behaviour and lie down at the same time" (The Dairy Group 2010: 12), so stress rises if space is not ample.

In the past, when farmers wintered their herds in small barns, the cows were in tie stalls or assigned stanchions to preclude fighting. (Cows quickly adapt to "their" stanchion.) Although outside exercise periods were pre-scribed when weather permitted, roomier loafing sheds or free stall areas can be healthier alternatives for cows. If they are not roomy enough, however, stress rears its head. Subordinate animals should always have ready access to space at mangers for feeding and free stalls for lying and resting. Veterinar-ians recommend ample space in confinement operations for cows to stand and socialize in their respective dominant or subordinate groups (Biagiotti 2013). When this author researched a report for *Hoard's Dairyman* (Scholten 1990a: 190) on rising European farmer concern for animal welfare, an elderly and well-regarded Bavarian veterinarian, Dr. Sambraus, stated in an interview that cows suffered stress in groups exceeding about 75. This is congruent with The Dairy Group's 2010 report suggesting 80 as the upper social limit for milk cow groups, which also confirms the experience of dairy farmers in Washing-ton State, who found that separating their 150 cow herds into two strings of 75 cows eased management.

Drawing on wide-ranging studies from the European Union and United States, The Dairy Group (2010: 13–14) presents a "typical daily time budget for [a conventional] lactating dairy cow" in six activities: (1) eating, 3–5 hours (in 9–14 meals per day); (2) lying/resting (12–14 hours); (3) social interaction (2–3 hours); (4) ruminating (7–10 hours); (5) drinking (0.5 hours); (6) non-housing time (milking, travel time) (2.5–3.5 hours).

America has led Britain in the genetics and management of intensive dairying, albeit largely based on Ayrshire, Guernsey, Holstein, and Jersey breeds developed over centuries of animal husbandry around the British Isles. In the United States, where cows produce even more milk than in the United Kingdom, a study by R. Grant (2009; in The Dairy Group 2010: 14) found high-yielding cows rested 14 hours per day—that is, two hours more than earlier studies found. Lying and resting time are crucial to cow welfare. In a sense—and considering the tremendous biological output of modern conven-tional and organic cattle—a cow is running when she's standing still. Higher yielding cows need more rest. The Dairy Group (2010: 10–11) suggests that, with yield rising a challenging 3 percent or so annually, modern cow comfort, health, and longevity depend even more than in the past on the "Five Free-doms" listed in the 1997 *Farm Animal Welfare Council Report on the Welfare of Dairy Cattle* (15):

1. Freedom from Hunger and Thirst—by ready access to fresh water and a diet to maintain full health and vigour.
2. Freedom from Discomfort—by providing an appropriate environment including shelter and a comfortable resting area.
3. Freedom from Pain, Injury or Disease—by prevention or rapid diagno-sis and treatment.
4. Freedom to Express Normal Behaviour—by providing sufficient space, proper facilities and company of the animal's own kind.
5. Freedom from Fear and Distress—by ensuring conditions and treat-ment which avoid mental suffering.

FAWC (1997) demanded that all livestock "must at least, be protected from unnecessary suffering." The Five Freedoms were offered as "ideal states rather than standards for acceptable welfare." In other words, they are meant to guide "the steps and compromises necessary to safeguard and improve welfare within the proper constraints of an effective livestock industry." The word "compromises" may have alarmed animal welfarists, but FAWC recognizes the fact that both farmers and animals sometimes engage with less than ideal conditions.

The Dairy Group (2010: 10–11) notes that in March 2011, the FAWC stated that farm animal welfare should move beyond the Five Freedoms to the quality of an animal's life to ensure an "animal, from its point of view, has 'a life worth living.'" The intuitive idea of a "good life" implies positive experiences and standards of welfare substantially higher than previous legal minimums. FAWC (2011) advised that these ideals inform "the design, construction, maintenance and management of all buildings and housing systems" for livestock.

Where greater yields place greater demands on cow health, the greater their need is to rest. This applies to both conventional cows (which spend most of the year in confinement) and organic cows (which spend winter months in barns when weather prohibits grazing). Bedding for lying down and resting must be comfortable (The Dairy Group 2010). If they are dissatisfied with bedding, cows approach it but stand chewing their cuds with only their front hooves in stall entries. In addition to slippery walkways, inadequate bedding can damage knees and forelocks, discouraging cows from getting up or lying down. Physically distressed cows may stand in alleyways ruminating when they normally would be lying down (FDA 2008: 3).

Welfare Definitions

Ethical concerns for the effects of stress, fear, and pain on animal welfare and well-being can trump other categories of debate in dairying. Terms of debate can be slippery, so it is helpful that the USDA offers lists of acronyms, definitions, and so on.

Apropos here is the USDA National Agricultural Library's *Agricultural Thesaurus* (2013) definition of *animal welfare*: "The sum or integration of an animal's past and present states of well-being as it attempts to cope with its environment; and human values concerning the social or ethical aspects of providing that environment." And *Animal well-being*: "The current state of an animal living in reasonable harmony with its environment." If ambiguity remains, the USDA definitions are at least points of reference. For simplicity's sake this book conflates the two terms into the phrase *animal welfare*. Most people endorse the use of animals for human needs, but many feel there are limits on that use, which should not compromise animals' welfare or well-being. What are those limits? One quantifiable parameter is longevity.

Jude Capper and Dale Bauman pursue research on the frontiers of intensive agriculture, working in what Lang and Heasman (2004) might call the "life sciences paradigm." Bauman and Capper (2011) and Capper and his coauthors (2009) note efficiencies gained between 1944 and 2007. In 1944 there were 25.6 million cows with a total milk production of 53 billion kilograms, and 63 years later there were 9.2 million cows with a total milk production of 84 billion kilograms. That is, around 65 percent fewer cows were producing around 50 percent more milk. Bauman and Capper claim yields were maximized "while emphasizing cow health and welfare."

However, an emphasis on welfare is not supported by data from the United States and other countries, if longevity is a proxy for cow welfare. Chapter 2 cited proceedings from the DeLaval supported Cow Longevity Conference in Sweden in 2013, where J. Rushen and A.M. de Passillé (2013: 3–4) stated that average culling rates in Canada were 30–40 percent. But the average culling rate in the United States seems to be even higher.

A number of articles in academic journals and trade publications were useful in establishing average annual culling rates for the entire U.S. dairy herd. In particular, articles published by Dr. Paul Biagiotti, DVM, a veterinarian with decades of practice in New England and in Idaho, drew from data from the National Dairy Herd Improvement Association (DHIA) in a report on culling rates in Idaho: "Every year, 44 percent of the cows and first-lactation heifers in an average herd are sold for slaughter, die or otherwise leave the farm" (*Progressive Dairyman* 2014). Biagiotti (whose book *Practical Organic Dairy Health and Management* is forthcoming with W.D. Hoard in 2015) notes it is a capital loss for farmers to cull a cow before she reaches her third lactation, and that dairy farmers may be reaching an era in which they cannot cull their way to profitability

Information from DHI Computing Service, Inc. in Provo, Utah, revealed national culling averages slightly lower than in Idaho. DHI-Provo kindly provided a table and chart of their "Rolling Herd Average - % Cows Leaving the Herd" for 2004–2013, based on their 1.6 million animal record (about one-sixth the U.S. dairy herd). This is a useful proxy for U.S. dairy trends (no identifier was given for conventional or organic herds). On a chart detailing half the states, the yearly culling rate steadily increased from a little over one-third (36.7 percent) of the national dairy herd in 2004, to within sight of one-half (42.7 percent) in 2013. It is worth noting surprisingly high culling rates among other states in 2013: Idaho (44 percent), Illinois (61 percent), New York (56 percent), Pennsylvania (39 percent), Texas (26 percent), and Washington (42 percent). Curiously, DHI-Provo (2013) documents relatively lower culling rates for herds it monitors in Columbia (24 percent), and Mexico (23 percent), but analysis of why is beyond the scope of this book.

Many factors influence culling rates. It is possible that herd removals in 2013 were motivated by high beef prices paid for culled cow carcasses, and low prices for replacement heifers (as the result of sexed semen which increases the proportion of females born). It may also be that the herds in this DHI (2013) sample represent some of the highest productivity cows in America (20,000–30,000 pounds yearly). For instance, high-production dairy herds of highly stressed cows in Idaho (also Illinois)—some with cull rates about 60 percent, and mortality rates of 10 percent—would skew both state and national culling averages upward, perhaps obscuring farms with culling rates of 25 percent.

*   *   *

Not only cows have noticed their decline in longevity. Public knowledge of declining welfare among conventional dairy cattle is suggested by the rising sales of organic milk. Many consumers doubt agribusiness prioritizes cows' health and welfare beyond their utilization for maximum profitability.

This negative public perception of agribusiness was addressed in a Bioethics Symposium held July 10–14, 2011, in New Orleans, Louisiana, and sponsored by Monsanto and Elanco, the principal firms involved with GMO dairy hormones. A subsequent paper titled "The Ethical Food Movement: What Does It Mean for the Role of Science and Scientists in Current Debates about Animal Agriculture?" by C. C. Croney and others, including Jude Capper, was published by the *Journal of Animal Science*. Croney and his coauthors (2012: 1570) note that "contemporary animal agriculture is increasingly criticized on ethical grounds" and that it is difficult for the industry to "reconcile concerns about the impacts of animal production on animal welfare, the environment, and on the efficacy of antibiotics required to ensure human health with demands for abundant, affordable, safe food."

Agribusiness actors are well aware that the ethical food movement portends legislation inimical to it. One solution, write Croney and his coauthors (2012: 1570), is for scientists to respond to public concern in terms of consumers' value systems and to clarify "misinterpretations of science" in the media. The outcome is uncertain because both agribusiness and the ethical food movement claim to have science on their side.

Agribusiness has softened the language used to reach consumers. The American Frozen Food Institute is a trade group that helped defeat California's ballot initiative Proposition 37 to label GMOs in 2012. At the AFFI Government Action Summit in Washington, DC, September 9–10, 2013, when Alexis Baden-Mayer, political director of the Organic Consumers Association (OCA 2013c) was allowed to present the case for labelling GMOs, she complained that the GMO food that Americans eat (many unaware) "has never been safety tested for human consumption, using reliable, independent long-term testing methods." The American Medical Association, she granted, says GMOs have been eaten "for close to 20 years, and during that time, no overt consequences on human health have been reported and/or substantiated in the peer-reviewed literature." Baden-Mayer questioned whether the AMA's view is proof that GMOs are safe, saying that when the AMA discusses "potential harm," the American Academy of Environmental Medicine (AAEM) talks of "probable harm." But both groups, said Baden-Mayer, support mandatory premarket safety testing. As mentioned before, the biotech industry has been avoiding such testing since the Reagan era, based on the U.S. government's doctrine of substantial equivalence between heirloom foods and GMO foods.

This infuriates greens, who see the public treated as virtual lab rats by the biotech industry.

With Baden-Mayer, a leader in the Organic Consumers Association campaign Millions against Monsanto, on the panel was David Schmidt, president and CEO of the International Food Information Council, which claims (IFIC 2013: 21) that the AMA, World Health Organization, and Food and Agriculture Organization of the United Nations "concluded that these foods [GMOs] are safe for human and animal consumption." To substantiate this sweeping statement, the IFIC cites the report on the Food and Drug Administration's review of the safety of recombinant bovine somatotropin (February 10, 1999). This is rBGH/rBST, which the Canadian government refused to license because "Monsanto's rBST submission failed to meet the standard data requirements for any new drug submission," and "Health Canada drug evaluators in the Bureau of Veterinary Drugs [determined] not all animal and human safety aspects were adequately addressed." Australia, the European Union, Japan, and New Zealand also refused to certify the drug based on human and animal health concerns.

Baden-Mayer (OCA 2013c) challenged David Schmidt and the IFIC on what she alleged was the Orwellian nature of its guide for the biotech industry "on what to say, and what not to say" on GMOs and other controversial topics when translating scientific research papers for public consumption. Baden-Mayer said the IFIC's manual *Food Biotechnology: A Communicator's Guide to Improving Understanding* contains a list called "Words to Use, Words to Lose." She adds, "The guide instructs readers to 'lose' phrases like 'not a direct danger to human health' or 'most research has not found an adverse effect' and replace them with 'safe, healthful, sustainable.'"

The IFIC website notes that the guides were "prepared under a partnering agreement" between the USDA Foreign Agricultural Service (FAS) and the International Food Information Council (IFIC). They are available in languages including Bahasa Indonesian, French, Vietnamese, Arabic, Mandarin Chinese, Russian, and Spanish, portents of the global diffusion of biotechnology exports from the United States.

Meanwhile, the OCA gathered up its cudgels for challenges to USDA politics at the spring 2014 meeting of the National Organic Standards Board, when Alexis Baden-Mayer would challenge what she saw as an industry power grab of the organic movement.

*   *   *

Animal welfare and well-being are popular issues (again, for simplicity this text conflates them in the phrase *animal welfare*). Animal welfare attracts massive attention from the public and celebrities. Consumers more often read *People* magazine and online feeds than, say, the *Journal of Animal Science*, so, as Croney and his coauthors (2012) advised, it is important to know who and what influence their opinions. The advocacy group Born Free USA lists endorsements from comedienne Joanna Lumley, rock musician

Bryan Adams, political comic Bill Maher, and comic writer Ricky Gervais in demanding more natural conditions for North American dairy cattle, as well as the African lions celebrated in the 1966 wildlife film that gave the group its name. They scorn confined feedlots as well as cages in zoos. Born Free has moved on from early baby seal campaigns, and now that it is over 40 years after the original Earth Day and past the hippie, boomer, generation X, and slacker eras, successful groups must appeal to millennials and the zeitgeist du jour.

Born Free USA (2013) condemns what it calls the "Destructive Dairy Industry" in which individual U.S. cows produce 250 percent of what they did in the 1950s, causing them to burn out much younger. The group claims the natural life span for cattle is 20–25 years, while the oldest cow on record lived to be 49 years old. In stark contrast to this natural state, Born Free USA claims 25 percent of dairy cattle are slaughtered before the age of three, and only 25 percent of the cows in the United States live more than seven years. As mentioned above, when milk prices are low, farmers succumb to the temptation to exceed reasonable stocking density. This results in more cows than there is shoulder room at feed bunks and a lack of stalls. Alternatively, if cows are fortunate enough to be on pasture, it might be overgrazed. Born Free points out that cows "confined to a barren fenced lot" face a litany of woes: "injury, illness, milk production lower than optimum, poor conception rates, and other factory-farming-induced health problems." Of course suboptimal milk production is no goal of any dairy farmer—which is exactly why such symptoms send so many cows to the knackers before their time. Born Free USA discusses the problematic physics of voluminous milk production. The bodily creation of upwards of 22,000 pounds (10,000 kilograms) of milk in one year is a remarkable physical process that takes a toll on even very sturdy beasts, such as Holstein-Friesians. High milk production entails mammoth udders, and these udders displace a cow's natural leg placement. Giant udders spread the cow's legs, affecting her pelvis, her spine, and ultimately her literal Achilles' heels—her hooves. As milk yields increased, many conventional farmers moved to rotational milking schedules of two and a half to four times a day, not only to collect more milk but also to ease the udder bulk that pulls on a cow's stomach muscles. This is also why robotic milking units, which can milk a cow whenever she wants, are selling well in the United States (in addition to the scarcity of farm labor due to delays in immigration reform).

The fulsome indictment of conventional dairying by Born Free USA is shared by many people in the organic community. But cows' litany of woes is lengthy: Mastitis of the udders affects about 33 percent of U.S. cows; this bacterial infection is exacerbated by high-energy diets and the added stress for the estimated 7–25 percent of the total 9 million U.S. dairy cows that are injected with rBGH/rBST to increase their yields, and in so doing enlarge their udders. Calves grown with preventive, or prophylactic, antibiotics (to spur growth) become cows whose clinical ills are treated with antibiotics before, arguably, early deaths. Of course, some argue that conventional cows simply serve a brief tenure as milk producer, before their ultimate role as beef provider, but the stress that disrupts cows' natural skeletal, reproductive, and digestive functions and routinely results in mastitis begs scrutiny. Is it any wonder that U.S. livestock, including cows, pigs, and chickens, receive 80 percent of the antibiotics sold in the country, four times more than the human population of 320 million people? It is now accepted by top Anglo-American health officials that the careless overuse of antibiotics in conventional livestock and human treatments have dangerously depleted medicine's arsenal against infectious disease.

Writing about antibiotics in *The Atlantic* (2012), Robert S. Lawrence, professor of environmental health sciences, health policy, and international health at the Johns Hopkins Bloomberg School of Public Health, believes the voluntary guidance touted by President Obama's Food and Drug Administration (FDA) in April 2012 is much ado about little. Lawrence notes that the FDA asks the agropharmaceutical industries to end the use of antibiotics to promote growth in young animals, including bovines. Only veterinarians should prescribe antibiotics, finally ending over-the-counter sales. But the FDA continues to endorse the "treatment, control, and prevention of disease" with antibiotics to "compensate for overcrowded and unsanitary conditions and prevent disease at the industrial operations that produce most food animals in this country."

Intensification and scale have concentrated disease vectors: in the last six decades, the number of livestock in America has doubled, while the farms that hold them are 80 percent fewer. (Conventional beef, pork, and poultry numbers rose, while dairy cow numbers were decreased due to their massive production amid stable demand in a relatively mature market.)

Lawrence says the danger of antibiotic resistance in human health is serious and increasing and that this is because the FDA still allows preventive or prophylactic use for disease prevention in concentrations of animals. The discussion above further reveals high levels of animal stress in conditions too crowded to allow them to express natural behaviors, making them vulnerable to infection. He observes that Denmark banned such misuse of antibiotics for livestock in 2000 despite fierce lobbying by the swine industry. The result was a temporary spike in mortality, which antibiotic proponents had warned of. However, after improvements, giving pigs more room, better nutrition, and delayed weaning to allow piglets to drink more of their mothers' immune system–boosting colostrum, things improved. After these meaningful reforms, mostly borrowed from the principles of organic farming, Denmark retains its position as the world's preeminent pork exporter.

Like the swine lobby in Denmark, conventional U.S. dairy farmers and veterinarians tied to the pharmaceutical industry decry what they perceive as misinformation about their practices. They insist, for instance, that periods of withholding milk after cows are treated with antibiotics are adequate and that dairy processors check stringently for antibiotic contamination to prevent any taint from reaching consumers. Their position is expressed eloquently by Dr. Angela King (*Hoard's Dairyman* 2013e), a Wisconsin veterinarian who serves 60 farms averaging two hundred cows, totaling 12,000 animals. She dispels the notion that veterinary care consists largely in administering antibiotics. On the contrary, there is a reluctance to prescribe antibiotics, as other interventions are often appropriate: "Sometimes cows need fluids, antitoxin, nutritional supplements, electrolytes, vitamins, anti-inflammatories, probiotics, aqua therapy, or even a magnet to solve their health issues." King explains farmers' financial incentives to avoid antibiotic use. First, "antibiotics aren't cheap," and second, "every time milk is picked up from a farm, a sample is taken and tested for antibiotics. If a tank tests positive, the entire load of milk is dumped."

Mainstream farmers who follow strict protocols are irked by careless charges that conventional milk is fraught with antibiotics. Not all farmers are so scrupulous. On Christmas morning near Crewe, England, in 1989, this author offered to help a local farmer milk. He mentioned one cow had mastitis. When he poured her milk in the bulk tank, I asked if she were not being treated with antibiotics. "Aye," he answered, "but they [processors] aren't checking in the holidays!"

Antibiotics Dwindle

Biological resistance has grown wherever antibiotics are used, and consumers are concerned that the list of antibiotics effective against common infections is dwindling. Tabloid readers in the United Kingdom—where penicillin was developed—read that only 10 percent of *Staphylococcus aureus* is controllable by penicillin today, compared to 95 percent in the 1940s (*Daily Mail* 2013). Germ evolution in this case is working against public health. About 50 years ago, agribusiness began using drugs, such as penicillin, streptomycin, tetracycline, and other antimicrobial additives, in feed for poultry and other livestock. Now medical professionals agree that the overuse of antibiotics in conventional agriculture is a factor in decreasing efficacy of the drugs in human health. MRSA and similar infections are said to result in five thousand unnecessary deaths per year in UK hospitals. Britain's chief medical officer Dame Professor Sally Davies calls the dearth of effective drugs a "ticking time bomb" for health—a potential catastrophe to rank with terrorism and climate change (*Independent* 2013). She warns of returning to the nineteenth century when a simple scratch can turn fatal.

With about five times the population of the United Kingdom, the United States counts similar death rates from superbugs, in MRSA-like conditions of antimicrobial resistance (AR). Conservative estimates by the Centers for Disease Control and Prevention (CDC) found 23,000 deaths annually (CDC 2011; *New York Times* 2013a). Estimates for AR-linked deaths had been as high as 100,000, but the CDC instructed researchers to eliminate cases in which AR may have been implicated but was not the prime cause of death. AR deaths totaling 23,000, still a disturbingly high figure, equates to more than 70 percent of the 32,000 automobile accident–related deaths in 2011.

Altogether, the CDC (2011) estimates two million Americans are infected with drug-resistant germs each year. So many cases turn serious when immune systems are impaired that an Interagency Task Force on Antimicrobial

Resistance was created in 1999 to coordinate a national response to AR. Cochairs with the CDC included the Food and Drug Administration and National Institutes of Health; participating agencies included the Agency for Healthcare Research and Quality, Centers for Medicare and Medicaid Services, USDA, Department of Defense (DoD), and Environmental Protection Agency (EPA). Policy changes regarding antimicrobial resistance (AR) practices were a high-stakes matter for agropharmaceutical companies. Lobbyists might press any political advantage in federal agencies or in congressional committees with oversight on antibiotic use.

Do Greens and Luddites bully Big Pharma? Following the development of penicillin in the first half of the twentieth century and the later discovery of streptomycin, millions of lives were saved from previously life-threatening injuries. This may have led to unreasonable public faith in the pharmaceutical sector—and not a little hubris by the industry itself. Evidence has accumulated against the overuse of antibiotics. There is a chance, however, that Big Pharma could silence critics with the equivalent of pulling a rabbit out of a hat.

At Oregon State University in Corvallis, Professor Bruce Geller heads a research team whose paper in the *Journal of Infectious Diseases* (Geller 2013) claims new drugs called PPMOs (peptide-conjugated phosphorodiamidate morpholino oligomers) offer a new approach to antimicrobial resistance (AR). In a press release from OSU (Oct. 15, 2013), Geller claimed: "The mechanism that PPMOs use to kill bacteria is revolutionary." Conventional antibiotics are often found in nature, but PPMOs are synthesized in the laboratory as analogs of DNA or RNA to silence the expression of specific genes. Geller explains: "They can be synthesized to target almost any gene, and in that way avoid the development of antibiotic resistance and the negative impacts sometimes associated with broad-spectrum antibiotics." If PPMOs succeed, they would be a victory for the life sciences paradigm favored by the pharmaceutical industry in competition with the ecological approach supported by organicists (see earlier references to Lang and Heasman 2004: 37–40).

Whatever promise PPMOs hold, antibiotics remain part of agriculture in the United States, United Kingdom, and many other countries. At a conference on Animal Health at Newcastle University in Northeast England (Nov. 27, 2013), students presented the pharmaceutical industry's advice on drying off cows between lactation and their next calving. Besides washing and teat dipping, farmers were advised to inject one tube of antibiotics in each quarter of the udders before the dry cow period. This is an apparent disconnect, when the UK's chief medical officer is cautioning against the routine use of the drugs.

Consumers are nervous that antibiotics are routinely administered to healthy cattle as MRSA becomes endemic in more hospitals. Lobbyists for Big Pharma protest that fears are misplaced. For example they claim antibiotics fed to poultry are different from those administered to bovines, and, as said, conventional dairy cows have periods of milk withholding after antibiotic treatment.

For decades after the Second World War the public trusted science to produce successors to penicillin. That trust is waning. Antibiotic angst has been one more trigger to consumer acceptance of the USDA National Organic Program (NOP), which prohibits their use in livestock products.

Singer's Animal Liberation

The contemporary animal welfare movement was pioneered by Australian philosopher Peter Singer, whose 1975 book *Animal Liberation* prompted many readers to take nonhuman creatures more seriously. When the book was published, animal welfare was absent from many more university syllabi than it is today. Academia was influenced more by B. F. Skinner's theories on behaviorist conditioning (1972) and inured to the effects of the mass testing of mice, monkeys, and other creatures in laboratories.

Before Singer, much of the philosophical analysis that reached lay readers implied that humans were utterly different from animals. Animals were assumed not only to be incapable of using tools, but bereft of the capacity for anything akin to human reasoning, fear, and other emotions. This led to the not-uncommon assumption that subjecting mice to laboratory electrical grids was ethically acceptable because *animals don't really feel pain*. Such attitudes seem antediluvian in retrospect.

Claiming he did not much care for animals himself, Singer nevertheless focused on the similarities between humans and nonhumans—and the ethical demands these implied for humans, who dominated such relationships. Singer (1975) dubbed the prioritization of human pleasure above sentient animals' pain as "speciesism." Prejudice against whales, chimps, or cows began to assume the hue of racism.

When his 1975 book was published, Singer judged dairy cows' lives better than animals subject to painful laboratory testing. But his ethical judgments became more critical over time. In his 2002 book *One World*, Singer deplores rulings by the World Trade Organization (WTO) that prioritize free trade over the interests of animals, workers, and the environment. He has always taken an ethical stance against causing animals unnecessary pain, but his reservations on eating meat do not extend to oysters because, lacking a central nervous system, it is questionable whether they experience pain. He is almost ambivalent about chickens because of their limited intelligence, and for this he has evoked ire among poultry advocates. He condemns the globalization of factory farming, saying cattle should be allowed to express natural behaviors, such as grazing. Over the decades, Singer developed a critique of intensive conventionalization in which most U.S. and UK dairy cows are confined away from green pastures. In the four decades since Singer's first book, many more universities address animal welfare (Scruton 1998), reflecting public concern for animal welfare.

High consumer priority for animal welfare was found in a study in the Center for Agroecology and Sustainable Food Systems (CASFS) at the

University of California, Santa Cruz. It was funded by a USDA grant to encourage sustainable agriculture on the Central Coast. Phil Howard (known for his graphical mapping of the corporate ownership of organic brands) and Jan Perez designed a 26-question survey on the relative importance of certain types of labels and mailed it to one thousand randomly selected households in San Mateo, Santa Clara, Santa Cruz, San Benito, and Monterey Counties. The survey response rate was 48 percent. They also conducted five focus groups, and Howard (2005: 2, 3, 4; also Howard and Allen 2006) found that "in the focus groups, the treatment of animals elicited the most emotion" and ranked higher than environmental impacts. Howard reported that, when asked to rank five potential "eco-labels," respondents were most enthusiastic about the idea of a "humane" label, with 30.5 percent of respondents citing it as their first choice, followed by "locally grown" (22 percent), "living wage" (16.5 percent), "U.S. grown" (5.9 percent), and lastly "small-scale" (5.2 percent).

While consumers become more sensitive to animal welfare, academic study is eroding the long-perceived dichotomy between animal and human feeling and intelligence. However, differences are still apparent in the blue state/red state polarization that afflicts U.S. politics from left to right. Anthropologist Barbara J. King's 2013 book *How Animals Grieve* combats the behaviorist notion that nonhuman creatures are motivated only by food, comfort, or procreation. She assembles evidence from ants to cats to elephants, suggesting that animal experiences after the death of a companion overlap with human experience and deserve human regard. When National Public Radio (Nov. 21, 2013) broadcast King's findings in a segment titled "The Ties That Bind Animals and Humans Alike," it is likely that some listeners on the political right associated King with a liberal leftist agenda promulgated by groups such as Compassion in World Farming (CWF) and People for the Ethical Treatment of Animals (PETA).

Some critics from the Judeo-Christian heritage reason that an obsession with animal welfare detracts from spiritual relationships between humans and God. In fact it is common to hear critics portray environmentalism as a new belief for hapless agnostics and atheists. Promoters of intensive livestock agriculture sometimes invoke Biblical texts such as Genesis 1:26–28: "Let us make man in our image, after our likeness; and let them have dominion over the fish of the sea, and over the birds of the air, and over the cattle . . . Be fruitful, and multiply, and fill the earth, and subdue it and rule over . . . every living thing."

Although this scriptural passage confers primacy on humans, it does not deny that animals feel pain or that humans have no duty to mitigate animal suffering. That cruelty to animals occurs is undeniable, but there is reason to believe it is more by sins of omission than commission. Few if any farmers this author has met revealed actual hostility to animals. (Although it is difficult for anyone to repress annoyance when a cow tail flicks one's face!)

Whether people lobby for animal *rights* or animal *welfare* may depend on their philosophical view of the sentience of animals. Those who speak of animal rights generally express human kinship with the sentience, emotion, even cognition of other creatures. Those lobbying for animal welfare acknowledge human/nonhuman differences while recognizing human ethical responsibility for creatures in their care.

The Guardian (2013d) reports that legal scholar Stephen Wise's Nonhuman Rights Project is suing to release chimpanzees from captivity in New York State. It is partly inspired by the work of ethicists Peter Singer and Paola Cavalieri, who established the Great Ape Project, litigating on several continents for the welfare of hominids, including humans, chimpanzees, gorillas, and orangutans. Quoting philosopher Richard Dawkins, Wise says, "We [humans] admit that we are like apes, but we seldom realize that we are apes."

Buttressing Wise's lawsuit are affidavits from primatologists stating that like humans "chimpanzees have extraordinarily complex cognitive abilities, and these include the ability to live their lives in an autonomous way." The argument is that such mental abilities accord chimps freedom from unnecessary constraint, akin to the Anglo-American right of habeas corpus. The aim is to remove chimps from close captivity in research or circuses to "primate sanctuaries where they have a limited version of 'the wild.'" Experiments on apes are already banned in New Zealand, the United Kingdom, and several European countries. The Nonhuman Rights Project claims to follow the same pattern of legal arguments that led to the abolition of human slavery that progressed in the nineteenth century. *The Guardian* (2013d) quotes Wise: "I think that people in the future will look back, and see this as the opening salvo in a sustained, strategic litigation campaign that led to a breach of the personhood barrier which currently divides humans from non-humans."

Of course bovines are not hominids. If the mighty aurochs was ever as intelligent as great apes, perhaps some of its cleverness may have turned to docility in the millennia of animal husbandry that begat modern dairy cows. But as Barbara J. King and many others, including Temple Grandin (1989), remind us, they deserve our respect. This book is not solely concerned with the politics of U.S. organic dairying and prospects for socioeconomic sustainability. People's altruistic impulses and empathy for animals also affect food chains from farm to plate. That is why it is important to examine the notion that small-scale family dairy farming is the best system for ensuring cow welfare. Few things are as uplifting as the human-animal symbiosis found on a healthy dairy farm. What is a healthy dairy farm? Let's take a brief look at the history of dairy expansion in the United States.

California or Bust: Westward Upscaling

Dairying first thrived in the colonial United States in the backyards of towns and the periphery of metro centers, such as Philadelphia and New York. Cities had hay markets for horses and cattle. New York is an example of urban growth, spawning the milk trains plying the Hudson River valley (DuPuis 2002). Dairying followed westward expansion, and, under the tutelage of

W. D. Hoard, it regenerated Wisconsin's soils that had been depleted by cereals monoculture (Osman 1985). Rainy green regions in New York, Pennsylvania, Wisconsin, and, a few decades later, western Washington State, led dairy production with their seemingly unassailable natural endowment of mild temperatures and three feet of rain per year watering green pastures.

Some of this has already been touched on in this book, but here the emphasis is on the march of technology as U.S. expansion moved dairying's center of gravity west. War and political economy altered the geography of dairy farming. It should not be forgotten how the Homestead Acts of the 1860s underpinned Manifest Destiny, thrusting dairying and mixed farming westward after the Civil War. Later, chemicals used for explosives and nerve gas in the First World War were turned into fertilizers and pesticides to increase yields on ground that previously relied on crop rotation and animal wastes for fertility. Elements of agricultural intensification (chemicalization, electrification, irrigation, mechanization, etc.) that advanced in the interwar Depression era were solidified after the Second World War.

The former Supreme Allied Commander in Europe, Dwight "Ike" Eisenhower, joined the Republican Party and became president, 1953–61. Despite public frustration with the bloody Korean War before Ike's election, a glorious American Century seemed underway—if voters could forget rural issues such as agricultural price supports problems debated in the 1952 and 1956 presidential campaigns waged by Democrat contender Adlai Stevenson against Ike. Stevenson's (1952) TV commercial titled "Let's Not Forget the Farmer" featured cartoons of cows in voting booths. To the tune of the oft-heard "Old MacDonald's Farm," singers compared mass farm foreclosures of the Great Depression in 1931 to the rural prosperity wrought by President Franklin D. Roosevelt's New Deal and urged voters to keep their farms prosperous with policies such as parity. The urbane and widely admired Stevenson, who enjoyed as much time on his Illinois farm as his national career allowed him to spend, urged the continuation of farm price supports and won rural votes. But it was insufficient to defeat Ike, whom a March 1952 poll declared the "most admired living American," and who vowed to end the Korean War and fix the "mess in Washington."

President Eisenhower initiated the interstate highway system, strategic infrastructure that was designed to disperse urban populations in the event of nuclear war. This highway system also stimulated tremendous residential, business, and farm expansion in the American West. Without the U.S. interstate highway system as a stimulus, it is hard to envision the present globalized food system. Eisenhower's freeways enabled gleaming stainless steel refrigerated tankers to rush southward on highways from Seattle processing plants, emblazoned with company names like Milky Way and Lynden Transfer Incorporated (LTI), and cartoon cows advising "Make Mine Milk!" or "Drink a Mug a Milk a Day!" But much of the liquid milk drunk in California was reconstituted from powder. So, the American fashion for teenagers to drink glass after glass of ice cold milk after school was ruined for some time in California by the odd gob of unmixed powder.

Californians wanted to change that, and the solution did not involve truck-ing milk 1,500 miles from Seattle. Truckloads of fresh milk and dairy products to Los Angeles were attenuated somewhat as local farmers and immigrants who had emigrated from the Netherlands applied knowledge from inten-sive dairying in their tiny homeland, where farmland was as prized as it was becoming in LA. As detailed in chapter 2, the model for periurban dairying in the suburbs surrounding Los Angeles, such as Bellflower, became one of little grazing or zero-grazing, where feed (some grown far away) was brought to cows in feedlots and their waste hauled away. Julie Guthman notes that "between 1950 and 1982, the number of milk cows [statewide] increased by 57 percent, although the number of dairy farms dropped considerably when dairy farmers started to practice intensive dairying" (Guthman 2004: 77–78; Jelinek 1979; Gilbert and Akor 1988).

Bellflower had a population of 45,000 people in 1960, but as urban pres-sure for residential real estate increased from central LA, Dutch, Portuguese and Japanese American dairy families from Artesia, Bellflower, and Cerri-tos relocated outward in concentric circles, especially east toward the Dairy Valley-Dairyland-Dairy City area. Today Bellflower is a dense suburban town with a population of about 75,000. A foray off Interstate 5 reveals fast-food restaurants more readily than vestiges of dairy farms.

New labor-saving building materials and technology, introduced in the 1960s, also spurred barn building for the covered feedlots and free stalls used in California-style confinement dairying.

Dutch Barn

The "Dutch barn" on the cover of this book was constructed over a period of months in 1951 by a contractor and crew of several men, including the farmer-owner and his brothers, in Whatcom County in Washington State. The labor intensive structure was an architectural monument of lumber cut from the dark, dense forests of the Cascade Mountains. Triangulation designed into the heavy-duty Douglas fir rafters and diagonal one-inch-by-eight-inch shiplap wood sheathing on the walls made the barn resilient in wind and rain. It might flex or swell, but it wouldn't permanently deform over decades of rough weather.

The wood-framed barn stands on a concrete foundation and guttered floor, with two longitudinally parallel rows of steel posts supporting six-inch-by-twelve-inch main wood ceiling beams. These support the lateral two-inch-by-twelve-inch wood floor joists and shiplap flooring of the haymow above. The roof is crowned by enough cedar barn shakes to cover a small field. Care had to be taken that hay was not blown in too damp, or it could spontaneously combust. A telephone ringing at midnight could be a neighbor's plea to help fork through a hay mound to relieve the heat before fire swept a barn.

Ready for winter, oats, grain, and wood shavings for cow bedding blown as high as the windows insulated the north gabled wall against wind roaring through Canada's Fraser River valley from the Yukon Territory. Mounds of chopped and baled hay rose past the windowed dormers, nearing the ceiling toward the south end. The only damage to the heavy building after 100 mph annual Arctic winter "northeaster" blizzards were a few dozen missing forty-inch-long cedar roof shakes. (The shakes were replaced by father and sons, most under age ten, which might breach contemporary health and safety regulations.)

California-style confinement dairies did not usually utilize such huge hay-mows after the 1960s. Hay bales were covered with tarps on the ground or in sheds accessible to tractor loaders. Instead of 40- to 60-foot-tall, cylindrical concrete, steel, or (outdated) wooden silos, grass and corn silage were stored in horizontal bunkers accessible to tractor loaders. Silos were problematic for several reasons: Wood, concrete, or metal silos were subject to rotting, chemical decomposition, or oxidization. Mechanical silo unloaders were expensive and often broke down, requiring specialist maintenance. Filling silos was dangerous because men and boys tramping the crop blown in through the top could suffocate in silo gas. According to the National Agricultural Safety Database, oxides of nitrogen and ammonia could affect farmers who entered silos up to two weeks after filling (NASD 2011).

There were good reasons to avoid silos, despite their pleasing church steeple–like aesthetic on farmscapes. Open air silage bunks and hay barns for larger rectangular or round bales minimized human labor and could be safer. Tractor loaders were reliable and fun to use compared to the chores of forking silage down a chute or bucking hundred-pound bales. Many owners of Dutch-style barns turned from silos to open bunkers for storage. Doing that sacrificed scarce flat space near the main barn, but the labor savings were significant.

New barn architecture became popular in the mid-1960s partly because it could be built quickly. Steel pole structures supported metal (or metal-and-wood-framed) trusses on two-foot centers that were strong enough to bear relatively low-pitched roofs, evading much wind force. In the Pacific Northwest and Great Southwest, Texmo Pole Structures were offered for a variety of applications from arenas to stables to loafing sheds for dairy cows (Texmo 2013). Progress in materials science and technology enabled new barn designs. Metal "galvanized tin" roofing that had rusted quickly in past eras was replaced with aluminum sheets with new long-lasting anodized coatings. Once the concrete foundations set, large buildings could be erected by a few workers and a crane in a matter of days.

Such architecture became the farm norm in California as its population increased toward its present-day 30 million people, and its economy reached the status of the world's seventh largest. When urbanization overran ruralities on the Central Coast between LA and San Francisco, much dairying moved to new, highly capitalized, multimillion dollar megadairies in the San Joaquin Valley, east of the coastal mountains. Many of these operations were and still are populated by thousands of cows. Unlike traditional farms built where the land can sustain their activities, these were farms irrigated by public water projects funded by taxpayers. Guthman (2004: 77) notes a wave of irrigation projects was promoted as a means to satisfy the agrarian dreams of small farmers, "but it was never successful in that regard." Agribusiness captured most of the gains from such schemes. The California Milk Advisory Board (CMAB 2013) boasts that "California staked its claim to the number one state ranking when it surpassed Wisconsin in total milk production in 1993." The Golden State's skills in wine and dairy production arrived with European immigrants. Women handled most of the dairying until the 1848 Gold Rush when commercial dairying took off. The Point Reyes area near San Francisco vied with areas of New York State as the nation's premier dairy region, and California made its name in cheese annals with Monterrey Jack. Most dairying was near Point Reyes and San Francisco, before the diversion of water to the southern part of the state. In 2013 CMAB counted over 1,500 dairy families and 1.82 million milk cows (20 percent of the U.S. total).

California-style dairying influenced the rest of the American West. The tables were turned on the temperate Northern states. As the twenty-first century proceeds, billboards boldly market California cheese in the soggy Seattle metropolitan area that once shipped trailer loads of milk and cheese south. Technology trumped the rain gods.

Taking grazing and pasture out of the equation meant dairying in California and much of the rest of America was typified by concentrated, confined-animal feeding operations—CAFOs. By and large, the cost of milk production per gallon or liter was less with confinement dairying (as long as cheap petroleum prices prevailed and farmers missed no mortgage payments on their loans when milk prices periodically fell). So CAFOs became the norm in conventional dairying. The trouble was, as their critics say, the negatives of this intensively scaled agroindustrial production system were externalized on the health of cows, the environment, and the livelihoods of smaller scale farmers who used more sustainable methods.

Steiner, Biodynamics, and Organics

Foreign countries have been regularly mentioned in this book. Indeed the growth of dairy systems in the United States is especially interlinked with Europe. The provenance of American organics runs from the Rodale Institute on the East Coast back to the British organic movement to the mid-twentieth century and to Austria, Germany, and Switzerland before that. National legislation can alter organic rules, but the methods, aims, and objectives are remarkably similar from Jersey to Japan and generally reflect the ideas of Rudolf Steiner.

Both conventional and organic dairy systems in the United States have roots abroad. Their primary breeds—Brahma, brown Swiss, Dutch belted, Guernsey, Holstein-Friesian, Jersey, and (water) buffalo—likewise have Indo-European origins. The provenance of American organics connects the Rodale Institute on the East Coast (see chapter 1) to the British organic movement led by Lady Eve Balfour (1899–1990) and botanist Sir Albert Howard

(1873–1947), back to the earlier work of Rudolf Steiner (1861–1925) in Germany and Switzerland. John Paull (2011) relates in the *Journal of Organic Systems* how at the young age of 22, the Austrian philosopher Steiner was so well regarded that he was tasked with organizing the eighteenth-century natural-science work of legendary polymath Johann Wolfgang von Goethe. Most pertinent here are eight lectures seminal to the organic movement called the "Agricultural Course," given by Steiner in Poland in 1924, one year before his death. Paull (2011: 1–2) notes that, even before the term *biodynamic* was coined, "Steiner set in train a process that led to the development, articulation, and naming of biodynamic agriculture, culminating in the publication of *Bio-Dynamic Farming and Gardening* by Ehrenfried Pfeiffer in 1938."

Strains of mysticism lace the writings of both Goethe and Steiner. The latter melded them into articles for the Anthroposophical movement, based at the Goetheanum study center in Dornach, Switzerland. As part of this movement, biodynamics is a spiritual-ethical-ecological approach to farming, food systems, and health. Positivist scientists discount the spiritual practices of biodynamics, such as planting seeds according to lunar cycles. But like organic leaders (such as Chuck Benbrook of the Organic Center in Oregon, and Professor Willie Lockeretz (2002), now retired from Tufts University Boston), Steiner extolled experiment over dogma. Paull (2011: 29, 32–33; Steiner 1924a; b; c) quotes Steiner extolling empirical experimentation to augment spiritual approaches to agriculture: "The aim of these lectures was to arrive at such practical ideas concerning agriculture as should combine with what has already been gained through practical insight and modern scientific experiment with the spiritually scientific considerations of the subject" (1924c: 9). The difference was that Steiner urged empirical experimentation on whole systems, while industrial agriculturalists highlighted gains in subsystems, sometimes to the detriment of surrounding ecologies.

Steiner's famous 1924 Agriculture Course did not use our modern lexicon, including the phrases *organic farming*, *biodynamic farming*, or *biological-dynamic farming* (Paul 2011: 32–33). But others, including his acolyte Pfeiffer, credit Steiner for inspiring these terms, based on the latter's key concept of the ideal farm as a living organism (Paull 2011: 32). Organic farming has many proud fathers and mothers, but Rudolf Steiner is primus inter pares, the first among equals when it comes to organic pioneers. What Steiner's lectures called his "hints" (for farmers to experiment with) agree with modern ecological tenets that organic farms aspire to be closed systems. These are in diametric opposition to conventional farms reliant on global outsourcing.

Steiner's ideas also tally with recent findings indicating that microorganisms in animal and human bodies, and in the microbiome in the soil around plants, play greater roles than previously assumed (*Scientific American* 2012; *Nature* 2013b). While our understanding of the microbiomes of living creatures is still in its infancy, the realization that bacterial cells in the human body outnumber human cells by a ratio of ten to one gives pause for thought on the damage to immune systems by broad-spectrum antibiotics. This is all the more reason for the bio-organic movement to promote health in conjunction with a living soil, obviating the prophylactic use of antibiotics, which they want reserved for emergency interventions.

Steiner's Anthroposophical movement diffused his ideas globally after his death in 1925. Paull (2011: 5) notes that by 1929, the Natural Science Section of the Goetheanum listed experimental stations throughout Europe and in Africa, Asia, Australia, and New Zealand, and America. Ehrenfried Pfeiffer was an effective apostle for Steiner's ideas. Pfeiffer's parents were Anthroposophists like Steiner, and by the age of 22 he worked closely with him in Dornach (*Biodynamics* 2008). Pfeiffer once said, "My innermost loyalty belongs to Rudolf Steiner. For him and his work I wish to continue to live." Pfeiffer's encapsulation of Steiner's ideas, *Bio-Dynamic Farming and Gardening*, was published at once in at least five languages, English, German, Dutch, French, and Italian (1938a; b; c; d; e; see also Pfeiffer 1938; Paull 2011: 29).

About 1,200 Steiner-Waldorf schools worldwide testify to Steiner's continuing influence. His ideas on ecology and social and economic sustainability are still evident in activities such as community supported agriculture (CSA) and so on, more of which below.

Steiner, Balfour, and Howard in Britain

Several times in his life Rudolf Steiner visited Britain, where Sir Albert Howard, Lady Eve Balfour, and others developed his "hints" on living soil in word and deed. Howard, who had worked a quarter century in India and across Asia as a soils and compost specialist, extolled the role of dairying in the tapestry of organic agriculture. Howard said that dairying was central to the virtuous circle of processes linking human, animal, and soil health.

Lady Balfour paid tribute to Rudolf Steiner as an ecologist who had great influence in the movement (even before Albert Howard or Ehrenfried Pfeiffer), in her address to the IFOAM International Federation of Organic Agriculture Movements (IFOAM) in Switzerland in 1977. "Towards a Sustainable Agriculture" was the theme of this, the first international research conference organized by the Forschungsinstitut für Biologischen Landbau (FiBL, Research Institute for Organic Agriculture).

Lady Eve Balfour fulfilled Steiner's mandate for experimentation on her own land. She began her pioneering Haughley experiment in Suffolk in 1939. The Soil Association, which she cofounded in 1946, assumed sponsorship the following year, for the next quarter century. The Soil Association (SA) was a charity established by farmers, scientists, and nutritionists who—contrary to the chemical reductionism of the day—believed farming practices, plants, animals, and people were interconnected in environmental health. One of Balfour's associates in Wales was Dinah Williams, who pioneered the non-chemicalized organic dairy farm that led to Rachel's Organic Dairy and twenty-first-century links with Dean-Horizon USA.

In her address to the IFOAM conference (1977), Balfour described the Haughley study, saying: "This pioneering experiment was the first ecologically designed agricultural research project, on a full farm scale." She described it:

> Three side-by-side units of land were established, each large enough to operate a full farm rotation, so that the food-chains involved - soil - plant - animal and back to the soil, could be studied as they functioned through successive rotational cycles, involving many generations of plants and animals, in order that interdependences between soil, plant and animal, and also any cumulative effects could manifest.

One result surprised even biodynamicists. Soil mineral levels fluctuated during the year in response to biological activity, and the finding was supported by a Scottish University study. Mineral fluctuation was highest in the organic section said Balfour (1977): "As much as 10 times more available phosphate has been recorded in the growing period of the year than in the dormant period. Potash and nitrogen followed the same general pattern." This reaffirmed the faith of biodynamicists, organicists, members of tilth societies, and Greens worldwide. They believe that rubbing tilthy organic humus between one's fingers feels livelier than chemically fed sludge under industrial crops. The bywords of biodynamics are *feed the soil, not the plant*. In the Haughley experiment, organic plants took longer to thrive but caught up with the nonorganic plants during the season and showed more resilience to drought and pests. Longevity and animal welfare and well-being were superior in the organic unit. Balfour (1977) remarked on "the longer working life of its livestock" and concluded that "with the livestock, the temperament of the animals composing the herds and flocks exhibited sectional differences, those belonging to the organic section being noticeably more contented."

Mad cow disease, or bovine spongiform encephalopathy (BSE), was first confirmed in the United Kingdom in 1986, four years before Balfour died. It peaked some seven years later, around 1993, after it had infected as many as 180,000 animals in the human food chain. In 1989, controls were introduced to stop feeding herbivore cows with the remains of other cows in the form of meat and bone meal. Claims remain that a mutation in prions caused BSE (instead of a switch from a traditional method of cattle rendering to a cheaper one), but feeding ruminants to other ruminants is indubitably the realm of industry rather than nature, and the carnivorous practice has not been exonerated (Whatmore 2002; Scholten 2007). BSE is a neurodegenerative disease that noticeably affects a cow's mobility, giving her a fearful mien. When BSE is transmitted to humans from eating infected beef, it results in a fatal condition called new variant Creutzfeld-Jacob disease (nvCJD), redolent of Alzheimer's disease. As recounted elsewhere in this book, the Soil Association (SA) expressed opposition to the industrial practices that contributed to the BSE crisis, and so the BSE crisis increased SA credibility for leadership in the struggle against GMOs. The SA leapt to preeminence in Britain's organic movement, working with protesters to veto live tests of GMO crops in the

United Kingdom (Scholten 2007; Reed 2006). The organization owns Soil Association Certification Ltd., the largest UK certifier of organic farmers and processors in the United Kingdom and worldwide.

In the spirit of Balfour's Haughley experiment, the Soil Association's director of innovation, Tom MacMillan, says the SA is establishing field laboratories to improve "productivity, quality and environmental performance in organic and low-input agriculture" in cooperation with Prince Charles' Duchy Originals brand (Soil Association 2013). Honoring scientific method, the Soil Association asks, How can animal welfare be improved if it cannot be measured? So its Innovative Farming programs include AssureWel, an assessment scheme that assurance schemes and producers can use to measure and improve welfare. Reducing antibiotic use on conventional dairy farms is another goal in its concern for animal welfare. Other programs address carbon capture, energy conservation, manufactured nitrogen on farms, and sustainable animal feed—all related to the question of whether organic farming can feed the world.

Pfeiffer and Rodale in the United States

The indefatigable Steiner directed an eight-hundred-acre experimental biodynamic farm in Loverendale, Holland, which performed studies for the movement's laboratory at the Goetheanum in Switzerland. In the 1930s Pfeiffer also spoke regularly in the United States. He was one of three European lecturers at the first Anthroposophical Summer School at Threefold Farm, in New York State in July 1933. Bill Day (*Biodynamics* 2008) writes that two of Pfeiffer's seven lectures bore the title "Dr. Steiner's Biologic Dynamic Agricultural Methods Practically Applied in Farming." In 1939 Pfeiffer was offered the means to start a model biodynamic farm and training program in Kimberton, Pennsylvania. So, Pfeiffer was able to devote the Second World War period to organic development, instead of returning to Britain, where the government was pressuring farmers to adopt chemically based farming in war preparations (Short, Watkins, and Martin 2007). Day notes that Pfeiffer also established the Biodynamic Farming and Gardening Association in 1938 and three years later its journal, *Biodynamics.*

Most serendipitous for the American organic movement was that, at Kimberton, Pfeifer met Jerome Irving Rodale, whose legacy, with his son Robert, is the Rodale Institute, widely seen as the birthplace of the country's organic movement, pioneering organic farming through research and outreach.

The title of J. I. Rodale's own magazine, *Organic Gardening and Farming*, resonated with that of Pfeiffer's widely sold 1938 book *Biodynamics.* Perhaps an American twist is discernible in a motto, inscribed on a chalkboard outside a Rodale farm shop, "Healthy Soil = Healthy Food = Healthy People." This is how Steiner's European ideas on living soil were telegraphed to American consumers. Rodale wrote in 1954 (Rodale n.d.): "Organics is not a fad. It has been a long-established practice—much more firmly grounded than the current chemical flair. Present agricultural practices are leading us downhill."

J. I. Rodale's son Robert established the thriving 333-acre organic center in Kutztown, Pennsylvania, named the Rodale Institute. Testimony by Robert Rodale and other Rodale officials helped persuade Congress to include funds for organic farming in the 1985 farm bill. Robert died in 1990, after which other family members stepped into the breach (Rodale n.d.). The institute, now a 501(c)(3) charity, continues to credit Rudolf Steiner and Ehrenfried Pfeiffer as forerunners in its commitment to "groundbreaking research in organic agriculture, advocating for policies that support farmers, and educating people about how organic is the safest, healthiest option for people and the planet."

IFOAM

When Lady Eve Balfour acted as a founding member of the International Federation of Organic Agriculture Movements (IFOAM) in 1972, the Anglo-American trend to chemicalized monoculture, intensive dairy farming, and confinement was already well underway. Her *Organics* is consonant with Steiner's idea of the farm as a living organism and with Ehrenfried Pfeiffer's use of the term *biodynamics*, perhaps without the optional elements of spirituality that accompanied earlier iterations of the movement, which still finds expression in various organic movements around the world. Balfour, for one, sounded permanently wedded to the spiritual components of organics when speaking at the 1977 IFOAM conference in Switzerland, claiming we cannot escape the "ethical and spiritual values of life" (Balfour 1977). She explained two motivations behind IFOAM's ecological approach:

> One is based on self interest, however enlightened, i.e. when consideration for other species is taught solely because on that depends the survival of our own. The other motivation springs from a sense that the biota is a whole, of which we are a part, and that the other species which compose it and helped to create it, are entitled to existence in their own right.

Four decades later IFOAM (2013: 4–5, 7) claims an "unchallenged position as the international umbrella organization of the organic world." Its flexible yet idealistic guiding principles bear quoting verbatim:

Vision: Worldwide adoption of ecologically, socially and economically sound systems, based on the Principles of Organic Agriculture.
Mission: Lead, unite and assist the organic movement in its full diversity.
Values: IFOAM acts in a fair, inclusive and participatory manner and highly appreciates the diversity of the Organic Agriculture movement.

Abbreviated versions of IFOAM's Principles of Organic Agriculture follow:

1. "Principle of Health: Organic Agriculture should sustain and enhance the health of soil, plant, animal, human and planet as one and indivisible."

Steiner's vision of the farm as an organism counters reductionism in conventional monoculture, urging farmers to bolster cows' immune systems rather than rely on antibiotics.

2. "Principle of Ecology: Organic Agriculture should be based on living ecological systems and cycles, work with them, emulate them and help sustain them."

In line with Lang and Heasman's (2004: 33–35) advocacy of the ecological approach, IFOAM questions the life sciences paradigm of the agropharmaceutical industries.

3. "Principle of Fairness: Organic Agriculture should build on relationships that ensure fairness with regard to the common environment and life opportunities."

Animal, human, and environmental stakeholders' welfare must be considered along with stockholders' quest for productivity. Principle 3 looms large in the discussion of IFOAM's (2006) St. Paul Declaration below.

4. "Principle of Care: Organic Agriculture should be managed in a precautionary and responsible manner to protect the health and well-being of current and future generations and the environment."

The precautionary principle requires assiduous scientific experimentation to evaluate new technologies that risk catastrophic damage to present and future creatures and biospheres. Examples include the risk of GMO maize trials in maize hearth areas of Mexico, and they also include the unrestricted breeding of (possibly genetically weak) cloned cattle with others. To the dismay of GREENS and organicists, the precautionary approach was abandoned by the Reagan-Bush administrations in favor of the USDA's present risk/benefit system, which neoliberals deemed more likely to boost the contribution of the biotechnology industry to the U.S. economy (Scholten 1990c: 8).

GMO advocates may be proven right on their profitability in the short to medium to long term, barring mishaps of Fukushima proportions. It is not, however, only American greens who are discomfited by Washington, DC, ramrodding of industrial agriculture. Asian and African buyers of U.S. grain and alfalfa have balked, at least temporarily, at imports of U.S. grain contaminated with GMOs. U.S. dairy and beef exports also suffered after 2003, when cases of mad cow disease were found. So the economics arguments cut both ways.

IFOAM recognizes our Malthusian dilemma, the need to increase efficiency and productivity in a world with finite agricultural area and rising population. New technologies are welcome, but, cautions IFOAM, they must be assessed carefully to avoid unintended consequences. This takes us back to the reliance that Steiner, the early biodynamicists, Balfour, Rodale, and contemporary organicists have placed on scientific experimentation buffered by the practical experience of real farmers. The way that foundational principles have been expressed globally has varied according to national and local conditions. Looking at such variation can illuminate the way forward for U.S. organics.

In the *Directory of IFOAM Affiliates 2013*, executive director Markus Arbenz lists 788 members, associates, and supporters in 117 countries, plus IFOAM self-organized structures, including regional bodies, sector platforms, and subsidiaries. Close ties are retained with the UK Soil Association that helped birth it.

IFOAM (2013: 6) is linked to four regional bodies (Asia, AgriBio-Mediterraneo, EU, and Latin America) and two national bodies (France and Japan). It has four sector platforms (Aquaculture, Amenity Agriculture, Animal Husbandry, Farmers). IFOAM is accredited by the United Nations, and its strategic partners include the UN Food and Agricultural Organization (FAO) and UN Conference on Trade and Development (UNCTAD) and FiBL, the Research Institute of Organic Agriculture of Austria, Germany, and Switzerland, which organized the first international IFOAM research conference in 1977 at which Lady Balfour spoke.

Its two daughter organizations are the Organic World Foundation (OWF, in Switzerland) and the International Organic Accreditation Service (IOAS). IOAS is registered in Delaware as a nonprofit NGO, independent because its income is derived from certification and paid studies. IOAS also offers accreditation regarding ISO/IEC standards, which are important in global organics trade.

Global reach gives IFOAM the capacity to collate streams of longitudinal data from its affiliates. For over a decade, editors Helga Willer and Minou Yussefi have compiled annual editions of *The World of Organic Agriculture— Statistics and Emerging Trends* at its headquarters in Bonn. Freely downloadable on the Internet, the compendium is a boon to researchers ready to peer above their national parapets and make international comparisons.

Key U.S. organizations affiliated with IFOAM figure in this book. They include California Certified Organic Farmers (CCOF Tilth), National Organic Coalition (NOC), and the Northeast Organic Farming Association (NOFA). NOFA, begun in 1971, housed the Northeast Organic Dairy Producers Alliance (NODPA), which began in a 2001 crisis when a processor summarily lowered milk prices, and farmers managed a group response. Lisa McCrory acted as coordinator, followed by Sarah Flack in 2004. NODPA became a separate entity in 2005, when Ed Maltby became executive director and webmaster of its lively Odairy listserv. Organic Consumers Association (OCA), Organic Trade Association (OTA), Organic Valley (OV/CROPP cooperative), and Rodale Institute are also affiliates of IFOAM.

IFOAM's International Conference 2006

IFOAM describes its mission as leading, uniting, and assisting the organic movement. This is complicated due not only to different climatic conditions worldwide, but also due to different traditions of animal keeping. The federation set its strategic policy on animal welfare at the first IFOAM International

Conference on animals in organic production at the University of Minnesota, in St. Paul, August 23–25, 2006. It was organized by Jim Riddle, who served on the National Organic Standards Board (NOSB) 2001–06. On his LinkedIn (Jan. 31, 2014) profile Riddle writes: "In 1997, I was incensed by the USDA's First Proposed Organic Rule, and wrote detailed comments, complete with replacement language." Thus the Big 3—GMOs, sewage sludge, and irradiation—were removed from permitted materials lists. As of April 2014, there is a less than subtle campaign by industry to increase the permitted percentage of nonorganic components, including GMOs.

The 2006 IFOAM conference was a pivotal moment in a delicate phase of U.S. organic pasture wars. Organic Consumers Association (OCA) had already begun its boycott against Aurora Organic Dairy and Dean Foods–owned Horizon Organic Dairy, and The Cornucopia Institute had filed legal complaints against Aurora and Horizon for neglect of pasture. As both Aurora and Horizon were among the sponsors of IFOAM's 2006 St. Paul conference, there was an apparent disconnect between the corporations figuratively dragging their hooves over an explicit endorsement of the 120 day/30 percent dry matter intake (DMI) grazing minimums demanded by Cornucopia and the OCA, when zero-grazing was banned in *The IFOAM Basic Standards for Organic Production and Processing Version 2005:* "5.1.3 Landless animal husbandry systems are prohibited. 5.1.4 All animals shall have access to pasture or an open-air exercise area or run, whenever the physiological condition of the animal, the weather and the state of the ground permit."

IFOAM essentially reprised its 2005 era standards in its 2012 *IFOAM Norms for Organic Production and Processing.* Notwithstanding the global organization's recognition of pasture as a defining component of organic animal husbandry, it was eight long months before May 2007, when Horizon Organic Dairy swung its weight behind the 120 day/30 percent DMI minimums. IFOAM, too, is an actor in the Pasture War. But in this battle, the USDA had the final say. The pasture paradigm of organic dairy farming was in the balance in the 2000s. It was likely that a USDA decision mandating more traditional patterns of extensive grazing through the growing season would help the Organic Valley cooperative woo more farm members. About this time a woman farmer on the Odairy online discussion list remarked: "The processors all know the consumers want pasture or they wouldn't show cows on grass on their cartons, or promote the benefits of grass fed without really having it."

The second term of President George W. Bush was, however, ill timed for the pasture movement to expect succor from Washington, DC, even if the president did spend vacations on a family ranch in Texas. In the capital Beltway, agribusiness consolidation was high on the agenda of lobbyists and politicians, who imagined this might generate more tax receipts, as the federal budget deficit ballooned during wars in Iraq and Afghanistan.

While the pasture rule was under consideration by the Office of Management and Budget (OMB), The Cornucopia Institute kept it in the public eye. When processors and retailers urged quiet to retain consumer confidence,

Mark Kastel was outspoken in saying that selling milk from cows not on grass made a sham of organic integrity. Kastel maintained that in 2004, when The Cornucopia Institute launched its Organic Integrity Project, it was already clear there was a perfect economic model for where the organic dairy industry was headed—the conventional dairy industry. Only a strong pasture rule could avert that dystopian dairy future.

Amid the legal and legislative doldrums came a little joy for pasture partisans in the summer of 2006, when IFOAM defended grazing at its St. Paul conference. Its perspective on welfare and well-being seemed more to come through the eyes of a dairy cow than an industrialist seeking to maximize profitability from milk production. That said, it should be recognized that the 2006 conference had a broad base of support with major sponsors, not only Horizon Organic and Aurora Organic Dairy (both being sued then by Cornucopia), but also Organic Valley, Newman's Own Organics, Whole Foods, FiBL, MOSES, Organic Trade, the Minnesota Department of Agriculture, and the National Center for Appropriate Technology.

A keynote speaker at the University of Minnesota conference was Temple Grandin, the animal behaviorist who credits the unique perspective of her own autism for insights on animal welfare. As a consultant, Grandin is estimated to have helped redesign as much as 50 percent of cattle corrals, water dips, chutes, and other elements of cattle transport and slaughterhouses in the United States. In her influential books, Grandin (1989; 2005; 2006) explains animals' sensory-based thinking, how cattle, like autistic people, process information in images not words. She maintains consciousness is possible without language. Seeking more dignity and less fear in the dispatch of beef and dairy cows, Grandin designed uncertainty out of walkways and lighting conditions in veterinary procedures and abattoirs. Her logical and passionate arguments won agreement from actors across the livestock sector, from small farmers, fast-food chains, and industrial agribusiness operations.

Antibiotics also figured in the IFOAM meeting's recognition of livestock as sentient beings, whose welfare and health management are important parts of organic food and fiber production, according to the Principles of Organic Agriculture. In the subsequent issue of *The Inspectors' Report* (2006: 13) one presenter concluded:

> Resistance to important antibiotics for human health—fluoroquinolones—had been developed by the highly pathogenic *Campylobacter jejuni* species of bacteria in 46 percent of conventionally raised chickens and 67 percent of conventional turkeys respectively, whereas less than 2 percent of the organically raised chickens and turkeys exhibited resistance.

Such observations on poultry had implications for cattle. A host of speakers showed how "organic livestock systems improve both animal and human health while protecting the environment." In this vein Grandin (ibid.) emphasized that "Organic Agriculture is positioned to create a new model of agriculture . . . diametrically opposed to industrial agriculture . . . Humans can no longer regard themselves as somehow separate from the ecosphere of which they are an integral part. In terms of organic standards and animal welfare we have to make sure that sick animals get treated in organic systems."

St. Paul Declaration 2006

After three days of deliberation, over 250 participants from two dozen (eventually 35) countries unanimously approved the St. Paul Declaration, which begins (IFOAM 2006b, *italics added*)

> We, the participants of the first IFOAM International Conference on Animals in Organic Production, recognize that *animals are sentient beings*. Rearing animals for production and domestic purposes is an evolving relationship that has spanned millennia and is based on ecological principles of *mutualism and interdependency, bound to culture and local circumstances* . . . The basis for organic animal production is the development of a *harmonious relationship among soil, plants, animals and humans. Organic animals should be provided with the conditions and opportunities that accord with their physiology and natural behavior.*

Analysis of the historic declaration finds, first, that organic-industrial monocultures are reproached by the phrase extolling a "harmonious relationship among soil, plants, animals and humans." Inputs should not be sourced from afar by truck. Ideally they are sourced on a farm working as an organism, as Steiner outlined. Phrases stipulating that "animals are sentient beings" were welcomed by animal welfarists who appreciated the reference to "mutualism and interdependency" connoting the respect for livestock demanded by Temple Grandin. Some wiggle room for confinement in animal rearing may have been interpreted in the phrase "bound to culture and local circumstances" around the world, but that interpretation was negated by this sentence: "Organic animals should be provided with the conditions and opportunities that accord with their physiology and natural behavior." The conclusion was inescapable that animals with the physiology of ruminants must express their natural behaviors by, among other things, grazing on pasture.

Agricultural ethics are seldom as black and white as Holstein-Friesian cattle. One example is that of a vertically integrated farm and dairy in Northern California, which produced, processed, and sold its milk and dairy products on-farm and in local outlets. It was established with a couple dozen cows several decades ago, certified organic in the 1990s, and later consisted of a few hundred cows on a square mile of ground. The operation was influential in state and national organic dairy movements. However, before the final pasture rule appeared, the family was distressed by rumors their farm might not comply with mooted regulations on stocking density. Waves of support for them appeared online and in newspapers. The contretemps blew over, but it illustrates how rules making can be a long process of fitting square pegs into round holes.

IFOAM's St. Paul Declaration can be read as an endorsement of traditional family-scale pasture dairying against megadairy confinement. The declaration condemned organic agribusinesses that denied cows their natural inclination to graze in pasture—or forest. On the other hand, at least one prominent confinement dairy business reportedly practiced natural breeding in animal husbandry, even before the USDA clarified its Pasture Rule in 2010. This is a more natural expression of instinctive cattle behavior, than the artificial insemination (AI) that is employed on some traditional pasture-grazing farms that introduce new DNA into their herd. When contemplating the variety of livestock traditions internationally, IFOAM remains sensitive to the fact that Rudolf Steiner offered *hints* not *prescriptions*.

This chapter began by claiming pasture as the common ground between cow welfare, longevity, fertility, lameness, mastitis, and the profligate use of antibiotics, whose overuse now limits the options available to sustain human health against infection. IFOAM's St. Paul Declaration comprehensively supports the role of pasture grazing in its biodynamic (i.e., organic) role binding animal, soil, and human health in a positive cycle. The declaration is a paean to extensive, connected organic agriculture and an indictment of unnatural industrial confinement that is bound to incur disease and more.

The following chapter recognizes, nevertheless, that issues are not entirely clear-cut regarding antibiotics and the relief of animal suffering. Much can be learned by exploring the Dutch dairy culture of Whatcom County.

5

Stewardship in the Northwest

Photo 5.1   Concepts of "stewardship" drive dairying around Lynden, Washington.
Photo Credit: Bruce Scholten.

Stewardship in the Northwest: Dutch Stewards, Vets, and Researchers Discuss U.S., Canadian, and European Rules

In Whatcom County, east of Puget Sound, on the U.S.-Canadian border, pastoralists have been squeezed by exurban sprawl from two prosperous cities: Seattle to the south and Vancouver to the north. Historically, this is one of the most productive dairy farming communities in the United States. As we will see further below, despite the economic pressures discussed throughout this book, family farming has remained viable due to the strong values held by Whatcom County farmers regarding stewardship of the land. Families on the small but growing number of farms belonging to the national Organic Valley cooperative were joined by consumers honoring the 2000s boycott of Aurora and Horizon industrial-organic products in dairy cases. Many people who were contacted by this author, on geographic fieldwork and subsequent activities from around 1998 through the present, reasoned that a strong ruling on pasture by the USDA National Organic Program would help preserve the agricultural heritage of the county, even if they weren't particularly fond of extant NOP rules.

Bellingham, on Puget Sound, is Whatcom County's largest city. The first non–Native American immigrants settled on Bellingham Bay in 1852, ten years after the first wagons crossed the continent on the Oregon Trail (Lynden Pioneer Museum 2007). It was once a lumber and plywood town known for the sour milk smell of its pulp mill, but stricter environmental regulations and foreign competition put an end to the old milky odor, and local politicians sought economic diversity beyond wood products. By the twenty-first century, Bellingham increasingly relied on Western Washington University to sustain its economy. Its students joined local residents in patronizing the small city's *two* farmers' markets, one of them a covered facility recently built to allow year-round market days.

Alternative food networks of organic and nonconventionally produced products are a vibrant part of the economy in Bellingham and around Puget Sound. One longtime actor, mentioned earlier in this book, is a woman we will call "Cath Illy" (pseudonym). In the 1970s she began baking organic bread in Seattle's Pioneer Square and then organic pizza in Bellingham's historic Fairhaven District, before growing organic fruit, vegetables, and flowers on plots near Mount Vernon. Cath is a vendor in Bellingham and at area farmers' markets when not farming with one or two apprentices, or working at the food co-op in Skagit County to the south. One could say Bellingham still fits into the green dream portrayed in Ernest Callenbach's 1975 novel, *Ecotopia*.

Lynden, 15 miles north of Bellingham and just 5 miles from Canada, is a dairy hub that punches above its weight. In the 1970s–1980s, farmer members of the Darigold cooperative claimed to own the world's largest conventional milk powder dryer, at the Darigold processing plant in Lynden (Scholten 1997).

In the 1950s–1960s Darigold was already a regional leader in the production of high-value butter, cream, cheeses, and ice cream, but for foreign policy and macroeconomic reasons, the USDA encouraged milk powder for export. In response Darigold made powder exports its core business, while licensing other companies to manufacture its consumer retail line. Co-op farmers prospered until U.S. powder became uncompetitive on the world market. As a consequence Darigold was unable to pay its unhappy farmers equity payments, a situation that continued till about 2012. With incomes languishing, farmers had unhappy choices. Some got on the push-pull treadmill, seeking economies of scale in confinement dairy farms with hundreds of cows. It was not uncommon to hear that Mr. and Mrs. So-and-So had retired from dairying because "they just didn't like seeing the cows off pasture."

A few farm families successfully established vertical operations, with the production, processing, and sales of milk, butter, and ice cream on the same farm and with select retailers. Their occasional unwillingness to participate in generic dairy industry promotional checkoff campaigns brought conflict with the Darigold farmers' marketing cooperative, but agreements were eventually reached.

Since strawberry and raspberry growing were already significant in Whatcom County's economy, some dairy farmers converted cow pastures to raspberry fields. Lucky ones on sandy river soil found raspberries thrived. It was a historic opportunity brought by misfortune in the former Yugoslavia. Due to the decline of production in the Balkans in the 1990s war, Whatcom County supplanted the Europeans' erstwhile world leadership in raspberry growing. Multinational corporations, such as Smuckers, contracted for local berries and stored them in refrigerated facilities next door to the Darigold facilities in Lynden, for transport to sale and further processing globally. People began to talk of Whatcom as a dairy and berry county.

Meanwhile, conventional dairy farmers around Lynden were aware of statewide interest in organic foods, as well as Whatcom County sales in supermarkets and farmers' markets in Bellingham and Seattle (Scholten 2007; 2011). Across the state in Pullman, Washington State University, which is known for teaching and research on intensive agriculture, including biotechnology, nevertheless responded to public interest in alternative food networks (AFNs). Melissa Bean, associate director of development for organic and sustainable agriculture and international research and agricultural development, explained that WSU was helping to develop the first four-year science-based organic agricultural systems undergraduate major in the country and the largest organic teaching and research farm on a U.S. land-grant campus. WSU's 30-acre Eggert Family Organic Farm integrates crops (wheat, etc.), organic beef cattle, and aquaculture (tilapia). Back in Whatcom County, red-blue political rivalry distinguishes the sympathies of rural residents, suspicious of liberal activities around Bellingham's university. This helps explain why climate change deniers write letters to *The Bellingham Herald* urging the city's development into a rail and ship megaport for massive exports of coal from Alberta, Canada.

One thing farmers have always appreciated about educated city consumers is their propensity to buy high-value foods. The Hartman Group, a consultancy in Bellevue, Washington, consistently found that high levels of education correlated with organic consumption, including dairy products (Hartman 1997; 2004; 2006). Seattle, a blue-voting city, boasted some of the highest per capita educational levels in the country, so it was no accident that organic producers trekked daily from Whatcom County to sell their homegrown organic fruit, vegetables, and flowers at Seattle farmers' markets (Scholten 2011: 124). Organics also sold well in urban Bellingham.

Near Ferndale in Whatcom County, Appel Farms (Scholten 2011: iii) subscribes to sustainable principles of animal and land care, if not USDA organic certification. Appel may be an example for some farmers who, frustrated by the burdens in time and money of USDA bureaucracy, go beyond USDA organic in their pursuit of sustainability mixed with traditional values. In the 1970s cheese making was begun in small but appropriately sized equipment imported from Europe by Dutch immigrant Jack Appel, and the family business is now in its third generation. Today Rich Appel manages the herd that has grown to 1,200 on extensive pasture. Rich's brother John Appel and the latter's wife Ruth manage cheese making from pasteurization to the distribution of cheddar, feta, gouda (the original specialty), and paneer (produced after requests from Whatcom County's growing community of people of South Asian descent, many of them berry growers). These and even more cheeses, plus quark and yogurt, have been sold in the farm store and local and regional supermarkets. By 2002 Appel Farm cheeses were prominently displayed in about 30 Haggen supermarkets in Washington and Oregon, where they caught the eyes of farmers considering organic conversion. As a local product, Appel Farms builds consumer loyalty in multiple demographics, for example, when its cheeses are displayed with toy Dutch windmills in supermarket cases, adjoining displays of Lummi Island Wild Salmon, a specialty of a local sustainable fishing cooperative.

Whatcom County farmers teetering on bankruptcy began to eye the approximate 25 percent price premium of organic milk in a market that was growing about 20 percent each year (which continued to grow until a dip during the 2008 recession, after which it recovered quickly to annual growth of 7–8 percent). They were aware of the segment of consumers that sociologist Melanie DuPuis said had a "*Not-in-my-Body*" (NIMB) attitude to milk produced with antibiotics and the GMO hormone rBGH/rBST, making organic milk "the fastest growing organic food segment in the United States, the 'star' of the organic foods industry" (Dupuis 2002: 285). One large-animal veterinarian, accustomed to treating cows on the growing number of megadairies, wrote an open letter to *The Lynden Tribune* offering to work with any farmers converting from conventional dairying (which allowed antibiotics) to organic dairying (which did not). In time, the Organic Valley/CROPP cooperative, growing from its Midwest origins, made organic dairying a realistic option for some in the Pacific Northwest (Scholten 2002; 2007; 2011).

South of the Canadian border, population pressure mounted on Whatcom County's 2,106 square miles (Quick Facts c. 2014; http://quickfacts.census. gov/qfd/states/53/53073.html). The population of Lynden tripled from about 3,000 to 9,000 citizens between 1970 and 2010, while Bellingham's population grew from about 35,000 to 80,000 in the same period. Total Whatcom County population increased from about 75,000 in 1970 to just over 200,000 in 2010. Urbanization was remapping this author's home county.

It was unsurprising that Nooksack Valley High School athletes were dubbed the *Pioneers*, because many of these rural dwellers had links to the pioneers of the late 1800s and early 1900s, whose lives were split between logging and farming. Whatcom farmers who carved their farms out of the woods with chainsaws and dynamite saw themselves as pioneers among Native Americans, although the latter lived mostly on the Lummi Island reservation. But if Abraham Lincoln's nineteenth-century parents cut their farms from virgin forest, Whatcom's twentieth-century farmers felled second-growth trees, left after the huge Weyerhaeuser lumber company clear-cut Washington State's virgin forests, after 1900 (University of Göttingen 2010).

On Front Street in the middle of town, the Lynden Pioneer Museum testifies to the community's identity (Lynden Heritage Foundation 2007). According to its "Lynden History Timeline," Phoebe and Holden Judson, of British Canadian and Presbyterian stock, followed the Oregon Trail to Centralia in the Washington Territory in 1853, motivated by the offer of free land from the U.S. government. The couple moved to what she would name the "Lynden" area in 1870; the town was incorporated in 1891, and Holden Judson was elected first mayor in 1894, not too many tears before he died.

About 1900 the first Netherlanders joined British, Canadian, German, and Scandinavian settlers in Lynden and the Nooksack River valley (Anderson 1957). The Dutch toehold was secured in the 1930s by a wave of Dutch Americans from the Midwest. Although the first church built in Lynden in 1887 was Methodist, the Christian Reformed Church (CRC) faith of many of the Great Depression-era Dutch American migrants, which mixed conservative values with a proclivity for livestock keeping like the Biblical Abraham, added a Dutch Calvinist flavor to Whatcom County farming.

Lynden's first creamery opened in 1888, a year before Washington statehood in 1889. The first cooperative creamery was formed in 1902, a harbinger of the Darigold co-op to come in midcentury (Lynden Heritage Foundation 2007).

The Calvinist Christianity of the immigrants gradually altered Lynden's church/tavern ratio; the latter dwindled to two by about 1980 and disappeared around 2010. The town was rumored to hold the world record for most churches per square mile. Although the Dutch Reformed church set the tone, Lynden served as a haven for other religious traditions where temperance and hard work won respect regardless of belief. St. Joseph's Catholic Church was sanctuary to residents of that faith and swelled with the seasonal influx of migrant Hispanic farm workers. A Sikh temple, or *gurdwara*, was located south of Lynden among berry farms cultivated by the South Asian diaspora, many of whom came via Vancouver, Canada. The nearest synagogue was in Bellingham, and Sol Lewis, the Jewish proprietor of *The Lynden Tribune*, was a respected community spokesman for decades. Bellingham hosted Baha'i, Islamic, and other places of worship.

Lynden city ordinances had long banned alcohol sales wherever dancing was permitted. Sports were encouraged, especially basketball, in which boys and girls in Lynden's public and private Christian high schools competed at the highest state levels. Most barn haymows had at least one basketball hoop for midwinter practice.

Community pride thrived on the dairying prowess of local farmers through the prism of Dutch dairy culture. Tree lovers could be nonplussed by the typical Dutchman's tendency to fell any tree that disrupted plow lines, and the county lost wetlands as compass-straight ditches drained marshes. Immigrants from "the old country" were expert at sustaining soil fertility and maximizing milk productivity through rotational cropping and grazing, although the introduction of chemical fertilizers and pesticides in the 1950s altered that pattern.

Lynden had long integrated Netherlands' themes into its Sinterklaas and Lighted Christmas Parade, and children's *klompen* (wooden shoe) dancing highlighted the Holland Days festival in May. The wooden shoe motif was anchored by giant semifunctional windmills on Front Street. These alternated with folksy murals depicting small-town life (tulips, boys plugging dykes, wives berating errant husbands) by local sign painter Bill Swinburnson. This civic facelift was led in the 1970s–1980s by Jim Wynstra, a local attorney and developer who observed how tourists flocked to Bavarian-themed Leavenworth, Washington, when city facades were redone as the half-timbered (*Fachwerk*) architecture of Alpine ski resorts. Compared to Leavenworth's makeover, Lynden's Holland dairy themes had stronger cultural provenance.

Although the number of Whatcom County's milk producers dwindled from about 3,000 in 1950 to 180 or fewer registered dairy farms in the mid-2000s, the county was the state's top milk producer, until megadairies in the central and eastern parts of the state challenged it near the turn of the century. Testifying to the speed of farm mergers, Dairy Farmers of Washington reported in 2014 that Whatcom County had just 123 dairy farms with 50,000 cows, while the Yakima Valley region of Benton, Franklin, Klickitat, and Yakima counties had 91 dairies with over 110,000 cows.

Back in the mid-1990s, Whatcom County farmers, Dutch or not, faced lower real prices for their milk. By then an 80-acre dairy farm was too small to feed a herd big enough to support a family and put kids through college (Scholten 2002). Hay and alfalfa trucked from eastern Washington was more expensive than formerly. Whatcom County farmers no longer had surplus pasture to scale up their operations. On the outskirts of Lynden, farmland that once hosted an international equestrian plowing match was now covered by new homes built by Wynstra's real estate firm Homestead Northwest. A surprising number of homes were owned by Canadians; some had emigrated from Hong Kong after reversion of that port city from British to Chinese authority in 1995. Farmers east of Lynden, which was more rural and less commercial than the highway between Bellingham and Lynden, wanted to maintain agricultural zoning, which Washington State taxes at lower rates than residential or commercial property. It was a controversial matter: Greens and town dwellers who valued pastoral farmscapes wanted to preserve this land for leisure-time recreation, and of course some aging farmers wanted to pass the property to a successor in their family's next generation. As ever, developers and builders sought land. Many farmers shared hopes that the dairy tradition would continue, but some loudly demanded their rights as property owners to sell their land to the highest bidder, be it a farmer or developer, in order to fund their retirement or pass a monetary inheritance to the children.

The battle line for zoning was identified by a leader of the Conservation District (personal communication). This was the Northwood Road running from the Nooksack River to the U.S.-Canadian border. If land to the east of this road were rezoned and lost its agricultural status, it could be overrun by a tsunami of sprawl. A border-area casino had also opened, despite opposition by locals. There was reason to worry, judging from the experience of farmers in nearby metropolitan areas. Since the 1970s and 1980s, much of the once overwhelmingly green area between Vancouver and the U.S. border had turned to suburbs and strip malls à la California. Similarly, south of Whatcom County, casinos, Wal-Marts, and other big box stores replaced the giant cedar stumps wide enough to drive a car though, which as late as the 1970s had lined the highway between Everett and Bellingham.

Coast Salish Sea Tribes and First Nations from both sides of the border now contest plans to use Puget Sound—which they call the "Salish Sea"— and points such as Bellingham for massive exports of coal or oil. The gambit portends political polarization and international litigation for the next few decades. The website Indian Country (Indian Country 2014) announced:

The Lummi, Swinomish, Suquamish and Tulalip tribes of Washington, and the Tsleil-Waututh, Squamish and Musqueam Nations in British Columbia stand together to protect the Salish Sea. Our Coast Salish governments will not sit idle while Kinder Morgan's proposed TransMountain Pipeline, and other energy-expansion and export projects, pose a threat . . . The Salish Sea is one of the world's largest and unique marine water inland seas. It is home to the aboriginal and treaty tribes of the Northwest whose shared ecosystem includes Washington State's Puget Sound, the Strait of Juan de Fuca, the San Juan Islands, British Columbia's Gulf Islands and the Strait of Georgia.

LeCompte-Mastenbrook on Dutch Stewardship

An anthropological study utilizing focus groups and interviews with dairy farmers in the Lynden, Washington, area by Joyce LeCompte-Mastenbrook (2004: 17) found that "the Christian notion of stewardship and its influence on farming practices as conceptualized by the Christian Reformed Church (CRC) . . . [is] influential in shaping community identity." That influence persists in the twenty-first century. A drive after church through Whatcom County reveals carefully groomed farms, but little labor performed between Sunday milkings. As Peter Moerland, a pseudonym for one of LeCompte-Mastenbrook's (2004: 47) retired farmers, explains:

> You might say that it's a religious thing to take care of the soil. We believe that all right. I think most anybody that has any common sense would think that. Sorry but . . . you take care of what God provides you. We believe that God provided it, and so you take care of it. It's still his world. But I would hope that my religion affects everything in my life, and that includes what I do, and what I think about what I do, and everything else. It isn't just the farming part of it, not at all. The only direct thing I can think that has to do with religion is that we didn't work on Sunday, you know. Well, we'd milk the cows and all that, but we didn't make hay. But my neighbor, he worked with the hay once in a while on Sunday, and he'd say, "Boy, that John Deere tractor never puttered so loud, except on Sunday." (47)

Before President Ronald Reagan's introduction of laissez-faire farm policy in the 1980s, dairy families were more confident of remunerative prices for their milk. This security stemmed partly from the principle of "parity" between rural and urban incomes fostered in President Franklin Roosevelt's New Deal programs in the Great Depression. The idea of parity was supported by Democrats, including John Kennedy and Jimmy Carter, in attempts to strengthen their rural-urban political base. Reaganism brought painful structural change. Its switch to reliance on market forces to guide agriculture simplified policy, but critics believed that considerations of animal welfare, environmental impact, and community disruption were often ignored in the breach.

National per capita milk consumption fell as baby boomers aged and sugary drinks filled the void. Worse, dairy farmers were caught in a "technology treadmill" of productivity increases forcing them to expand production in order to maintain income (Geisler and Lyson 1991: 562; Hinrichs and Lyson 2007). Based on data from the USDA Economic Research Service (ERS), the Congressional Research Office (CRO 2013: 5–6, 26) confirms that productivity gains "have resulted in agricultural output tending to expand faster than demand. As a result, farm prices declined in real terms steadily from the late 1940s until 2006." The farm share of U.S. food expenditures in the period 1950–2006 for ten major food groups for at-home consumption—that is, retail purchases of "beef, pork, broiler, eggs, dairy products, fats and oils, fresh fruits, fresh vegetables, processed fruits and vegetables, and bakery and cereal products"—fell from 41 percent to 15.5 percent. When grain prices soared due to the U.S. biofuel program in the mid-2000s, Chinese demand for U.S. soybeans, and drought, dairy farmers were hit harder than most farmers because outsourced feed and grain were a major share of dairy budgets. Much can be inferred about the pressure that agribusiness appropriation places on farmers and the inherent rivalry that farmers have with industry and traders—one of this book's themes—by an observation in the same report (4):

> The farm value of a retail food product is the portion of the farm share that actually stays with the farmer. In 2011, ERS estimated the farm-value share at 7.9 percent. The remaining 7.6 percent (of the farm share of 15.5 percent) went to agri-businesses and marketing industry groups that furnished inputs to the farm production process to produce the raw farm-gate commodity.

The upshot of these trends for Whatcom County was that family farms that had been successful for decades went to the wall. Megadairies with hundreds of cows, some close to 1,000 cows, were formed by merging neighboring farms in attempts to achieve better economies of scale on the technology treadmill. The most fortunate farms had enough acreage to raise their own hay, silage, and grain, but few had enough pasture for lactating cows. Talk on the street, or after worship services, was gloomy. It was heartbreaking when respected farmers, who had been "big farmers" in their heyday with 50 or even 100 cows, on learning a journalist or academic was researching their plight, asked for advice "because farming is dying" (personal communication). Some indebted farmers auctioned land and equipment for scant return, leaving them with little to offer the market but their labor, in a process akin to what that Marx called "proletarianization" (Glassman 2006: 610). One small but important comfort to Whatcom County farmers forced to merge their holdings with another farm is that most of these mergers were by local farmers or neighbors, not by outside corporate agribusinesses.

Most previous social science studies of the dairy transition referred to above were quantitative in character, by agroeconomists and rural sociologists. LeCompte-Mastenbrook (2004: 1, 16) chose a qualitative approach, drawing on ethnography and political ecology. These research tools are suitable for exploring how local values of "Christian stewardship" interact with external pressures, such as USDA policy, currency rates affecting U.S. export prospects, and local resource conflicts (e.g., the Dairy Nutrient Management Act negotiated to attenuate nitrate pollution that imperiled Native American fisheries). LeCompte-Mastenbrook (2004: 13) applied Fikret Berkes (1999: 6) conception of "place-based knowledge" (PBK) to learn the Northwest dairy community's

> way of knowing and behaving . . . based and rooted in the land, and passed on from generation to generation informally via cultural transmission. Berkes suggests that the concept of PBK extends beyond knowledge of the local environment to include "practice . . . in the way people carry out their livelihood activities . . . [and] belief in people's perceptions of their role within ecosystems and how they interact with natural processes."

LeCompte-Mastenbrook (2004: 16) quotes a dissertation by Burton L. Anderson, *The Scandinavian and Dutch Rural Settlements in the Stillaguamish and Nooksack River Valleys of Western Washington*, which focuses in part on the Nooksack Valley in Whatcom County. Anderson (1957: 148) is sensitive to the cultural norms of the Christian Reformed Church (CRC) and sister churches descended from the Dutch Reformed Church in the Netherlands:

> A common phrase encountered when speaking to farmers is "Man is but a steward for the Lord." The manner in which a Hollander cares for his Maker's property is regarded as a measure of his reverence. This attitude, in conjunction with socially inherited knowledge of methods of maintaining soil fertility, has made the area prosperous . . . Since both the philosophical attitude and tradition are engendered by the Church, it seems entirely plausible that religion has been a paramount influence in shaping . . . the agricultural practices in the Nooksack Valley.

Anderson's words on stewardship triangulate with LeCompte-Mastenbrook's study and this author's life experience. Even many non-Christians are familiar with Matthew 25:20–21 in the Digital American Standard Version Bible (DASV):

> [20] The one who received the five talents came and brought five more talents, saying, "Lord, you gave me five talents. Look, I have gained five more talents."[21] His lord said to him, "Well done, good and faithful servant."

A right-wing political interpretation of this parable could accept it as a paean to stockholder capitalism, an injunction to maximize the boss's earnings in on-farm productivity increases. But the meaning of the parable is hardly "greed is good." Greens and those of left environmental convictions might not reject the pervasive profit motive (few Dutch Americans would, speaking as one myself!), but they might parry that argument in the following way. The parable depicts a good steward's management practices over the owner's absence of many years, possibly equating a seven-year organic crop rotation. It must also be considered that the early Christian era predated the advent of chemical fertilizers and pesticides by two thousand years; therefore a good steward would not have been able to make crops, herds, and vineyards on an estate prosper except by honoring ecological principles and organic practices, such as fallow, crop rotation, and grazing when animals fertilize the land.

Environmentalists sometimes air misgivings over the Deity's command in Genesis 1:26–28 (DASV) to "fill the earth, and subdue it and rule over the fish of the sea, and over the birds of the sky, and over every living thing that moves on the earth." Granted, this passage depicts humans as the crown of creation, but Judeo-Christian ethics, a pillar of community ethics in Whatcom County, are greener than sometimes thought. Scripture offers ample passages limiting the scope of human activities to those sustaining biodiversity, the environment, and social responsibility. To begin with, farmers are instructed to let their land lie fallow with regularity in Exodus 23:10–12 (DASV):

> [10]For six years you shall sow your land and harvest its crops.[11] But on the seventh year you shall let it rest and lie fallow; that the poor of your people may eat and what they leave the animals of the field may eat . . .[12] Six days you shall do your work, but on the seventh day you shall rest; so that your ox and your donkey may have rest, and your female slave's son and the foreigner may be refreshed.

What is remarkable is not that slavery existed in Old Testament times, but that concern and kindness for servants and foreigners—for stakeholders— was part of mainstream ethics. After all, the Israelites had been foreigners in Egypt, so empathy with outsiders was part of the religion. In the context of Whatcom County then, Christian stewardship can be understood as "sacred ecology" in the words of Berkes (1999). In everyday terms this means duty toward animal welfare, environmental sustainability, and social justice.

Animal welfare is the topic of Deuteronomy 25:4 (DASV): "You shall not muzzle an ox when it is treading out the grain." In other words, be kind to farm animals and do not unnecessarily encumber them. The inclusion of animal welfare within general morality is unmistakable in Proverbs 12:10 (DASV): "The righteous cares for the needs of his animals, but even the kindness of the wicked is cruel."

Concern for the environment as well as social justice is implicit in Leviticus 19:9–10 (DASV): "[9]When you reap the harvest of your land, you should not reap right up to the edges of your field, neither should you gather the gleanings dropped during your harvest.[10] You must not glean stripping your vineyard bare, and you must not gather the fallen fruit of your vineyard. Leave it for the poor and the foreigner." The first verse connotes the importance for biodiversity of buffer zones around crops, which leave crop wastes to enrich the soil. The second verse commands concern not only for the poor in one's country people, but also foreigners. This duty was quietly honored by those Lynden farmers who looked beyond their own racial-cultural profile to build better housing for migrant workers, to foster care children, or to visit county jail prisoners on some mornings before other citizens reached church or the golf course. The reasonable conclusion is that when it comes to stakeholders—animals, the environment, farmhands, and other people in the community—the primary value of Christian stewardship is symbiosis, not exploitation.

In her research from August 2003 to May 2004, LeCompte-Mastenbrook (2004: 5, 6) found the "majority of dairies" in Whatcom County were herds of 100–500 cows. This was medium to large on the U.S. national scale (and a world apart from 1950 when the county had three thousand farms averaging 11 cows!). At the time there were few if any megadairies over 1,000 cows. Most herds were "housed full-time in free stall barns, a practice that is commonly associated with industrial-style milk production," but she found that there remained "a substantial number of farms that pasture their animals or practice intensive rotational grazing" (5). A decade later this remains the case; a few grazing herds still ship to the conventional Darigold cooperative (2014 personal communication with Whatcom Farm Friends). Most of the county's registered dairies (from 120 to 180, depending on source) remain in family ownership. LeCompte-Mastenbrook notes that during her study, "Northwest Dairy Association (NDA) was the cooperative, and West Farm Foods, also owned by co-op members, processed the milk and marketed it under the Darigold Brand name." This text uses Darigold as shorthand.

In her ethnographic approach, LeCompte-Mastenbrook (2004: 9, 94) asked informants over one hundred questions, in semistructured interviews on how traditional values of Dutch Christian stewardship interacted with the dairy industrialization that was underway. In late 2003 the cost of milk production exceeded the farm gate price by 15 percent, so farms that had been enlarged were hemorrhaging money. On farm or in town, she heard the opinion that dairying was "dying." Several prominent farmers had taken advantage of Cooperatives Working Together (CWT), a herd buyout organized in 2003 by national cooperatives with assistance from the editors of *Hoard's Dairyman*, to sell their cows and take a minimum five-year timeout to decrease supply and allow prices to stabilize.

This was not the first buyout program in the cyclical dairy sector. In 1986 the USDA managed a herd buyout that sent cows to slaughter or to China (Scholten 1989c: 10). One of the participants was Joe Verdoes, whose analysis of the dairy sector (based on his business and history degrees) led to his selling of his family dairy farm near the Skagit River and becoming a successful fisherman on Puget Sound in the 1990s.

Many farm incomes, in the adjacent Skagit and Whatcom Counties, were supplemented by off-farm activities in teaching, retailing, and so on; on-farm pluriactivity was also attempted (Scholten 1997; 2002). Several farmers responded to the national low-fat diet trend by raising ostriches. Memories of the giant birds grazing on pastures persisted after the ostrich bubble collapsed. Farmers complained that the birds were poor milkers anyway.

*   *   *

It is not uncommon to hear conventional dairy farmers expressing resentment of organics and proclaiming the safety and nutrition of their milk to be *every bit as good as organic,* in everyday conversation, or in this author's presence in forums such as the National Dairy Leaders Conference (2001). In a red-blue political divide, they suspect elite urban consumers of seeking yet another expensive way to perform their superiority. (LeCompte-Mastenbrook cites Scholten's [2003; 2011: 193–94] idea of "merit-badging" as relevant here.) Yet most large dairy farmers seem to quietly yearn for the sight of cows grazing on pasture, as in their grandparents' day. Whatcom County Farm Friends and the Whatcom County Ag Preservation Committee held meetings exploring options for the many farmers considering exiting dairy, trying to find a way to remain by establishing direct sales with the public or converting to organic. In 2004, 2005, and 2006, three Whatcom County farms began shipping to Organic Valley (a few more farms joined, but reverted to conventional when the economy weakened). Conventional farmers who avoided bankruptcy in this period were thankful to receive $4/hundredweight (cwt) more for their milk. But there was no question that many farmers had looked into the abyss—Farmageddon—and did not like what they saw.

Responses to LeCompte-Mastenbrook's questions (2004: 42) connected the dots on several issues; replying to question 82, "Do you use milk production hormones, such as rBST?" one farmer replied: "The whole BST thing. We're just so sorry that that cat ever got out of the bag . . . Part of the reason our milk prices jumped phenomenally is because Monsanto has . . . some problem with production in Austria." This was the same story heard among megadairies in California. Farmers who had angrily demanded the right to apply biotechnology to dairying later blamed rBGH/rBST as part of the technology treadmill that generated price-killing surpluses. Ironically, they found that when the dairy GMO became scarce and too expensive to buy, milk production dropped barely 1 percent, far from the 15 percent drop feared by some intensive farmers (ibid.: 41–42). Similar to the experiences of farmers in California and Denmark mentioned above, some intensive Lynden dairy farmers found that improving animal welfare and comfort by giving cows more space for socializing and lying down, along with careful attention to nutrition, were most crucial to milk yield—making rBGH/rBST superfluous.

LeCompte-Mastenbrook (2004: 18) suggests: "The stewardship ethic continues to be influential despite the fact that it was not overtly stated to me." There was consensus that the land and water must be sustained to provide for succeeding generations. After decades of red tides that closed Native American fisheries in the Nooksack River estuary and Puget Sound, some dairy farmers continued to question whether their rogue nitrate and manure effluents were responsible for fecal coliform contamination that annually shut down the Lummi Island shellfish beds. When 2013 passed without a fisheries shutdown, a longtime conservation district official told this author (2013 personal communication): "That's because farmers finally learned not to spread their fertilizer on the river banks and follow the Dairy Nutrient Management Act."

In her conclusion, LeCompte-Mastenbrook (2004: 53–58) suggested that the "ethnic identity of Whatcom County's dairy farmers has contributed to their being the most productive dairy farming community in the United States today."

The concepts of stewardship among Dutch Americans influenced by the Christian Reformed Church remained vital in the vocation of farming as a mission to feed humanity. But stewardship evolved in the industrial transformations that occurred after the 1950s, especially in the 1980s, when the Reagan-Bush administrations rejected the New Deal politics of the past, including farm-friendly policies of "parity" between rural and urban incomes. The Reagan-Bush administrations ushered in the era of laissez-faire economics and rejected dairy farmers' pleas for supply management (i.e., quotas) to stabilize farm gate prices in the cyclical dairy industry that was characterized by boom and bust (Scholten 1989a; c).

When the author grew up in Whatcom County in the 1950s and 1960s, small herds of 20–40 were common, and cows and bulls had names as familiar as one's aunts and uncles. Many herds had favorite cows aged ten or older. By the mid-1960s, cow talk in coffee shops or at the cattle auction frequently questioned expert advice to cull cows as soon as their yield fell or they approached the age of seven. But while it sounded like advice to sell granny, to some farmers, it was good management to others. Since the 1960s, herd sizes have upscaled by a factor of ten, and, following the industrial practices of California, the slaughter age for cows has become closer to four than seven years. The thought of slaughtering a cow so young violates the sensibility of farmers, who know milk cows do not reach their production plateau until the third lactation. Stewardship of the land is closely linked to cow welfare and animal husbandry on small farms. LeCompte-Mastenbrook (2004: 53–58) notes that "farmers, particularly older ones, feel a deep sense of loss when they see the decline of small farms in the county." Her study, like this book, uses number of lactations and age at culling as proxies connoting animal welfare. She suggests that, "as output maximization becomes the focus, one of the consequences may be that their cows live shorter lives." This source of ethical uneasiness is a sore spot in rainy Whatcom County as it follows arid upstart California's industrial model. Conventional breeders have begun to admit longevity problems (Biagiotti 2013; 2014). Until the average culling age of herds rises to those of their great-grandmothers during the 1950s, industrial dairying is vulnerable to consumer rejection on this basis.

A decade ago, animal welfare figured in some farmers' objections to rules of the USDA National Organic Program (NOP). LeCompte-Mastenbrook (2004: 42) writes that "the primary argument against organics is the antibiotics issue. Most farmers feel it is ethically wrong to not be able to treat a sick animal with antibiotics." An interviewee with the pseudonym Jake Huizenga made the following remarks that resonate so strongly with my own experience of Whatcom County that they are related at length (ibid.: 42):

> At this meeting with these organic officials, they were telling us all the rules, what we can and cannot use with our cattle, we can't use any artificial drugs, and I said to this guy "only things found in nature can be used on the cattle right?

Like garlic, aloe vera, Echinacea, and I said, by the way, where does penicillin come from?" "Well, from mold, it grows." "Well then I can use penicillin, right?" "Well no, you can't." So we just went around and around in circles. So I said, "Okay, so my cow gets sick, and I've got to kill it?" And he says "No, no. We don't want your cow to suffer. If your cow is suffering and you can't cure her with any of these natural things we want you to use the drugs, make her better, and then you can sell her to the neighbor or to somebody else for conventional milk production." And I say, "Well okay, then what we have here is a two class system. If it's not good enough for us, then we're going to make the cow better and we're gonna sell it to the peasants. The lower class people who can't afford our product."

Care for other creatures is mixed with the politics of outraged egalitarianism in this Dutch American farmer's response. The next chapter shows that his perspective on matters such as antibiotics can be found across the country. Ultimate solutions are not simple.

6

Antibiotics and Health in the Northeast and Beyond

Photo 6.1 Jerseys graze on the University of New Hampshire Organic Dairy Research Farm.

Photo Credit: Bruce Scholten.

Antibiotics and Health in the Northeast and Beyond: Experts on U.S., Canadian, and European Rules

About 2,460 miles as the crow flies east of Lynden, Washington, lies the first organic dairy research farm established at a U.S. land-grant university. Located in rural Lee, seven miles from the main University of New Hampshire campus in Durham, the New Hampshire Agricultural Experiment Station bought parts of the former Burley-Demeritt and Bartlett-Dudley farm properties comprising dairy, sheep, poultry, and crop farms in 1969.

The NHAES organic dairy farm has been established for a decade. Of its 300 acres, 120 are in woods. White pine shavings compost well and are useful as bedding for the cows. Several New Hampshire towns, from Durham to Albany in the north, are experimenting with renewable energy systems fueled by wood shavings or pellets (drawing partly on Scandinavian technology). A heat exchanger that is part of the experimental aerobic composting facility was installed to preheat water for the step-up four-stall milking parlor on the Burley-Demeritt dairy farm in fierce New England winters.

With his ideal of a farm as a single organism, Rudolf Steiner would appreciate the farm's operation as an integrated agroecosystem of biological, physical, and human-related components. State budget cuts of 48 percent in 2011 have crimped its ability to hire enough labor to harvest all its own hay and other fodder for now. Meanwhile, university research conducted is relevant to both conventional and organic farms. There is interdisciplinary research, much of it supported by the NHAES, with scientists from the College of Life Sciences and Agriculture and the Institute for the Study of Earth, Oceans, and Space. Recent work includes studies of kelp or flaxseed as dairy nutritional supplements, to improve pasture productivity across the growing season (Antaya et al. 2013).

On a warm day in May 2013, farm manager Nicole Guindon and faculty fellow Anita Klein kindly show me the farm, facilities, and animals. The well-being of the 45 or so Jerseys is evident by their calmness and friendliness. It is supported by impressive statistics on the somatic cell counts (SCC) of their milk, a very low 77,000 bacteria per cubic centimeter, against a standard of 400,000 (formerly 750,000) in USDA rules. Mastitis is happily rare in this herd.

Disease prevention is becoming more important in the world dairy trade, as the European Union seeks to lower its present rate of antibiotic use in conventional herds by 75 percent. This is a challenge for herds in America's warm, moist Southeast, while this Northeast herd meets the SCC standard more easily.

Biosecurity is the byword. Global epizootics, such as foot and mouth disease (FMD) and severe acute respiratory syndrome (SARS), have taken some of the joy out of visiting new herds. They are petting zoos no more. We wear clear plastic moon boots to prevent contamination from our shoes and refrain from touching the animals. It is hard to resist scratching the ears of the dozens of dark-eyed Jersey heifers quartered near the main barns, eating from mangers in sunshine. They enjoy visitors who, in turn, are intrigued by their ear tags. Besides the customary bar code, the tags show each heifer's name. One bright spark is called "Bonanza." We imagine "Bo" responds when we call her name.

Walking up a cow lane to a section of pasture, Dr. Klein, who has a PhD in biochemistry and teaches in the College of Life Science and Agriculture, says pasture alone does not guarantee high animal welfare: "We measured stress levels, and cows grazing on hot days were more stressed than those inside." To maximize nutrition and welfare, she explains that managed intensive rotational grazing (MIRG) is practiced on the 40 acres of pasture. Introducing the herd to a new paddock after each milking not only excites the herd, it also regenerates grass at a faster rate. Klein mentions that Guindon assumed management of the farm even before the completion of her master's degree. The double load of graduate studies and dairy farming could daunt anyone, but she exudes capability. Guindon recalls growing up on horseback "all my life." Obviously, some of her equine passion has turned to bovines, and she seems born to run a dairy farm.

Taking a coffee break from paperwork in the farm office, Klein and Guindon discuss animal welfare, cow longevity, and antibiotics. Rumors come up of local farms where the average age of organic cows at culling is just four, compared to five for conventional cows. This counters intuition and the, albeit incomplete, data so far encountered. Many factors besides cow health affect culling rate, including the size of facilities and the number of heifers ready to replace them, which is often in surplus given proper animal husbandry. Later it turns out the rumors are based on anecdotal evidence from the university's conventional Fairchild Dairy. Most sources suggest organic cows generally outlive conventional ones, depending on conditions. But this chat reminds us that any intervention with antibiotics bans a cow permanently from USDA organic certification. If a conventional farm does not buy her (note: organic Jerseys generally produce less milk than conventional Holsteins), the next stop is the abattoir.

Antibiotics are prohibited in the USDA National Organic Program (NOP), and cows treated with them are banned from organic certification. In the European Union (EU), some antibiotic treatment has been allowed for organic cows, with a subsequent withholding of milk for a certain period. The same is true in Canada, but with a longer withholding period. Guindon says, "There are some things I really like about organics, and some things I really don't like." We discuss other aspects of animal welfare, and she adds, "I wish we'd adopt the European system of using antibiotics, but with the penalty of three times the withholding period of that of conventional cows in the U.S."

Guindon is not alone in this opinion. It matches those of some farmers in my home town of Lynden, Washington (see LeCompte-Mastenbrook 2004, as referenced in chapter 5). Why should farmers deny cows life-saving interventions they would not deny their children or themselves? (This author has heard this argument in Whatcom County since at least the 1990s.) Why must a cow permanently lose organic certification after the use of antibiotics? Later in this chapter is analysis of this dilemma with one of the country's foremost veterinarians.

IFOAM on Antibiotics

Just as IFOAM philosophy, influenced still by Rudolf Steiner's Anthroposophy, prioritizes feeding the surrounding soil rather than the plant itself, its principles of animal husbandry stress improvement of environment and nutrition to strengthen an animal's immune system, rather than radical interventions with pharmaceuticals to address disease after illness strikes and produces visible symptoms.

At the same time, IFOAM standards and norms have been careful to prohibit farms and veterinarians from allowing unnecessary pain or suffering by farm animals. That is so even if antibiotics are necessary and their use entails loss of organic certification, either temporarily or permanently according to national legislation. National spats on rules are easier to analyze when put in the international context, so IFOAM standards and norms are quoted at length here.

The 2005 IFOAM Basic Standards for Organic Production and Processing (2005: 34, 41) state:

"5. 7.2 If an animal becomes sick or injured despite preventative measures that animal shall be treated promptly and adequately, if necessary in isolation and in suitable housing . . .

"Producers shall not withhold medication where it will result in unnecessary suffering of the livestock, even if the use of such medication will cause the animal to lose its organic status. An operator may use chemical allopathic veterinary drugs or antibiotics only if . . . preventive and alternative practices are unlikely to be effective" and if they are "used under the supervision of a veterinarian," and "withholding periods shall be not less than double of that required by legislation."

These IFOAM rules recall the humility of Rudolf Steiner who made suggestions, rather than prescriptions, on organic solutions. They underline the principle that ideology, even organic philosophy, must not hinder the relief of animal suffering.

"5.7.3 Substances of synthetic origin used to stimulate production . . . are prohibited."

Synthetic stimulants, such as the GMO dairy hormone rBGH /rBST, are among drugs prohibited in this section. As we have seen, such synthetic hormones increase the stress on conventional cows using them, which may lead to early culling.

IFOAM 2012 Norms for Organic Production and Processing (2012: 18–19) state:

"7.2 Health Care. Livestock production:

"Organic animal management systems follow the principles of positive health, which consist of a graduated approach of prevention (including vaccinations and anti-parasite treatments only when essential), then natural medicines and treatment, and finally if unavoidable, treatment with allopathic chemical drugs."

Relevant to our concern with antibiotics, this section reiterates the prioritization of animal welfare and the avoidance of pain over the retention of organic status. This increases motivation for farmers to preclude illness by improving their animals' immune systems.

A later more detailed section on IFOAM (2012: 49–50) norms states:

"5.7 Veterinary Medicine: General Principle:

"Organic management practices promote and maintain the health and well-being of animals through balanced organic nutrition, stress-free living conditions and breed selection for resistance to diseases, parasites and infections.

"Requirements: 5.7.1 The operator shall take all practical measures to ensure the health and well being of the animals through preventative animal husbandry practices."

Practically speaking, measures to honor the veterinary general principle include the "selection of appropriate breeds" for farm environments and "animal husbandry practices" conducive to an animal's natural requirements, including exercise and "pasture" in rotational grazing in comfortable stocking densities, to sustain their immune systems.

These IFOAM requirements are congruent with the efforts of organic pastoralists managing intensive rotational grazing systems (MIRG), as opposed to confinement strategies to maximize milk production. IFOAM could hardly be clearer on the importance of pasture areas and comfortable stocking densities to reduce stress for cattle.

Many small-scale organic farmers also interpret IFOAM requirements to include other species, such as poultry. Mobile units of truly free-range, longer-lived organic chickens can aerate and fertilize pasture soil in an ecologically sound manner, while improving weed and insect control, which in turn improves the welfare of cows in summer. Hens, cows, and people win.

Research at Pennsylvania State University found pastured hens, compared to battery-caged hens eating commercial feed, lay eggs with double the vitamin E and 2.5 times greater amounts of healthy omega-3 fatty acids, according to Scrambled Eggs, a major report by The Cornucopia Institute (2010: 10, 11, 12).

"5.7.2 If an animal becomes sick or injured despite preventative measures, that animal shall be treated promptly and adequately." Natural approaches such as Ayurvedic medicine, acupuncture, and homeopathy are recommended.

Paragraph 5.7.3 states that the use of synthetic drugs, such as antibiotics, results in the loss of organic status but specifies that farmers must not withhold them if that would result in unnecessary suffering. However, compared to USDA organic regulations, IFOAM's broad church approach has scope for regaining organic certification after using antibiotics, if the farmer can "demonstrate compliance with 5.7.1, and natural and alternative medicines . . . are unlikely to be effective" or are unavailable, and drugs, such as antibiotics, are given under a veterinarian's supervision. To limit moral hazard, a penalty is exacted: withdrawal of a cow's milk from organic supply for double the withdrawal period of conventional milk supply or at least two weeks. The IFOAM requirements prohibit repetition of such an exception more than three times for any animal.

A crucial requirement, divorcing IFOAM 2012 norms from those of conventional agriculture, is found in paragraph 5.7.4, prohibiting prophylactic use of antibiotics or other synthetic allopathic drugs. Critics of conventional agriculture wish it would adopt such an approach, to reduce the danger of antibiotic resistance to general public health.

Finally, paragraph 5.7.6 permits vaccinations only when endemic disease cannot be controlled by organic methods or when vaccination is required by law.

Standards in IFOAM 2005 and norms in IFOAM 2012 documents show unbridled concern for animal welfare. Point 5.7.3 allows limited retention of organic status with some penalties, so the total U.S. ban on antibiotics is draconian, while the guidelines hint that farmers who resort to antibiotics to relieve animal pain and suffering have failed in their duty of care to establish farm environments that strengthen livestock immune systems. Point 5.7.5 also prohibits recombinant dairy hormones, such as Elanco's (formerly Monsanto's) GMO Posilac, as unsuitable for organic dairying.

It is not easy to keep large herds without resorting to antibiotics. Moreover, the skills needed to raise calves are different from managing productive milk cows. Even farmers who subscribe to enlightened practices of feeding newborn calves their mothers' colostrums in order to kickstart their immune systems, and raising them initially in separate hutches, or perhaps with one other calf, before including them in small groups, may lose some calves to disease. Nor is the science absolutely firm on this. For example, the Hoard's Dairyman working model farm at Fort Atkinson, Wisconsin, is currently switching from individual calf hutches to small groups.

As British cattle nutritionist Paul Robinson illustrated in an earlier chapter, highly productive cows are a special challenge against mastitis and infection when calving. The next section draws on another veterinarian who has committed decades of his career to bovine health in both conventional and organic herds.

Hue Karreman on Antibiotics

No one in America understands better than Dr. Hubert J. Karreman, DVM, how controversial antibiotics are in the organic sector. His professional life provides lessons in organic politics. In 2013 he accepted a post as veterinarian with the Rodale Institute, saying he felt like he was coming home because "Rodale is the pinnacle of the organic community" (McMinn 2013). He is best known for his work with dairy cows, although at Rodale he also works with organic pigs, goats, and poultry. His regard for Rodale's commitment to scientific experimentation is clear in discussion with colleagues (Odairy Feb. 11, 2014): "Rodale Institute has been doing their Farming Systems Trial for the last 30 years and has real data which irrefutably state that organic [crop] yields match or outperform conventional in the long run and especially drought years (Rodale Institute 2011 *The Farming Systems Trial: Celebrating 30 years*)."

Karreman referred to the Rodale (2011 FST: 1–13, 3) report celebrating 30 years of the farming systems trials around Kutztown, Pennsylvania, concluding that in organic crops for humans and livestock "organic yields match conventional yields. Organic outperforms conventional in years of drought. Organic farming systems build rather than deplete soil organic matter, making it a more sustainable system. Organic farming uses 45 percent less energy and is more efficient. Conventional systems produce 40 percent more greenhouse gases. Organic farming systems are more profitable than conventional."

Karreman now represents Rodale at conferences and develops research and teaching programs that serve both conventional and organic farmers raising poultry, swine, other animals, and his specialty, bovines. With such impeccable organic credentials, it is surprising that in correspondence he says he once "committed political suicide" in debates on antibiotics.

Karreman's CV reflects two decades of top-level input in organic policy making, alongside hands-on veterinary work. Since graduating from the University of Pennsylvania as a veterinary medical doctor (VDM) in 1995, his career has intersected with many of the people and events recorded in this book. Karreman served on the USDA National Organic Standards Board 2005–10, which overlapped or followed the terms of pioneers such as Goldie Caughlan, George Siemon, Jim Riddle, and William Lockeretz. In 2007–10, Karreman chaired the NOSB Livestock Committee, writing recommendations on pasture and other matters for regulatory implementation by USDA. He participated in the first IFOAM International Conference on Animals in Organic Production in 2006, which made the historic St. Paul Declaration on animal welfare. He also attended the second IFOAM conference in Hamburg, Germany, in 2012.

Over 1995–2009 Karreman was a full-time veterinarian in Lancaster County, known as "Pennsylvania Dutch country," with about 1,850 conventional dairy farms and 150 certified organic, many of them Amish. He was responsible for emergency, routine, and preventive medicine, surgery, reproduction, and obstetrics on about 85 certified-organic grazing family dairy farms, plus 10 conventional farms that grazed cows. Amid all this activity, he presented educational seminars for veterinary schools and farmer organizations. Besides contributing to peer-reviewed publications, such as the *Journal of Dairy Science*, Karreman is well known for his 2004 book, *Treating Dairy Cows Naturally*, which has become one of the bibles of the organic movement. Major U.S. organic companies vouch for the book, and one sent this author a copy of the 2007 edition in its corporate information pack. Enthusiasm for animals rather than careerism seems to propel Karreman's efforts. Hence his 2011 book for organic farmers is *The Barn Guide to Treating Dairy Cows Naturally*. In that year Karreman founded his company Bovinity Health, providing nonantibiotic treatment for infectious diseases and nonhormonal treatment for infertility via botanical derivatives, biological immunomodulation, acupuncture, low-dose therapeutics, and hands-on physical therapy.

Many people in the organic community know Dr. Karreman as "Hue," a longtime participant on Odairy, the organic email list moderated by NODPA's Ed Maltby. Generous with his time, Hue advises faraway farmers worried about sick cows or calves. He is well regarded by organic leaders in North America as well as Europe, where he attended Certified Cow Signals Training sessions in the Netherlands in 2012 and 2013.

Over 2007–09 Karreman was affiliate assistant professor in the Department of Animal and Nutritional Sciences at the University of New Hampshire, home to the Burley-Demeritt organic dairy farm discussed above. It was natural to invite correspondence with Dr. Karreman on topics such as antibiotics and longevity, and he kindly granted permission to include the edited exchange below:

Author: Intuitively, cow longevity and culling rates are proxies for animal welfare. So, it was surprising when University of New Hampshire Organic Farm managers cited figures from farms where the average culling age for organic cows is 4, a year younger than conventional cows at 5. Conjecture tells me that if organic cows do go to slaughter earlier, it's likely due to acidic diets on organic-industrial megadairies that do not graze.

Karreman (personal communication, Aug. 31, 2013, 1:38 p.m.): It depends on what kind of culling you're talking about, doesn't it? Voluntary culling is good [for herd performance] in a sense whereas involuntary culling is not good. There should be numbers for longevity, I agree. In a small study we did here in Lancaster County 2006–2008, we saw that organic animals live on average about 50 months and conventional animals 48 months. Didn't do statistics. But I was surprised. However, many of those farms were recently transitioned. I think one would need to look at blocks of farms which have been organic for different periods of time, say 15–20 years, 10–15 years, 5–10 years and less than 5 years. Just an idea.

Author: The U.S., Canada and European Union (EU) organic dairying programs have varying rules on antibiotics. None are permitted in the USDA National Organic Program (NOP). Canada permits them with a long withholding period, while the EU has a shorter withholding period. Which is best?

Karreman: Ah, the perennial question of antibiotics and animal welfare . . . my specialty in a sense due to the permanent removal of livestock from the organic sector due to NOP rules. I used to hate that rule, like when I was on the NOSB and took the request up to write about it in a magazine article, with some sidebar input of a friend of mine who is in the Canadian organic system. I basically got my head chopped off for even hinting that an animal treated with an antibiotic could stay in the organic sector, even a very young animal two years away from milking. It is a mindset here and it has been established long enough that people simply take it as one of the "commandments" of U.S. organic livestock agriculture.

Author: Engraved in stone in the U.S.?

Karreman: That said . . . I was at the 2nd IFOAM Livestock Conference in Hamburg last September, 2012, sponsored by the Animal Welfare Institute. The panel was to discuss the use of antibiotics in organic livestock. Initially we were to take up the opposite viewpoint of what our country's official position was. But that was kind of difficult for most (although I would gladly have) so we basically discussed antibiotics from our own experiences in our own countries. The panel was made up of a German veterinarian who strongly defended their use of antibiotics, an Australian beef producer who basically is an NOP adherent (with beef, not too much gets sick anyway), a Namibian (ex-pat from Germany), myself and I think one other person (probably from Britain).

Author: How do the U.S., Canadian and European organic systems compare?

Karreman: After having "committed political suicide" from the article I wrote while I was on the NOSB raising the possibility of using antibiotics for an emergency situation, I realized that the NOP rule had (and continually has) stimulated me to come up with alternatives to antibiotics—and that whenever antibiotics are easily used (as with EU organic regs), the use of alternatives will never be truly tested and developed from actual clinical use. Moreover, the whole notion that "antibiotics = animal welfare" is completely mistaken, in my opinion. Yes, indeed, there can be individual cases, at times, of [organic] animal neglect due to not using antibiotics—but that also happens in the conventional world . . . I think animal welfare is a 24/7/365 type process, not only to be raised in the specter of when one animal, that otherwise has enjoyed a great life in the organic system, is in distress. Additionally, my petitions to allow the use of xylazine, butorphanol, and flunixin to relieve pain and suffering for use in organic livestock here in the U.S. (allowed as of Dec. 12, 2007) speaks much more to helping animals feel better, and ability to do humane surgery and other procedures anyway. I maintain that antibiotics do NOT equal animal welfare when considering what I've written here. That said . . .

Author: What's your preference vis-à-vis antibiotics in U.S., Canadian and EU organics rules?

Karreman: I really like the Canadian system: 30 days milk and meat withhold when an antibiotic or hormone is used therapeutically. In a nutshell, it is the best of both worlds. Why? Because the penalty of using an antibiotic is severe enough to give pause to consider and truly try alternatives (and thus enhance the clinical application and trial of alternatives), yet it is not so severe as to never use an antibiotic. It is the solution that I brought to the panel in Hamburg last September. I really believe that it is the moral high ground as well, and will stimulate the use of alternatives to antibiotics. Since we don't have even this in the U.S. organic sector—only the permanent removal of an antibiotic treated animal—I truly believe that the U.S. organic livestock sector is leading the way globally to finding . . . effective alternatives to antibiotics. Everyone thinks that the EU is very naturopathy oriented—with humans, yes; with animals, no—the organic sector [is] allowed to use antibiotics with a simple twice the label withholding time, which often translates to less than a week out of the milk tank. No incentive there to try alternatives—see what I mean?

Author: So, the European and Canadian systems are more flexible for organic animals in emergencies. But the USDA NOP's prohibition of antibiotics is forcing development of alternatives—a boon to the world?

Karreman: I was asked at the panel if I'd be willing to bring this Canadian paradigm up in the U.S. I would, actually. I feel much more confident in my role in the sector, having un-plugged from political aspects three years ago. It is the one topic (antibiotic use in organic livestock) that I am willing to debate—with anyone, anywhere. I have lived through countless cases in the trenches with the No Antibiotics rule hanging above my head. And, by clinical trial and error, I have sifted through numerous alternatives to antibiotics and hormones and have come up with about a half dozen treatments that work across the spectrum. Have I had steam coming out of my ears when farmers make a conscious decision to withhold antibiotics when they are truly needed? Yes. But I must say they are few and far between.

Author: You believe penalties are not strict enough to deter overuse of antibiotics in European organics, and prefer the Canadian system with three times the conventional milk or meat withholding, to deter the moral hazard of overuse, while protecting cows in emergencies without banning cows from their herds. Meanwhile, U.S. rules have forced you and others to develop alternatives. How effective are they?

Karreman: Unfortunately, many of the alternatives are not officially approved by FDA, or Health Canada, or the British, Dutch, German regulatory authorities. This makes it difficult from a completely different perspective. There are professors that are real "sticks in the mud" and will point out the supposedly "illegal use" of such materials (even though they are allowed within the organic system regs). However, when a veterinarian has a valid veterinary client patient relationship (VCPR) . . . The FDA farm health inspectors tend to give such things a pass.

Author: You seem to have tackled all aspects of antibiotics in the welfare of conventional and organic cows.

Karreman (Aug. 31, 2013, 15:44): I honestly believe that I'm the only one that has grappled with the NOP "No Antibiotics" rule for years in the trenches as a licensed provider of animal health sworn to reduce pain and suffering as per my veterinary licensure—and come out the other end a better alternative health provider as well as more informed at various levels . . . Certainly more so than EU folks asking that the NOP allow antibiotics, as well as farmers in the U.S. that wish the rules were changed. I'm here to help educate and teach, from direct experience in the trenches as well as knowing the political realities, and also knowing the reams of science behind alternatives, about this very delicate but mighty important subject.

Dr. Karreman summed up U.S. cow welfare and antibiotics rules vis-avis Canada and Europe (personal communications, Aug. 31, 2013):

> Though the antibiotic prohibition has indeed been a mother of invention for alternatives here in the U.S., and I argued that at least organic animals have a much better daily life than conventional animals here on a 24/7/365 basis—the EU organic animals have a much better life and there is no punishment when trying to save their lives by any and all means. So, in a sense it trumps the U.S. organic sector if comparing them. But the U.S. organic sector trumps the U.S. conventional sector (and that is essentially what U.S. organic farmers care about it seems—not how they stack up against Canadian or EU organic farmers).
>
> The antibiotic prohibition makes for good trade protection for the U.S. certified organic livestock sector since not many, if any, foreign livestock producers care to go the extra mile to get NOP certified. That said, Organic Meadow in Canada is trying to go the NOP route. Also, there is stirring already among the Danish and Dutch organic dairy sectors to do completely no antibiotic organics. That is why the Danes are coming . . . to check out . . . U.S. farmers' attitudes here about not being able to use antibiotics. I have also given a seminar, at the vet school in Utrecht, Netherlands a few years ago on this topic. The idea of reducing antibiotic use in livestock is hot in Holland and there is a small cooperative of organic dairy producers checking into the no antibiotic route   (yet not getting certified to NOP).

Conclusion

There is consensus that antibiotics have been chronically overused in industrial agriculture, reducing their effectiveness as life-saving interventions for animals and humans alike. Overuse has dramatically increased the routine contamination of poultry and other meats, while MRSA has become endemic in U.S. and UK hospitals. Government officials warn that medicine is in danger of losing effective antibiotics, so it is not surprising that many U.S. organicists have zero tolerance for antibiotics in their sector. Individual cattle cases, such as cows exiled from organic production after interventions with antibiotics, are sad. But they may play a sacrificial part in an overall solution.

A deciding factor is that there are alternatives to antibiotics. Much can be learned by perusing pre-1940s veterinary journals, when infections were treated with a variety of plant-based tinctures. The fact that U.S. organicists are required to work without antibiotics has, as Dr. Hue Karreman (personal communications) observes in his own veterinary work at the Rodale Institute, been a mother of invention for alternatives.

There are also marketing tactics to consider. A historic contingency that bootstrapped organic milk sales in the United States was the introduction of rBGH/rBST in 1994. In what Melanie Dupuis (2002) termed the "not-in-my-body"(NIMB) movement, many consumers reacted in horror at the flood of GMO milk by seeking organic milk. Ergo, part of organic milk's product differentiation was its status as non-GMO milk. Barely a decade later, several conventional U.S. supermarkets and processors asked their farmers to stop using GMOs, so it is less a headline issue than before. However, public anxiety about antibiotics is widespread. The strategic need to maintain unique selling points for organic dairy products is a compelling reason to exclude antibiotics from organics when the public worries about MRSA hospital infections.

From statements on its website, one can surmise that Aurora Organic Dairy (AOD) supports the USDA NOP prohibition on antibiotics. The company has in some instances recommended a study that supports the NOP ban on antibiotics. Readers may wish to access a research study comparing U.S. conventional to organic herds in a variety of herd health topics by Pamela Ruegg and her coauthors (2013) at the University of Wisconsin–Madison, indicating that mortality and culling rates were similar in organic and conventional herds.

Ironically, as mentioned in previous chapters, the USDA NOP has allowed the continued use of antibiotics such as streptomycin and oxytetracycline in—of all things—apple and pear production through 2014 (Granatstein 2011; WSU 2014). Fire blight can devastate an orchard in days, necessitating the razing of all trees. The threat of onslaughts of fireblight, which has been described as arboreal gangrene, understandably prompted growers in susceptible areas to lobby the NOSB for exceptional uses of antibiotics, and the extensions persisted for two decades. But many parents believing fruit to be nature's purist food for their children would blanch at realizing their apples and pears were treated with antibiotics. The 2014 phaseout voted by the National Organic Standards Board (NOSB) in Portland, Oregon, in April 2013 will ultimately generate alternatives.

Yet, the veterinarian's oath and the IFOAM principle that animals never should suffer unnecessary pain argue for the occasional emergency use of antibiotics in extremis. Dr. Hue Karreman (Jan. 25, 2014, on Odairy) made a sobering observation in a recent email exchange over an organic cow whose calf was delivered, dead, after a caesarean section in which some antibiotics were administered: "Why not do more antibiotics? She almost definitely needs them; I can say that even sitting here at a distance . . . I've only seen one cow make it through a C-section without antibiotics . . . She's no longer 'organic' anyway, so you might as well treat her to the maximal extent."

If additional antibiotics help her recover, is it not a sad anticlimax if she cannot rejoin her organic herd after a suitable timeout? It requires the wisdom of Solomon to resolve such dilemmas. Fairness requires consideration that MRSA, the dangerous increase in resistance to antibiotics in hospitals, is not due to the occasional cow receiving antibiotics. The blame for MRSA is largely due to overuse on factory farms, according to the Organic Consumers Association (OCA 2013). That is the inevitable result of government policies and regulation allowing overuse of prophylactic antibiotics in mass animal confinement—and the concomitant marginalization of family-scale farms that can produce more hygienically on pasture.

Weighing all factors, the best way forward for the USDA National Organic Program is to continue the exclusion of antibiotics. But the price in animal welfare would not be worth paying if leading individuals and organizations were not developing alternatives to antibiotics.

The rise in numbers of conventional confined-animal feeding operations (CAFOs) since the 1950s has removed cows from farmscapes. Confined bovines now produce about 250 percent what they did then, but longevity is less. For decades milk yield rose around 3 percent per year with traditional breeding methods, without the intervention of the dairy hormones rBGH/rBST (Scholten 1989; *Hoard's Dairyman* Mar. 10: 183, 194). Stressed cows live shorter, meaner lives than their forebears in the twentieth century. IFOAM's St. Paul Declaration suggests it is time to prioritize animal welfare over productivity in conventional dairying. There is also room for improvement on some organic dairy farms, one of the themes of the following chapter.

7

Family Farms and Megadairies

Photo 7.1   Aurora Organic Dairy in Colorado: thousands of cows on this megadairy.
Photo Credit: The Cornucopia Institute 2006.

7

Family Farms and Megadairies: Effects on Cows, Land, and Society

Leo Tolstoy, whose novels subsidized farming throughout his life, wrote that all happy families are alike, hinting they are quite boring (Bartlett 2008). Fortunately, people's comments in a survey conducted for this chapter show the "family" of big and small farmers, processors, and traders comprising the U.S. organic sector is rather interesting.

Respondents compare organic-industrial scale megadairies to family-scale farms with reference to animal welfare and sustainability issues. This discussion is back-dropped by the well known Food and Agricultural Organization (FAO-UN 2006) report *Livestock's Long Shadow* by Henning Steinfeld and others claiming that livestock contribute about 18 percent of global greenhouse gas (GHG) emissions, surprisingly more than transport. Steinfeld et al. claim that when land use and changes such as deforestation are factored in, the livestock sector accounts for 9 percent of anthropogenically-produced CO_2, and a disturbing amount of even more harmful greenhouse gases such as nitrous oxide, methane, and ammonia.

A study by Maurice Pitesky, Kimberly Stackhouse, and Frank Mitloehner (2009) sets the figure of human-related livestock emissions closer to 3 percent. Some observers claim the FAO overestimated the GHG contribution of deforestation involved in the planting of soybeans for cattle feed, and expect the actual amount to be somewhere between the low estimate of Pitesky et al. and the high of Steinfeld et al. (FAO-UN 2006). Accurate measurements of dairy and livestock emissions are needed to inform policies responses to climate change. Pastoralists extol the role of pasture as a helpful carbon sink for climate mitigation, and condemn the emission of greenhouse gases (GHGs) by confined animal feeding operations (CAFOs) in their outsourcing of fodder. Advocates of industrialized dairying such as Jude Capper (2009) claim that intensive systems are more efficient and in the aggregate, emit fewer GHGs than cows in grass-based systems. Life cycle analysis (LCA) of intensive and extensive systems has been conducted by researchers such as Charles Benbrook (2014) at The Organic Center (more of which below).

Emissions from biofuel systems, shale gas, and fracking are also in the context of current debates. There is evidence that the dedication of U.S. cereals to the biofuel boom increased the price fluctuations of cattle grain so much that many organic dairy farmers seek to acquire additional land, in order to accommodate their grain needs (Scholten 2010a). As regional and global food systems gear up to feed a forecasted 9.1 billion people by the year 2050, it bears asking whether the USDA certified National Organic Program will last till then and whether small farms managed by nuclear families will still exist.

Organic actors from farm to plate generally share a vision of a better, more sustainable world. However, just as birth order affects the psychology of family members variably, the politics of organic dreams vary with an actor's position and power in the food chain. To understand how political power functions in national and global food systems, it helps to go back in time. Decades before Ronald Reagan and Margaret Thatcher presided over the application of neoliberal market economics to their respective countries' dairy policies, neocorporatist political blocs played greater roles in negotiating national farm policy.

Political scientist John T. S. Keeler (1981; 1987) found precedent for such dynamics in his study of neocorporatist politics in France, 1947–70. Keeler compared the power of corporatist entities like the NFU in the United Kingdom, the National Farmers Union in the United States, and the powerful German Farmers Association (Deutsche Bauernverband). Keeler describes how France's main farmers' organization, the National Federation of Agricultural Workers' Unions (FNSEA), oversaw the formation of a post–Second World War rural consensus of farmers across scales. Large-scale farmers fomented a "we are all in this together" alliance with peasants, in negotiating farm policy with the national government in Paris. One national political economic outcome that pleased President Charles De Gaulle was that, after the establishment of the Common Agricultural Policy (CAP) in 1957, France garnered a lion's share of crop subsidies from Brussels in the European Economic Community. German farmers eventually topped the surplus EEC Butter Mountain as chief recipients of CAP dairy subsidies (Scholten 1989a). But France did well itself—much to the complaints of the United States and Cairns Group countries, such as Argentina, Australia, and New Zealand, which scorned the European farm subsidy regime. CAP crop subsidies incentivized French farmers to apply chemical fertilizers at rates double those of their North American counterparts, polluting waterways as they made the country the world's top wheat exporter. Unfortunately for French peasants, their share of EEC subsidies was comparatively scant, despite the blandishments of better capitalized farmers and agribusiness.

Henry W. Ehrmann (1983: 30–31) cites Keeler in his account of French farm politics: The "vision" was of farms that were not the "tiny, inefficient units predominant in the past, nor the giant agro-business concerns found in the United States, but rather a compromise . . . medium-sized family farms which are socially and humanly viable." The vision faded, writes Ehrmann (1983: 31): "Traditional individualism has won out in many instances and has prevented more cooperative forms of farming from spreading." While French smallholders focused on their nuclear families, Ehrmann notes: "Improved productivity has mostly benefited the large farms whose share in acreage and total production has risen substantially."

Farm politics in France are a smaller rendition of the "get big or get out" policies of conventional U.S. farming. Despite neoliberal rhetoric about cutting subsidies in the 1994 Uruguay Round Agricultural Agreement (URAA 1994) in establishment of the World Trade Organization (WTO 1995), subsidies remain a formidable element of commodity programs for wheat and other export crops (USDA 2006).

The Environmental Working Group describes itself as a public interest group dedicated to protecting public health and the environment. From USDA statistics gathered through the Freedom of Information Act (FIA), the EWG (c. 2013) compiled a "Table of USDA Subsidy Program Recipients/ Amounts 1995–2012" that reveals the preponderance of government support to large-scale and corporate agribusiness. From 1995 to 2012, $292.5 billion in subsidies were paid in commodity subsidies ($177.6 billion), crop insurance subsidies ($53.6 billion), conservation subsidies ($38.9 billion), and disaster subsidies ($22.5 billion). The top-heavy nature of subsidies is plain in this array: 62 percent of farms in the United States did not collect subsidy payments (according to the USDA); 10 percent collected 75 percent of all subsidies, totaling $178.5 billion over 18 years. The top 10 percent of farms garnered $32,043 on average per year between 1995 and 2012. The bottom 80 percent received just $604 on average per year.

EWG (c. 2013) data reveal the pecking order in farm subsidies. Keeping in mind that *organic dairying* received very little in nominal dollars or relative support, *dairying as a whole* ranked only ninth in the top ten list of USDA support 1995–2012. Rounded off to the nearest billion dollars:

1. corn subsidies ($84 billion) were by far the biggest subsidy program with 1.64 million recipients 1995–2012;
2. wheat subsidies ($36 billion) went to 1.37 million recipients;
3. cotton subsidies ($33 billion) accrued to 265,000 recipients;
4. The Conservation Reserve Program ($32 billion) paid out to nearly 925,000 recipients;
5. soybean subsidies ($28 billion) were paid to over one million recipients;
6. disaster payments ($22 billion) were disbursed among 1,384,956 recipients;
7. rice subsidies ($13 billion) went to 70,033 farmers;
8. sorghum subsidies ($7 billion) went to 615,810 beneficiaries;
9. down the list in ninth is dairy, in which dairy program subsidies ($5 billion) were shared by about 161,000 farmers 1995–2012.

Careful readers will note that, while there may have been 161,000 U.S. dairy farms in 1995, that is no longer the case. Since 1995, when the decline in U.S. dairy farm numbers was already steep, the total has fallen by more than half to fewer that 50,000—another reason for writing this book. With relatively low levels of support, it is understandable why so many families said "Bye-bye!" to dairying (Scholten 1997b). Other farm families sought a price premium in organic sales of milk produced in more environmentally sustainable and animal friendly settings.

Economists with their eyes on gross domestic product (GDP), rather than equitable distribution of wealth (Gini coefficient), might defend the decimation of family-scale dairy farming as another version of the decline of the buggy whip industry that attended the switch from horses to automobiles. It is, they may say, a harsh but necessary form of the *creative destruction* that Austrian economist Joseph Schumpeter claimed gave capitalism its vigor. But if the switch to large-scale megadairies has brought unacceptable levels of point pollution to the environment and worsened and shortened the lives of cows, this creative destruction becomes a question of environmental sustainability and the ethics of animal welfare—not to mention the livelihoods of family farmers. Richard G. Wilkinson and Kate Pickett, the authors of *The Spirit Level* (2009), combed data from many other researchers to conclude that equal societies almost always do better. From here it is not a great leap of faith to imagine that a Jeffersonian democracy of many prosperous rural farmers, rather than a few very rich ones, weaves a more sustainable economy for all.

Returning to the Environmental Working Group's list of subsidy program recipients and amounts (EWG c. 2013), please note that these 1995–2012 programs were approximately 99 percent for conventional ag programs, not organics. (Organics are low priority for federal subsidies. As mentioned in a preceding chapter, the 2012 farm bill extension earmarked money for organic research, but it was 20 percent lower than 2008 farm bill levels. This was followed by 2013 in the doldrums, a year of no USDA funding for stranded research projects. Comprising just a small, but growing, percentage of the total food system, the organic sector has been a poor shirttail cousin to Big Ag. Fortunately, organics garnered more support in 2014.) The previous data establish that corn (maize), wheat, cotton, and soybeans are the top four commodity crops, although the Conservation Reserve Program gets more funding than soybeans. Program subsidies were received by an estimated 161,463 conventional dairy farmers amounting to $5,334,467,679 over 1995–2012, about $1.28 billion more than subsidies shared by livestock (and beef) farmers. The meaning of corn topping the list is manifold. Farmers have always grown corn in the western hemisphere, but U.S. production accelerated in the 1970s as more dairy cows were fed a higher-energy mix of totally mixed rations (TMR) in confined-animal feeding operations (CAFOs). This was also true of beef finishing, which became almost totally conducted in huge feedlots. Humans also ate more corn directly as ingredients in fast-food restaurants. The biggest boost in corn production was the launch of President

George W. Bush's biofuel program in 2005. The effect persists, although it is somewhat tempered by the U.S. boost in energy production via fracking.

Greens critical of President Obama's and USDA secretary Tom Vilsack's support of biotechnology in agriculture were less dismayed by some aspects of his farm policy. *The New York Times* (2014; see also Riddle c. 2014) claimed that close examination of the 2014 farm bill showed it to be "more whole grain than white bread." The newspaper found: "While traditional commodities subsidies were cut by more than 30 percent to $23 billion over 10 years, funding for fruits and vegetables and organic programs increased by more than 50 percent over the same period, to about $3 billion." The farm bill increased funds to support the transition from conventional to organic farming from $22 million to $57.5 million. Funds for oversight of the National Organic Program almost doubled to $75 million over five years. This could help NOP staff retention and improve monitoring and enforcement of dairy pasture rules. Bipartisan support was key. Laura Batcha, executive director of the Organic Trade Association (OTA), explained how organic lobbying efforts bore fruit, "We kind of over performed with younger new members of Congress on both sides of the aisle." Because Democrat and Republican parents alike are worried about childhood obesity and diabetes, they are attracted to the "farm-to-table movement promoted by the first lady, Michelle Obama, and other national figures." Ferd Hoefner, the policy director of the National Sustainable Agriculture Coalition. "Even the most ag-centric member of the Agriculture Committee knows that is what helps sell the bill when it gets to the floor."

Further good news was that the $16 million allotted to the Organic Agriculture Research and Extension Initiative (OREI) in the 2012 farm bill extension was, after no funding in 2013, increased to $20 million. The National Sustainable Agriculture Coalition (NSAC 2014) happily announced that "Congress reauthorized OREI and provided $100 million over five years ($20 million annually) in mandatory funding." The outlay can pay for research, education, extension, and conferences.

There was a slightly greener sheen to the 2014 farm bill than the last one. But it did not obscure the dominance of GMOs in commodity agriculture and the potential for crops, such as GM alfalfa, to contaminate fodder for organic dairy cows (*Guardian* 2013b). The official USDA line was that conventional, GMO, and organic crops could "coexist," but so many examples of cross contamination were known that public meetings were scheduled to discuss this thorny issue of biodiversity in 2014. Fortunately, organic milk product sales continued to increase, but the high price of fodder sourced off-farm prompted farmers to seek more cropland—or sell out (Scholten 2010a).

*   *   *

Michael Pollan (2001; 2003) famously called attention to the emergence of an organic-industrial complex that advanced corporate agribusiness and that he described as against the original organic dream of cooperative members and others in alternative food networks. For a decade, Philip H. Howard (MSU) has illustrated the propensity of corporate agribusiness to acquire small-scale organic entities and incorporate them into their own structures. Answering questions on the corporate ownership of organics is Howard's (2014) graphic "Organic Industry Structure: Acquisitions and Alliances Top 100 Food Processors in North America," which traces the organizational changes that challenge traditional meanings of *organics*. Julie Guthman (2004) describes the process as the "conventionalization" of organics. To her credit Guthman's analysis is anything but naïve, and she points out that much of the conventional methods adopted in Californian organics were summoned by small farmers from below, not originally imposed by big agribusiness.

This text has applied the term "appropriationism" to instances when highly capitalized entities have entered the organic sector (for definitions see Guthman 2004 and Fine et al. 1996). There can be huge positive outcomes from large-scale investments in the sector; it is probable these have attracted attention from the USDA in regulating the sector and also convinced processors, distributors, and supermarkets to make room in their operations for organic. In other words, without the participation of highly capitalized agribusiness in the USDA NOP, it is difficult to say whether or not it would have gotten established in the first place and, second, whether its annual sales would have exceeded the current $32 billion per annum.

Over the decades it has become apparent to this author that there is some truth in the claim that farmers are "all in this together." Big and small farmers share hostility to urbanization, and they can imagine walking in each other's boots. But they are also competitors, and if some actors skimp on traditional organic methods (such as pasture) or add nonorganic ingredients to processes (e.g., synthetics or inputs made with GMOs) that challenge the integrity of USDA organic certification, it can evoke ire.

U.S. Organic Dairy Politics Survey in 2013

There was a time, in the 1950s, when two-hundred-cow dairy farms seemed unimaginably big to most family farmers. Things change. In 2007 a report for the USDA on conventional dairying by James MacDonald and others (USDA 2007: 5) found "the number of dairy farms with fewer than 200 cows is shrinking, while the number of very large operations, with 2,000 or more cows, doubled between 2000 and 2006." MacDonald's study did not treat costs and farm sizes of organic dairy farms, but it illustrates the rush to scale from which so many organic dairy farmers fled, in order to preserve livelihoods on family-scale farms.

In the organic sector, attention to animal welfare, environmental sustainability, and social justice are apt to be prioritized differently, depending on the positionality of the actors. Despite communal exhortations to work together to make the U.S. organic sector thrive for the good of all, different perspectives inevitably bring conflict in organic politics. The National Organic Program's first decade featured a Pasture War that was apparently resolved by the final 2010 Pasture Rule mandating that cows graze on pasture a minimum of 120 days per year and consume a minimum of 30 percent of their dry matter intake (DMI) on grasses and so forth. This Pasture War pitted small-scale family farmers against large organic-industrial dairies, which the 2010 rule obliged to desist from the confinement of lactating cows. The monitoring and enforcement of pasture rules lacks transparency according to critics and awaits resolution.

Survey Construction and Methods

To better understand political dynamics within the organic sector, this author constructed an electronic test instrument via SurveyMonkey in late 2012. This was followed by piloting among colleagues and revision and addition of questions. A colleague mentioned, "I [usually] select *Agree*, not *Agree 100%*. I don't know how many others do the same." This writer often duplicates that practice himself. Alas, interpreting responses is a subjective task. But it was decided to retain the wording. An advantage is that, when respondents' opinions are not ambivalent, they can indicate them unambiguously by responding *Agree 100%*.

The "U.S. Organic Dairy Politics Survey" was advertised in early 2013 at the Association of American Geographers (AAG) conference in Los Angeles, at the April 2013 meeting of the National Organic Standards Board (NOSB) in Portland, Oregon, on the Odairy email discussion list, and in the newsletters of farmers' organizations such as FOOD farmers, MOSES, NODPA and WODPA. Suggestions to advertise the survey on Facebook were rejected; although this could have attracted more respondents, it was suspected that few would be versed in organic issues beyond blog headlines. A demographic of people participating in some aspect of organic dairy or food chains was sought. Respondents were encouraged to ask others to take the survey in snowball distribution. By the end of 2013, respondents numbered 65, representing a wide range of people working with or concerned about different scales of farms, processors, traders, and consumers.

The electronic SurveyMonkey (http://www.surveymonkey.com/s/83 QSVHP) was designed to be completed in seven to ten minutes, or longer if participants made comments in the space provided.

Farm Models

Among the issues mentioned in the survey, pasture and confinement issues are clearly queried, along with other environmental issues. Comments usually hinge on the respondent's relationship to farm scale. To visualize these relationships on a virtual map, we begin with descriptions of four cases of generic model farms dubbed West Family Farm (WFF), West Megadairy (WMd), East Family Farm (EFF), and East Megadairy (EMd).

These models do not represent individual farms; they are composites of dairy operations that could exist in their geographical contexts, like fictitious farms in a novel. A useful generalization is that organic herd size decreases from historically newer, larger dairy farms in the West to older, smaller farms in the East. It is common for small family-scale farms with small herds to raise some or most of their replacement heifers, bulls, and feed crops. Larger farms and megadairies tend to contract with other businesses to raise their heifers and source much hay, grain, and other inputs off-farm. A former Washington State conventional dairyman explains that his dairy success did not come without difficulty:

> Right before leaving the dairy business for fishing in 1986, the herd was about 425 cows including dry animals, but not young stock. There were about 400 young animals of various ages. But I don't have a clear idea of my culling rate because it varied upon the stage of my business growth. That is, when I first moved into the remodeled facilities I had room to grow and culled few because there was room. As young stock matured I started culling more and also sold some young stock. It helps if you know how to raise calves without killing them—a skill not all farmers possess. Systematically giving newborns colostrum and isolated individual pens help . . . But I suspect the remaining farmers are better than previous generations. Anyway I can only estimate that the average animal lived to be about six years old. Culling reasons were first of all reproductive (slow or late conception), lack of production, injury/health problems from calving, and other health problems. A guess as to the cull rate is about 10–18 percent initially and, as young stock numbers rose, increased dramatically to 20–25 percent thereafter. Remember, if 50 percent of calves are heifers and 85 percent of those survive to become replacements you have 43 percent replacement animals available per year.

This former farmer's comments also highlight themes regularly discussed in this book, those of herd expansion and of culling. Current culling rates on U.S. conventional dairy farms are about double the cull rates given above; organic farms that do not achieve better breeding results than that are probably prioritizing productivity over cow health Dr. Paul Biagiotti, DVM, writes in *Progressive Dairyman* (Biagiotti 2014) that in Idaho "Every year, 44 percent of the cows and first-lactation heifers in an average herd are sold for slaughter, die or otherwise leave the farm. This suggests that the average cow has only about a one-in-two chance of completing one full lactation." The national culling average is close to Idaho's, nothing to be proud of.

Prior to the 2010 USDA Pasture Rule benchmark rule, organic megadairies often illustrated the conventionalization thesis (Guthman 2004), in which cows that had been transitioned to organic in a former confined-animal feeding operation (CAFO) dairy were pastured infrequently and fed organic rations in a feedlot. Some early organic megadairies in the Northwest and Southwest began on previously established conventional CAFOs with at least four thousand cows on each. This followed the Dutch American model of packing the property with milk cows and buying hay, grain, and pregnant heifers off-farm. This description is corroborated by William McBride and Catherine Greene (2009: iii) in the summary of their report to the USDA:

> Between 2000 and 2005 . . . certified organic milk cows on U.S. farms increased by an annual average of 25 percent, from 38,000 to more than 86,000. To meet the growing demand, the organic production sector has evolved much like the conventional sector. Along with primarily small, pasture-based organic operations located in the Northeast and Upper Midwest, larger organic operations, often located in the West, that use more conventional milk production technologies have increased in number. Economic incentives, driven largely by lower production costs, are behind much of this change. Proposed changes in . . . NOP . . . seek to clarify and stiffen pasture requirements.

Splitting survey models into West and East archetypes does not deny historic aspects of Midwestern conventional dairy farms. Before the boom in confined battery pork production, farms often raised grain, kept chickens, and fed pigs extra milk. The Midwest is also important for the birth of the CROPP/Organic Valley cooperative, which sought to retain the mixed farming tradition. Midwest farms are subject to Mother Nature's extremes, when snow in the winter and soaring temperatures in the summer can keep organic cows in the barn for more months of the year than in temperate microclimates, like the Pacific Northwest.

As mentioned above, the survey analysis imagines four archetypes to represent organic dairy farms nationally. The "family" subgroup for both West (WFF) and East (EFF) farms covers smaller family-scale farms with herds under two hundred cows, which a nuclear family can manage. This category includes larger farms based on evidence that an extended family group (e.g., adult children who expand their parents' dairy and buy neighboring farms) can successfully manage an organic grazing herd up to one thousand cows. This limit is based on the author's correspondence with sources, including a key figure who sustainably ran an organic dairy herd approaching one thousand cows via managed intensive rotational grazing (MIRG).

In this chapter's definition, an organic megadairy has more than one thousand cows. Such large herds are, in many informants' opinions, difficult to graze according to the 2010 USDA Pasture Rule. It is physically impractical for two or three thousand cows to walk from pasture to barn, especially if milked more than twice daily. To address this problem, more U.S. megadairies could adopt the European practice of machine milking in fields, but with a hypothetical addition of robot milkers, which allow cows to choose the time of milking. A megadairy that complies with USDA NOP rules may be a constellation of hub-and-spoke models, that is, multiple barns and milking parlors set within a local area. Subunits can benefit from economies of scale in sourcing building services, equipment, veterinary care, organic hay, and grain from afar. More discussion on this follows in the concluding chapter.

Here the reader is reminded of approximations of gross income for organic milk, based on The Organic Center's (Benbrook 2012a: 19) model averaging

$30 per hundredweight (cwt) for organic milk and $20 cwt. for conventional. Assuming 20,000 pounds milk produced annually, an organic cow would gross $6,000, and a conventional cow, $4,000.

West Family Farm (WFF organic): A USDA farm typology by Doris J. Newton and Robert Hoppe (USDA 2001d; also Scholten 2011: 118–20) disaggregates conventional farms in a pattern applicable to organic farms. "Small family farms" have limited resources with fewer than one hundred cows, gross incomes under $100,000, and net incomes under $20,000. "High sales family farms" might have up to two hundred cows and gross incomes up to $249,000. "Large family farms" have more than two hundred cows and gross incomes of $250,000 to $499,000. "Very large family farms" have up to one thousand cows and gross incomes of $500,000 or more. The difference between such a farm and the category below is their ownership. Note: projecting these 2001 patterns to 2014 must factor in higher prices for off-farm grain and fuel.

West Megadairy (WMd organic): Borrowing from the same conventional typology, organic megadairies have at least one thousand cows and gross incomes over $500,000. Such farms can, with well-trained employees, function like family megadairies, but because most organic WMds have been corporate owned (e.g., "organized as non-family corporations or cooperatives, as well as farms operated by hired managers" [USDA 2001d]), this category excludes family ownership. New corporate megadairies built since 2010 reportedly average two thousand cows, half the size of some former WMds. As discussed in earlier chapters in this book, larger operations, both organic and conventional, have a higher rate of cow burnout—that is, cows so stressed that their milk production and health decline at an age younger than expected. This requires a proportionally larger pool of replacements (pregnant heifers) than smaller organic family farms. But cow longevity on large organic-industrial dairies may now be better than it was on earlier facilities with higher stocking densities and less pasturing of lactating cows, before the USDA 2010 final Pasture Rule. Multiple sources on WMds report better cow longevity than on comparably sized conventional megadairies (with the latter averaging cull rates of 40–50 percent, and a few hyper farms around 60 percent). Health emergencies occasionally prompt thoughts of intervention with antibiotics, but proactive veterinary practices have diminished the incidence of such emergencies on organic farms compared to conventional megadairies. Managers of WMds publicly support the USDA NOP prohibition on antibiotics, noting the importance of consumer expectations.

East Family Farm (EFF organic): East Family Farms follow the West Family Farm pattern, but they average smaller herds and lower gross incomes depending on milk prices paid by processors (USDA 2001d). Herd sizes could be closer to 50–100 than 200, although mergers in recent years have driven herd numbers upward. Maryland, New York, and Pennsylvania have many family organic dairies of long provenance. New England has a vibrant organic network, with Vermont claiming many family organic dairy farms. Some farms with sufficient acreage have stopped feeding much grain and rely on grass to lower overhead and boost healthful omega-3 fatty acids in the milk. The Southeast has many family farms but often struggles with heat stress and mastitis.

East Megadairy (EMd organic): Based on the USDA farm typology of Newton and Hoppe (USDA 2001d), organic megadairies average fewer cows than a West Megadairy (WMd) with lower gross incomes, depending on farm gate price. Harsher winters in parts of the East require more wall cladding on barns for wind and snow protection, while simple roofs suffice for climatic conditions on some Californian WMds. East Megadairies can be "organized as non-family corporations or cooperatives, as well as farms hired by professional managers." Like many dairy farms, corporate facilities may keep a pet cow in the ten-year-old range. They may be the exception, not the rule, but cows on organic megadairies have a better chance of living six score months than cows in conventional operations.

Now we turn to the survey results. Where specific farms have been mentioned, their identities are anonymized by use of the above four categories. As we will see, these representative actors in the organic sector—farmers, activists, and consumers—reinforce the picture presented in earlier chapters. They have myriad concerns and varied opinions, reflecting the many factors involved in dairy farming today.

Survey Questions and Comments

Almost 85 percent of the 65 respondents disagreed to some extent that organic and conventional dairying had the same environmental effects (Question #1).

A prominent veterinarian responded that the two modes were not utterly different due to "diesel fuel usage in tractors for tillage, etc. Exhaust coming from tractors is pollution and can make one sick. Extraction of oil to make diesel fuel is a global problem: societally, internationally, etc." Then the vet added an observation reminiscent of "repeasantization" activities around the world described by Dutch researcher Jan van der Ploeg (2009) and observed by the author of this book, involving draft animals by mostly part-time farmers in several developed countries: "If all were horse powered, then, yes, organic would be a 100% solution to environmental problems."

A British geographer, who develops equestrian tourism in Scandinavia and happens to be familiar with West Coast U.S. dairying wrote: "Cows still

Question #1 Environmental effects of organic and of conventional dairying are about the same.

Agree 100%	Agree	Unsure	Disagree	Disagree 100%	Total
4.62%	6.15%	4.62%	35.38%	49.23%	
3	4	3	23	32	65

emit methane." A Southwest farmer of Dutch extraction with experience in the conventional mode commented: "Manure is organic." Both observations are largely true, but the amount, composition, and presence of synthetics in manure vary with diet. Gene Logsdon, author of *Holy Shit* (2009: 37), suggests liquid manure from farms big enough to dig lagoons makes "farmland soils harder, not soft and mellow like bedding manure does." Manure decomposing anaerobically in lagoons emits more methane than the same amount deposited by cows in fields.

News reports worry the general public that fish and frogs are undergoing gender realignment, and many organic farmers suspect this is caused by synthetic estrogen mimics used in conventional farms and wider commerce. An organic farm owner in Vermont observed: "Organics don't use chemicals used on conventional farms."

In New York State, a husband and wife active in dairy policy lamented the "toxic pesticides, pharmaceuticals, and hormones that conventional farmers dump into the environment."

A Nebraskan agronomist wrote: "Feeds used, hormones used, processing methods are but three reasons [to reject conventional]. Grass versus confinement is a major factor."

The higher the stocking density, the more point pollution affects flora and fauna. A trainer for a major West Coast consumer cooperative assailed: "Artificial drugs and hormones to promote production, which have no testing threshold or withholding period. Confinement (versus outdoor access or pasture) in conventional encourages an unnatural diet that eventually would kill them." Sadly, this statement is supported by culling and longevity statistics referred to earlier in this book.

In Thousand Oaks, California, an insurance adjuster who trains race horses part-time writes: "By feeding organic hay and grain it reduces the amount of harmful pesticides in the environment directly on the fields, also reducing or eliminating pesticides that end up in milk and the bodies of those who drink it. The use of antibiotics also affects the environment as more people are becoming resistant to 'super bugs' like MRSA."

Not all respondents were so critical of conventional dairy farming. One anonymous New York farmer who admitted disapproval of some organic-industrial dairies wrote: "Environmental effects can be very different, depending upon the individual organic and conventional farms being compared."

In New England, a molecular biologist wrote: "It is too easy to paint all conventional dairying with the same paint brush. I don't believe rigorous studies have been done."

The productivity of conventional intensive dairy farms and reduction of the industry's environmental footprint were praised by a prominent agricultural journalist based in the Midwest: "Organic milk production is less efficient because of the lower levels of milk production. Maintenance cost is spread over fewer pounds of milk." One conclusion is that, while it is right to pose ethical questions on animal welfare and longevity, costs in labor, land, and materiel benefit from careful life-cycle analysis (LCA). The degree of appropriationism involved in outsourcing globalized dairy inputs is so high that analysis is difficult.

Nearly 77 percent of survey takers disagreed that animal welfare was commensurate in organic and conventional confinement dairying (Question #2).

A histology technician in Seattle who taught 4-H classes in animal care noted: "Organic raised cows have to be treated by the organic guidelines and put on pastures 30 percent of the time during a year, which is better for the animals than being 100 percent in confinement." Pasture is easier on a cow's hooves and musculature than confinement on concrete, and grass offers less exposure to pathogens than most confined facilities.

A New York organic couple accustomed to winters on the Canadian border joked: "Grazing is a great life for cows but the winter could be easier on them with conventional dairy practices." Of course, successful organic farmers barn their cows in inclement weather.

Taking a health tack on individual cow welfare was a leading dairy writer and consultant: "It is a welfare problem to not be able to give antibiotics to cows with infections."

Moderating that view was a Southwest dairyman who extolled proactive conditions that fortify cows' immune systems: "Take care of cows and they take care of you."

A UK colleague questioned terminology: "I am wondering about the word 'confinement farms.' It might work with those in the know but to ordinary people it sounds like torture. You might want to say something like large-scale agri-industry farms." This reasonable-sounding suggestion was rejected for two reasons: First, the term "confined-animal feeding operation" and acronym "CAFO" appear in nonspecialist U.S. publications. Second, the term "large-scale agri-industry farm" is so ambiguous that respondents might believe the herd usually grazed on pasture.

Question #2   Animal welfare is about the same in organic or in conventional confinement dairying.

Agree 100%	Agree	Unsure	Disagree	Disagree 100%	Total
3.08%	9.23%	10.77%	44.62%	32.31%	
2	6	7	29	21	65

Question #3   Dairying is as sustainable on large-scale confined feedlot farms as it is on family-scale pasture grazing farms.

Agree 100%	Agree	Unsure	Disagree	Disagree 100%	Total
4.62%	9.23%	12.31%	26.15%	47.69%	
3	6	8	17	31	65

Almost 74 percent of respondents disagreed with this assertion, though nearly 14 percent agreed (Question #3). A key to interpretation is understanding how the respondent prioritizes sustainability regarding animal welfare (e.g., longevity) or the environment or people and communities.

The husband and wife owners of a West Family Farm (WFF) with 160 cows and replacements, on rich river bottom land in the Pacific Northwest, asserted that conventional megadairies were lucky that consumers did not understand their breeding practices. They said consumers roundly rejected GMO dairy hormones, but if they knew about drugs given to cows on megadairies to synchronize breeding, more would buy organic milk.

A Californian woman showed uncertainty about the final 2010 USDA Pasture Rule but was sensitive to boredom afflicting animals in CAFOs: "Currently [confinement dairying is not as sustainable]. I think with innovative ideas and technology that create a more natural environment, for cows in a large-scale operation, that it's possible for it to be sustainable, if they are fed organically and given access to sunshine and grass . . . but a feedlot style farm with constant close contact perpetuates the chance of disease and the farmers 'need' to feed antibiotics which end up in the food supply."

The cycle between land and animals envisioned by pioneer organicists is broken by confinement, according to an agronomist from North Bend, Nebraska, who adds: "Knowing your land and making it better is not happening on confined lots."

A career military airman, with a degree in international politics, who was familiar with pasture and confinement dairies wrote: "I think cows should be able to graze 100 percent of the time." That is possible in parts of his native Washington State, but many climates require housing cows in the winter. His comment hints at a consumer assumption that cows belong on pasture as much as possible (Question #4)—otherwise, why do marketers flaunt images of cows on grass on milk labels?

The 4-H teacher near Seattle asked: "Why wouldn't it be easier to keep them grazing most of the year?" The answer is that farmers in many geographic areas say it is easy to pasture cows most of the year. Some wish USDA organic grazing minimums were 240 or more days per year, rather than the current 120 days.

Nevertheless, a farm consultant from Wisconsin who suspected big producers were not honoring the 2010 Pasture Rule wrote: "Regional grazing

Question #4 Organic dairy cows should graze on pasture as much as possible during the year.

Agree 100%	Agree	Unsure	Disagree	Disagree 100%	Total	Average rating
61.54%	30.77%	4.62%	3.08%	0%		
40	20	3	2	0	65	1.49

rules should be adapted over one national rule. Too many climate and soil variations exist around the country."

On a positive note, the agronomist from North Bend, Nebraska, extolled the human nutrition benefits of pasture-based milk, revealed in Anglo-American studies (Benbrook 2012a; Benbrook et al. 2013): "Better feed [produces] healthier Omega-3 fatty acids versus corn high in [unhealthy] Omega-6 fatty acids."

The next question evoked much uncertainty, 52.31 percent, in respondents (Question #5). As discussed in previous chapters, some of America's most critical advocates for small family farms are unsure how closely the USDA NOP has been able to monitor and enforce the final 2010 Pasture Rule. For expert observers, the greatest uncertainty on pasture compliance concerns West Megadairies (WMds), including some in mountain states. Websites of some leading megadairies show cows on fields, but the details of farm and paddock layouts, distance to milking barns, milking schedules, heifer replacements, culling, and longevity are not transparent. It follows that nonexperts are even more uncertain of the status of USDA grazing regulation. (More discussion follows in chapter 8.) The few respondents on either side of this bell curve—that is, the 7.69 percent who *Agree 100%* with the proposition that the NOP strictly enforces the Pasture Rule and the 6.15 percent who *Disagree 100%*—arouse curiosity on whether their responses are based on insider knowledge or personal political economic interests.

A husband and wife in New York, active in regional dairy politics, praised a Midwest farmer advocacy group for litigation that prompted the USDA to investigate pasture violations by WMds. Because of a lack of transparency on pasture regulation, the outcome is still in doubt. But the couple predict the outcome will "set the future of the 50–100 cow organic farms that were driven out of the conventional sector" by confined operations (CAFOs).

Question #5 The USDA National Organic Program (NOP) strictly enforces the Pasture Rule that cows must graze pasture at least 120 days in the grazing season.

Agree 100%	Agree	Unsure	Disagree	Disagree 100%	Total
7.69%	21.54%	52.31%	12.31%	6.15%	
5	14	34	8	4	65

Question #6 The USDA National Organic Program (NOP) treats large and small dairy farms fairly.

Agree 100%	Agree	Unsure	Disagree	Disagree 100%	Total
3.08%	20%	44.62%	24.62%	7.69%	
2	13	29	16	5	65

This question is similar to the previous one on pasture but wider in scope. Expressing the frustration and uncertainty, 44.62 percent, of many observers (Question #6) is an anonymous New York EFF farmer who offered this criticism: "Lax enforcement overall of all dairies, and especially for certain rules, e.g. pasture requirements."

An organic inspector from Oklahoma added a request to the survey, addressing a major concern of watchdog organizations: "Please put pressure on the NOP to address the Origin of Livestock provision they have been promising to do. Thanks." According to gadflies, loopholes allowed organic-industrial megadairies to boost profits by buying conventional pregnant heifers, which had been fed on cheap conventional feed through the second trimester of their calves' gestation. The inspector also remarked: "At the organic dairies I see I would not characterize animal welfare as any different as the conventional dairies I used to work with. The main difference is the access to pasture." This statement is memorable, but without a quality yardstick, such as longevity, we are left wondering exactly how good or how poor welfare was on those farms.

A full 80 percent of respondents were adamant (*Agree 100%*) that GMO foods be labeled, abetted by another 13.85 percent who agreed (Question #7). High-profile political campaigns have fought this issue in California and Washington, with more political contestation on GMO labeling in New England and Oregon (see next chapter).

In New Mexico, a couple in the Dutch American tradition commented: "Organic was created because of GMOs." It is true that public wariness of biotechnology and GMO dairy hormones drove demand for USDA organics, but impetus also came from food scares, such as mad cow disease, and demand for systems perceived as kinder to animals, biodiversity, and family farming (Whatmore 2002; Scholten 2007). The New Mexico couple was skeptical of organics until they noticed keen consumer demand and the price premium paid by a national cooperative. When the co-op could not accept milk from more than five hundred cows on their West Family Farm (WFF), the couple multiplied their herd into a very large operation that could be called a family West Megadairy (WMd) and established manufacturing and retail outlets for added-value product sales. Their income multiplied to gross about $3 million annually, though net income fluctuated according to the price of any outsourced organic feed.

Negative responses to the question of GMO food labeling deserve consideration. A microbiologist in New England disagreed 100 percent and wrote: "GMO plants and animals are tested for safety. Conventionally bred plants and

Question #7   Genetically modified organisms (GMOs) should be labeled in food.

Agree 100%	Agree	Unsure	Disagree	Disagree 100%	Total
80%	13.85%	0%	0%	6.15%	
52	9	0	0	4	65

Question #8   It doesn't matter whether food is labeled "Natural" or "USDA Organic."

Agree 100%	Agree	Unsure	Disagree	Disagree 100%	Total
4.62%	4.62%	3.08%	26.15%	61.54%	
3	3	2	17	40	65

animals aren't. There are a number of examples where conventional breeding has introduced new allergens or toxins into crop plants." It is true that plants and animals bred with traditional methods epitomized by Gregor Mendel can have toxic or other unfortunate properties, but these methods are generally better understood by scientists than anthropogenically designed organisms that may borrow genetic material across biological kingdoms. The extent that "GMO plants and animals are tested for safety" is debatable because the Food and Drug Administration (FDA) relies largely on company testing to determine a GMO's "substantial equivalence" with natural forms, and there is skepticism of these procedures in many countries (Smith 2003).

A strong 61.54 percent *Disagreed 100%* and 26.15 percent *Disagreed* with the notion that labeling food "Natural" or "USDA Organic" is an inconsequential matter (Question #8). "Natural could mean almost anything," writes a plumbing salesman and amateur ornithologist from the Midwest, now living on Puget Sound. This man and his wife were aware of incidents when organic dairy and soy companies were bought by corporations that switched from organic to cheaper nonorganic ingredients and continued marketing them in similar packaging to consumers who did not notice the change.

Food products labeled "Natural" are sometimes processed with GMOs, heavy metals, and irradiation, which the USDA included in organic rules proposed in 1997. That plan was met by 275,000 largely hostile comments before the USDA published National Organic Program (NOP) rules without them in 2002.

Conclusions

This chapter began by exploring the National Organic Program's relatively minuscule size in USDA budgeting compared to conventional commodities programs. This relationship was compared to the unequal alliance between French peasants and large-scale farmers in that country's postwar farm policy during 1947–70. And while the NOP continues as a junior partner to conventional and GM farming in the USDA budget, its status as a growing sector merits attention. Mark Keating, a former NOP livestock specialist notes (personal communication):

> As a portion of the whole, organic funding at USDA remains small, but in raw numbers, the 2014 Farm Bill is the best that organic agriculture has ever done,

by far—$11.5 million per year for organic cost share, $20 million per year for OREI, and the Organic Data Initiative. Someone from the Organic Trade Association said, "We over-performed on the Farm Bill and they are correct. USDA is clearly on-board with organic."

More research funds for organics do not, however, erase unsustainable practices in other sectors. Secretary Tom Vilsack's shepherding of new genetically modified crops, such as GMO alfalfa, to certification is a threat to organic dairying, which relies on organic fodder that is not adulterated with GMOs. Relevant to conventional farming, President Barack Obama has presided over recent increases to the permitted parts per million (PPM) of the GMO pesticide glyphosate in the environment. The USDA speaks of "coexistence" among organic, conventional, and GMO crops, yet hard-core environmentalists and deep greens suspect this is a delaying tactic by biotech firms intent on contaminating the U.S. biosphere to the point that recombinant DNA is so ubiquitous that its presence is a fait accompli.

The 2013 U.S. Organic Dairy Politics survey explored attitudes to animal welfare and sustainability on conventional and organic dairy farms. Most respondents displayed support for pasture dairying, although one respondent pointed out that cows in some parts of the United States would be up to their ankles in mud much of the year if put on pasture. The same source claims conventional intensive dairying is more efficient in resource use—and a lesser source of climate-changing greenhouse gases (GHGs) than organic methods. Some respondents hinted that cow welfare was better in organic methods, but another maintained that a prohibition against antibiotics was a welfare issue.

There was strong support for keeping organic cows on pasture when practical but significant uncertainty that the final 2010 Pasture Rule was being strictly enforced. There were demands for stricter enforcement of all organic rules and a call to effect an origin of livestock provision.

Nearly 88 percent of respondents found it important to distinguish between so-called *natural* and *organic* labelling. More discussion on how natural labeling can blur the lines between organic and conventional food is found in the next chapter, as well as interesting political developments inside and outside Vermont.

8

Conclusions and Outlook

Photo 8.1 Horizon Organic megadairy in Idaho in 2009.
Photo Credit: The Cornucopia Institute 2009.

8

Conclusions and Outlook:
Agribusiness, Cooperatives, and
Power to 2050

Dairying in the USDA National Organic Program has thrived partly due to negative public perceptions of industrial practices on conventional farms. But as organic dairying comes to resemble conventional farming, dangers exist of disenchantment with it. This chapter summarizes the book's arguments on the appropriation of traditional methods by agribusiness, cow longevity, antibiotics and biotechnology, implications for policy on GMO labeling, and an update on the Pasture War of the 2000s.

Many consumers sought organic milk after losing trust in the safety of intensive conventional dairying. As earlier chapters detailed, a series of transatlantic food scares involving *E. coli*, listeria, and salmonella roused disquiet. The late 1980s Alar scare in the Pacific Northwest apple industry about possible carcinogenic effects of this chemical had a ripple effect. In 1994, biotechnology entered conventional dairying in the form of rBGH/rBST injected into dairy cows. In 1996, the UK government admitted a link between mad cow disease (BSE) and new variant Creutzfeldt-Jakob disease (nvCJD) in humans. Anxiety about conventional food was linked to a rise in vegetarianism in young people, especially girls (Atkins and Bowler 2001). Animal welfare groups alerted consumers to the less pleasant aspects of intensive livestock keeping, and shoppers increasingly perceived a gap between the rural idyll and the reality of industrial livestock operations. Food scares combined with fear of biotechnology prompted by a so-called turn to quality in a move back to nature manifested by the consumption of organic foods (Murdoch and Miele 1999; Whatmore 2002; Scholten 2007a; b).

During the planning and writing of this book, this author was astonished by a relative dearth of data on cow culling and average longevity in conventional dairying. A decade ago insiders on state-of-the-art California megadairies with multiple thousands of cows confided, with some gloom, that their culling rates approached 40 percent. This seemed much higher than rates in

the 1950s and 1960s, so it was assumed to be an anomaly of these particular farms. Initially it was suspected that weakened body condition and cow
burnout resulted from the improper use of GMO dairy hormones (rBGH/
rBST can avert culling by stimulating a cow's milk production, but its effect
on reproductive performance is often negative). Aside from GMO hormones,
a high energy diet and life on concrete were also factors. The pasturing of
cows seems to be such a benign influence on cattle welfare that one is tempted
to see it as a cure-all. Yet, good longevity rates, on some Midwest farms that
rarely pasture, show grazing is not the silver bullet to herd health. The mys-
tery culprit may be stress, which comes in different forms (bullying by larger
cows, overcrowded facilities, poor diet, injuries, and so on) that weakens
cows' immune systems, increasing their vulnerability to mastitis, and hoof
and reproductive ills. However, the verdict of experts familiar with longevity
rates on conventional and organic farms is clear: bigger is not better when it
comes to dairies. A corollary to this rule is that the bigger the herd, the harder
it is to pasture all the cows well.

While this book cannot address all aspects of longevity, it is a call to
improve cows' welfare and breed longer-lived animals—Methuselah cows, if
you will—that live as long as cows did in the 1950s. But there is evidence that
early culling and poor longevity are more the result of inappropriate diet and
crowded conditions than genetic heritage.

As mentioned in chapter 4, a recent article by veterinarian Paul R. Biagiotti
circulating among experts in America suggests the national average culling
rate is within sight of 50 percent. Biagiotti calls this a "culling crisis" in large
Idaho dairies where the annual death rate has hit 10 percent. Biagiotti (2014)
writes that "data published by the DHIA shows an average 44 percent culling
rate for Idaho dairy herds. . . . the average cow has only about a one-in-two
chance of completing one full lactation." Nor is this pattern restricted to Idaho
(see chapter 4).

These topics are not simple. It is hard to get a handle on longevity because
so many factors are involved. Nonspecialists should realize that culling and
turnover in herds can, ironically, be complicated by positive factors, such as
successfully raising calves to be first lactation cows with calves of their own.
This dynamic, which increased with "sexed semen" producing more female
than male calves, pressures farmers to sell or slaughter older cows if the stocking density is already high. As we have seen in this book, this is a factor on
both conventional and organic dairies, where overcrowding negatively affects
cow welfare, including health. Herd size is, of course, a factor in the farmer's
marginal ability to graze cows.

Consumers suspect confined-animal feeding operations (CAFOs), which
have become standard in conventional farming, induce stress in cows, with
mobility and reproductive ailments limiting longevity. Not all confined
operations are unhealthy, but megadairies with thousands of cows suffer an
uncomfortable proportion of cows that burn out and go to slaughter before
their fourth birthday. Many stressed cows suffer bouts of mastitis requiring antibiotic treatment. The U.S. Centers for Disease Control warn that the

overuse of antibiotics in conventional heifer and cow management has, as in battery poultry and swine operations, led to the loss of drugs vital to maintain animal and human health and contributes to as many as 23,000 human deaths annually (CDC 2011). Although the USDA has announced reforms, they seem feeble (*The Atlantic* 2012; BBC 2013c) because they allow the use of antibiotics if disease is likely to occur without them—a tautology because disease is more likely to occur in CAFOs than in spacious pastures. There are hopes of new classes of antibiotics, including one rumored to attack MRSA (BBC 2013c), but such evidence makes no case for allowing the prophylactic use of them in livestock agriculture. Fortunately, as chapter 6 relates, organicists, such as Rodale Institute veterinarian Hue Karreman, are developing proactive protocols to raise cows' immune systems, as well as alternatives to antibiotics in emergencies.

*   *   *

The official USDA story, of a beneficent 250 percent productivity boost per cow since 1950, seems an environmental boon because fewer resources seem to be used in the production of each unit of milk. Consumers who enjoy meat can hardly complain that culled cows are sacrificed to their pleasure. Chapter 2 notes that animals have been consumed for millennia, but some consumers perceive a shattering of an ancient human/nonhuman covenant—that is, that we protect domestic animals from predators and care for them well until their demise. Denying cows pasture seemed a violation too far of animals' needs to perform natural behaviors, such as grazing, as articulated in the St. Paul Declaration on animal welfare (IFOAM 2006).

However, this pasture-centric view of conventional dairying is critiqued by Steve Larson, until 2013 the managing editor of *Hoard's Dairyman*. The Hoard's Farm in Wisconsin features a number of Guernsey cows at least eight to ten years old, topping national lists as lifetime leaders in milk production with at least six to eight lactations. Jerseys are being introduced into the herd, cared for by trained professional herdswomen in roomy, well-designed barns, where cows are free to feed or lie down. They do not graze on pasture, due to frequently muddy fields, but exude well-being. It is apparent that animal welfare is multifaceted.

Across the world in India, where about 40 percent of the population is vegetarian and dairy products are a mainstay, small-scale dairying is shifting from roaming groups, tended by cowherds in forests and on common land, to tie-stall (tethered) barns, as agricultural land is squeezed by the urban population. Even so, with close human-animal interaction, milch cows can live many years. An agribusiness analyst at *The Hindu Business Line* writes (personal communication), "I think 10–11 years is normal." And Joseph A. Purathur, deputy general manager for cooperative communication, at Dudhsagar Dairy, one of India's largest co-op processors in Mehsana, Gujarat, writes: "One of our experts in the animal health department says the longevity of cows in cooperative dairying averages 17 years." These answers point to a range from about three to five times the longevity of cows in U.S. conventional confinement dairying (Biagiotti 2014; *Hoard's Dairyman* 2014e). The Indian rates seem even better than those for U.S. high-production organic cows.

It is one thing to extol model instances of cow care in confined operations, such as the Midwest Guernseys above, but how good is cow care in general? Not good enough if U.S. national cull rates hover around 40 to 50 percent, connoting cow longevity of barely four years from birth to slaughter, with too many cows not birthing a second calf (DHI-Provo 2013). Pasture covers a lot of sins. Meaning what? Cows' feet and udders are healthier on pasture than in poorly designed confinement. Falls are not as injurious. That knowledge is shared by many consumers and, as a member of the Washington State Dairy Federation told a government panel before the certification of GMO dairy hormones, "to the consumer, perception is reality" (personal communication 1990). That is why shoppers who dislike the idea of cows injected with synthetic hormones and confined on feedlots buy milk from cows on pasture.

Recapitulating briefly, The Cornucopia Institute (2006a; Feb. 2014d), initiated what it called its Organic Integrity project with a Dairy Scorecard rating what it considered ethical family farm producers, and listing other farms that did not meet its standards. (Some of them were still suspected of violating National Organic Program rules on grazing in 2014.) Cornucopia cooperated with other activist organizations, such as the Organic Consumers Association, to uncover violations of USDA rules. Horizon and Aurora were not the only companies suspected of rules infringement but attracted complaints from consumers in Seattle and other markets due to their prominence in the dairy case. Partly as a result of the OCA milk boycott and legal complaints filed by Cornucopia, organic-industrial megadairies got the message in the final 2010 USDA Pasture Rule that cows must be pastured. Earlier chapters in this book noted that, in 2007, the USDA decertified 3,500 organic cows confined alongside 6,500 conventional cows in the 10,000-cow split herd of Vander Eyck megadairy near Fresno, California—and ordered Horizon Organic Dairy to make other major changes. Later that year, Aurora Organic Dairy was ordered by the USDA to pasture lactating cows and decrease stocking levels, before further legal challenges by consumers (*The New York Times* 2007).

Pastoralists rejoiced that pressure from citizens and organic activists was liberating more cows from CAFOs to organic farmscapes. But pessimists among them feared the USDA delayed a final ruling on access to pasture until 2010 because—in light of food and oil price inflation and competition for space from biofuel—it was analyzing the aggregate cost of pasture-based dairying with an accountant's eye. If the USDA concluded that grazing was no longer affordable, the National Organic Program could have turned into an industrial husk of what pioneers had dreamed. Opposing that view with science-based evidence is The Organic Center, headed by Dr. Chuck Benbrook (2010, 2014) with analytic tools such as the "Organic Calculator."

One research paper that gave everyone pause for thought was published in *Proceedings of the National Academy of Sciences* in 2008 by Jude Capper of Washington State University, with coauthors Roger A. Cady, of Elanco Animal

Health (Elanco bought the GMO hormone rBGH/rBST from Monsanto), and Dale E. Baumann, of Cornell University. Titled "The Environmental Impact of Recombinant Bovine Somatotropin (rbST) Use in Dairy Production," Capper and her coauthors (2008) conclude that CAFOs utilizing GMO hormones have a lighter environmental impact than organic pasture dairying. As to charges that CAFOs emit more greenhouse gases (GHGs) than pastured herds, proponents of industrial-scale confinement predict that research on fodder and genetics will reduce enteric methane missions in a "cow for the future" (Knapp et al. 2011; *Hoard's Dairyman* 2013c). This prospect is argued by Jude Capper and Dale Bauman (2009). In other articles Capper and colleagues (Capper, Cady, and Bauman 2009; Capper 2012a; b) tout productivity increases of the U.S. dairy herd during the period 1944–2007, arguing that confinement dairying with high-producing Holsteins using GMO hormones is more environmentally sustainable than organic pasturing.

Taking the opposite view are Chuck Benbrook at Washington State University and colleagues at the Organic Center in the nation's capital. With a grant from the Packard Foundation and financial support and technical assistance from Stonyfield Farm, Horizon Organic and WhiteWave Foods, Aurora Organic Dairy, and Organic Valley, Benbrook created the *Shades of Green Dairy Farm Calculator* (Oct. 2010). In an SOG update called *A Deeper Shade of Green* (2012a: 1, 18), he mentions the quest by proponents of the Cow of the Future Project to alter "diets and animal nutrition, rumen function, genetics, and herd structure" to cut methane emissions 25 percent by 2020 (Knapp et al. 2011), while ignoring cow health and longevity. Benbrook likens the lower-producing Jerseys favored by many organic farmers and the high-volume Holsteins favored in conventional operations to the tortoise and the hare. Energy corrected milk (ECM) calculations reduce bias in comparisons of Holsteins producing over 70 pounds of milk per day to Jerseys producing under 50 pounds of milk per day, when Jerseys have better nutritional quality in protein and fat and live longer with more lactations.

With the SOG calculator, Benbrook assessed four scenarios. Before we turn to them, please note that the four scenarios analyzed by Benbrook at the Organic Center (2012) are configured differently from the four generic organic farm models that this author constructed for chapter 7. However, there is some overlap: Scenarios 1 and 2 share aspects with the West Family Farm (WFF) and East Family Farm (EFF). Scenario 3 overlaps with the East Megadairy (EMd), which may stress their cows somewhat. Scenario 4 is a nonorganic megadairy, which likely stresses its cows so much that culling exceeds rates on organic-industrial operations that, in turn, are criticized for stressing their cows so much that they cannot raise sufficient replacement heifers for their burned-out mothers.

At the Organic Center, Benbrook (2012a: 5) assessed these four scenarios:

Scenario 1. Double J Jerseys Farm in Oregon (playing the role of *tortoise*), producing an average 40.5 pounds of milk per day, relying on "home-grown pasture and forages year round" to "minimize stress and disease and maximize health and longevity."

Scenario 2. California Cloverleaf Farm, an organic, grazing-based opera-tion milking crossbreds and Jerseys, producing an average 41.5 pounds per day. New Zealand–style, the cows are milked seasonally, so there is no milking in midwinter.

Scenario 3. Hypothetical organic farm managed to minimize methane. Crossbreds eat mainly forages, grain, and protein supplements. Ergo milk production averages 50 percent more than the two grass-based organic farms. Manure is managed to minimize methane.

Scenario 4. Hypothetical megadairy (playing the *hare*), based on U.S. aver-ages (NAHMS, 2007). Holsteins given GMO hormones average 73.4 pounds milk daily; fed total mixed rations in a feedlot (NAHMS, 2007). Cows stand on concrete or rest in free stalls with little if any pasture. Cattle wastes are flushed with water to an anaerobic lagoon.

Please note that the SOG calculator assumed reasonable averages of $30 per hundredweight (cwt.) for organic milk and $20 cwt. for conventional. Therefore conventional (40.5 pounds of milk per day × 365) divided by 100, and multiplied by $30 =

14,782.5 lbs. = 147.825 cwt. x $30 cwt. = $4,434.75 gross income per cow annually
[Multiply each cow × herd count] Ergo:
50 organic cows = gross income $221,737.
100 organic cows = gross income $443,475.

In the key findings, Benbrook (Nov. 2012: 4, 19, 18–22) notes: "It is com-mon for conventional dairies like those modeled in Scenario 4 to require 40 percent to 60 percent replacements annually, compared to about 20 percent to 30 percent on farms with long-lived cows." (Mention of 60 percent culling gives pause for thought.) The two years needed to raise a replacement rep-resent significant amounts of "feed, nutrient excretions, and methane emis-sions." These are resources that are saved by long-lived cows. In Scenario 2, California Clover Leaf Farm emitted just over one-third the manure methane, "0.006 kilograms of manure methane per kilogram on energy corrected milk, compared to 0.017 from cows in Scenario 4" (18). Total methane in kilograms per year of productive life was 277 kg. in Scenario 1, 189 kg. in Scenario 2, 239 kg. in Scenario 3, and a weighty 539 kg. in Scenario 4. By this metric, the Jersey "tortoises" on year-round grass in Scenario 1 averaged 6.3 lacta-tions over 8.5 years and won the race to sustainability, beating the confined GMO-enhanced conventional Holstein "hares" that averaged only 2.3 lacta-tions in 4.3 years. Scenario 4 cows had such high emissions because archaea microorganisms release methane in the anaerobic conditions of megadairy manure lagoons. That is not a problem on airy pasture.

In a March 18, 2014 webinar for Shades of Green (SOG) calculator users, Organic Center staff explained how analysts studying the same farms could conclude opposite sustainability assessments (Benbrook 2014): A literature review of previous "environmental footprint models" found they were based on "a year in the life of a cow" instead of a "cow's productive life," overlooking externalities of intensive "management systems on cow health and longevity, soil quality and productivity, and overall, lifelong economic returns to a lactating cow." In other words, life-cycle analysis of all factors finds organic pasture-based dairying has a lighter environmental hoofprint than intensive confinement dairying. The webinar resonates with the themes of this book: "Cow health and longevity, and in particular, reproductive performance, are critical variables in determining a dairy farm's environmental footprint."

GMO Revolving Doors in Washington, DC

Visitors to the USDA cafeteria during the 2000–08 Republican era of President George W. Bush found a shrine to Big Ag with examples of every fast-food restaurant that populates Main Street, USA. Major retailers of foods made with GMO milk and grains, high-fructose corn syrup, and trans fats, such as McDonald's, KFC, Pizza Hut, and Taco Bell, reflected what Eric Schlosser (2001) called the Fast Food Nation's prowess at marketing the highly processed foods that encourage obesity around the world (the latter three are leading brands among the 39,000 restaurants owned by YUM Brands Inc. in over 130 countries).

The election of President Obama in 2008 held promise for greens and organicists. As a candidate, the senator from Illinois waxed lyrical on sustainability and labeling GMO foods. After the election, greens suggesting organic pioneer Jim Riddle as the new USDA head were disappointed by Obama's appointment of Tom Vilsack. They were slightly mollified when Kathleen Merrigan, an academic from Tufts University who'd helped draft the Organic Foods Production Act (OFPA 1990) was named deputy to Vilsack. Many greens saw Vilsack as a poster politician for biotechnology who, as two-term governor of Iowa, won awards from that industry. But he made conciliatory gestures to greens after accession to the USDA, for instance by imposing moratoria on the introduction of certain GMO crops. Later, Vilsack quickened the pace of GMO crop introductions, such as alfalfa. All of this was decried by nonprofit organizations, such as The Cornucopia Institute (2013; 2014) and the Organic Consumers Association (2013; 2014). A chronic matter of speculation is complicity between the USDA and major organic-industrial processors and retailers vis-à-vis synthetics permitted in organics. Synthetics with dubious health effects, such as carrageenan, and a type of fat chemists call docosahexaenoic acid (DHA) have been approved against the objections of critics. Greens claim DHA approval was given without adequate consultation with the National Organic Standards Board as stipulated in the OFPA 1990. There are signs that ignoring the NOSB as originally constituted—effectively marginalizing farmers and consumers in favor of industry and traders—is the new modus operandi at the USDA. It is worth noting that carrageenan was allowed on the NOP-certified synthetics list, even though the Organic Valley/CROPP cooperative announced plans to switch to safer alternatives after 2013. The OV cooperative's stated goal, along with continued consumer disquiet over carrageenam, may have prompted competitor WhiteWave to restate its own position in 2014 (Odairy 2014e).

Meanwhile, Secretary Vilsack seemed happy to support product differentiation for USDA-certified organics as long as it was the fastest growing sector of agriculture. But greens wonder how committed he and his boss, President Obama, really were to organics. The administration appeared more interested in the potential of biotechnology patents and intellectual property to bolster the country's global trade balance. When Merrigan, who guided the Pasture Rule to a successful conclusion in 2010, resigned her post as deputy to Vilsack early in the second term, greens took it as a grim bellwether on farm policy. Some suspected Vilsack's zeal for GMOs was behind her resignation, but Merrigan's views are nuanced. At a meeting of the Ohio Ecological Food and Farm Association, *The Plain Dealer* (2014) reported her saying: "I've never been anti-GMO . . . but the marketplace is demanding it." Merrigan backed compensation for organic farmers from GM pollen drift and said that despite recent legal defeats, plenty of hope remains for organics. She also urged people to comment on pending federal food safety regulations: "There was a rule that farmers wouldn't spit or chew gum. The government can do real harm if the regulations don't really fit the needs." Merrigan's next post is the executive directorship of George Washington University's Sustainability Institute, and it will be interesting to see how GMOs and organics fit into its 40 academic programs.

Canadian writers see bipartisan support for GMOs in Washington, DC. A month before the November 2012 U.S. election, the Centre for Research on Globalization, an independent nonprofit organization based in Montreal, published an article by Josh Sagar (CGR 2012), suggesting Monsanto expected federal support even if Barack Obama won reelection and certainly if Mitt Romney (who helped design Monsanto strategy at Bain Capital) won the presidency. Sagar noted that in the previous electoral cycle, "Republicans in the legislature have taken $226,000 from Monsanto Co., while Democrats have taken only $90,500." Sagar wrote, "The Republican Party is based in the center of the country and the south, much of which is dependent upon farming." That explained why, according to him, "Republicans are far more politically friendly towards agribusiness than the Democrats and more likely to support companies like Monsanto." Sagar went on to say Romney took $4,075,531 in campaign donations from agribusinesses, about three times the $1,377,503 taken by Obama, although the aggregate total including Super PAC dark money is unknown. Sources of such money became more obscure after the Supreme Court's 2010 Citizens United ruling that special interest groups, such as corporations, unions, and other associations, shared certain First Amendment rights with human citizens. As a result, Republican strategist Karl Rove's group Crossroads GPS and industrialist David Koch's group Americans for Prosperity could legally finance negative media attacks on left-wing candidates, despite efforts by Priorities USA, organized by allies of President Obama (ProPublica 2011). In this political climate, Republican President Dwight Eisenhower might be considered left.

Monsanto need not have worried about President Obama, whose first term, beginning in 2009, utilized the revolving door like his predecessors. Greens were appalled, in Obama's second term, when USDA secretary Vilsack called for—in what came to be known as the Monsanto Protection Act—the doctrine of substantial equivalence to be extended to virtually all GM-derived crops, without the minimal observation and testing accorded previously certified glyphosate-resistant seeds. It would block court review of environmental impacts of GMOs from most agencies except the USDA. Despite at least 250,000 email protests, in March 2013, President Obama signed House Resolution 933, which contained the Monsanto Protection Act. GM salmon were also given the go-ahead despite protests.

According to Sagar, three key first-term appointments from Monsanto personnel in 2009 were Michael R. Taylor, Roger Beachy, and Islam Siddiqui (CGR 2012; also MacMillan 2002; Nestle 2003: 101). Taylor's complicated chronology as a Monsanto attorney, vice president, and lobbyist, and tenure with the Food and Drug Administration in the 1990s (he was responsible for policy on the GMO dairy hormone rBGH/rBST) is mentioned above. Greens were stunned when the Obama administration named Taylor the FDA's first deputy commissioner for foods in January 2010. The fact that the new FDA food safety czar was a veteran Monsanto insider incensed greens. This and anger by the GMA's 2012 defeat of Proposition 37 to label GMO food in California fueled the international March against Monsanto in 2013. Ronnie Cummins, of the Organic Consumers Association (OCA 2014a), claims one million people marched against GMOs in cities across the United States.

Other key appointments by the Obama administration featured Monsanto alumni. Roger Beachy went from directing the Monsanto-linked Danforth Plant Science Center to be director of the USDA's National Institute of Food and Agriculture, its grant-writing division. At the NIFA, Beachy holds sway on the allocation of farm research grants. This disappoints organicists who echo calls from Steiner to Rodale for more scientific research, outside the laboratory on real farms, to improve productivity and sustainability. They expect little help from Beachy.

Islam Siddiqui, a lobbyist for Monsanto, became Agriculture Trade Representative. The Obama administration gave him responsibility for promoting American export crops, such as corn and soybeans. It is likely that he promotes overseas sales of rBGH/rBST, which Monsanto sold to Elanco a few years ago.

The nomination and appointment of Michael R. Taylor to the Food and Drug Administration topped these potential conflicts of interest. While Taylor was at the FDA in the early 1990s, he oversaw policy on Monsanto's recombinant bovine growth hormone (rBGH/rBST). Marion Nestle (2003: 101, 400) found "revolving door" networks among Monsanto, the USDA, and the FDA, although she notes that the famous case of Taylor (who went from a position as counsel for the FDA to work for an Iowa firm representing Monsanto to a return to the FDA) was, perhaps implausibly, judged by the General Accounting Office not to be a conflict of interest in 1994. Many greens suspect Vilsack is single-mindedly promoting GMOs as a U.S. industrial champion domestically and globally. After all, Monsanto and other biotech companies have been major contributors to political candidates for decades, and the Supreme Court Citizens United decision according corporations the same First Amendment rights as people suggests that corporations are in the driver's seat of American democracy. Protests that Citizens United also grants labor unions equivalent First Amendment rights, for instance to fund political campaigns, should note that unions have lost their proportion of membership in the population since the onset of Reaganism and neoliberalization in the 1980s.

Jeffrey M. Smith, consumer advocate and author of the anti-GMO books *Seeds of Destruction* (2003) and *Genetic Roulette* (2007), was one of the first popular writers to document irregularities in the North American certification of the dairy hormone rBGH/rBST and other anomalies in the politics of biotechnology. He questioned the nomination of Taylor as FDA food safety czar (Smith July 23, 2009), claiming Taylor had suppressed safety warnings from government scientists over dairy and crop GMOs and promoted the risky principle of substantial equivalence between GMO and heirloom varieties, with the policy outcome that GMO milk was not required to be labeled as such. Most galling, wrote Smith, was that "Monsanto used Taylor's white paper as the basis to successfully sue dairies that labeled their products as rBGH-free."

Greens saw Taylor in the FDA and Vilsack heading the USDA as varmints guarding the henhouse. In this GMO horror show, one victory salves the wounds of animal welfarists. Since the unequivocal USDA Pasture Rule of 2010, certified organic cows must enjoy at least 120 days grazing on pasture during each year, with a minimum of 30 percent of their dry matter intake coming from pasture—and more if the growing season allows. But activists will be vigilant: the thrust of USDA livestock policy in conventional dairying under Vilsack still favors the confinement and pushed production that have sent so many dairy cows down the lane to early slaughter, since intensification of dairy farming was promoted by USDA secretary Earl Butz in the 1970s.

Farmers Lose Political Clout

Neoliberalism, in forms from Reaganism to Tea Party libertarianism and the procoal energy activities of the Koch brothers and the American Legislative Exchange Council (ALEC), does not stop at disempowering urban labor unions (*The Nation* 2011). It also seeks to whittle away at the power of U.S. farmers' cooperatives and the agricultural activists mentioned throughout this book. The economic contexts are different, but the dynamics of India's dairy sector in which liberalizations were introduced in 1991 are parallel to the neoliberal drive to delete New Deal farm policies (such as parity between urban and rural incomes) from U.S. farm policy. Congressmen Newt Gingrich and Dick Armey's 1994 Contract with America, which eventually brought a government shutdown before breathing new life into Bill Clinton's

presidency, produced the 1996 so-called Freedom to Farm Act. This neoliberal legislation eventually disempowered many of the farmers who originally
supported it (see David R. Harvey, *Food Policy* 1998). Recent U.S. census data
show such policies have not brought prosperity to ruralities. On the contrary
they show declining rural population and country towns with fewer amenities
than once enjoyed. That was also my impression when visiting a corporate
organic megadairy in Maryland. A supermarket in a nearby town revealed
dominance by heavily packaged processed food and such a poor selection of
fresh fruit and vegetables that the town's food supply seemed tantamount to
the food deserts that afflict inner city areas with high unemployment.

In these food deserts are the oases of deeply organic organizations committed to animal welfare, environmental sustainability, and social justice, such as
OCA and Cornucopia. The Rodale Institute and creative veterinarians such as
Dr. Hue Karreman are doing research that can help avert the global threat of
MRSA. But it remains unclear whether programs such as the UK Royal Society's "sustainable intensification" (Basu and Scholten 2012) truly encompass
the care for animals of the St. Paul Declaration (2006 IFOAM).

In twenty-first-century agriculture, the processes of agribusiness appropriationism have given the patented and intellectual copyrights of chemical
and seed companies far more power than conventional and organic farmers
alike. Agribusiness leads the consumers astray where foods and ingredients
marketed as "natural" blur lines between industrial products and unadulterated food. When the products of established organic brands are repackaged
as natural, with nonorganic ingredients under the same logo, it amounts to
the asset stripping of organic integrity. It is not only dishonest to consumers,
it endangers the livelihoods of organic farmers.

Critics claim that another type of marginalization of farmers is evident
in a campaign, by members of the Organic Trade Association (OTA), such
as the manufacturers General Mills and Kellogg's and retailers Safeway and
Wal-Mart, to mandate a percentage "checkoff" from organic farm gate earnings to fund the promotion of organic products. This seems like a reasonable idea, but organic dairy farmers resist the checkoff, complaining of the
persistence with which processors and retailers press farmers to agree on it.
Processors and retailers are more likely to gain from the fund, not farmers.
Tension was palpable at the NOC's premeeting before the National Organic
Standards Board rules meetings in Portland, Oregon, in April 2013. The pressure intensified. Jim Riddle, the esteemed former chair of the NOSB, resigned
from OTA over what he perceived as unfair marginalization of farmers (Cornucopia 2013b). An eastern farmer commented in a discussion group that the
mandatory checkoff was legalized robbery, showing disregard for producers.

NODPA (2014c) officials noted that costs to processors and handlers have
historically been exacted from farmers, as in the conventional dairy sector's
"Got Milk?" campaign. The NODPA led dairy farmer resistance against the
checkoff with OFARM organizations—including the Buckwheat Growers
Association of Minnesota, Kansas Organic Producers Association, Midwest
Organic Farmers Co-op, Montana Organic Producers Co-op, NFOrganics,

Organic Bean and Grain, and Wisconsin Organic Marketing Alliance—fighting with them.

Some claimed the Organic Trade Association had wrongly told Congress that farmers backed the checkoff. Few are the issues that arouse such ire as a checkoff that farmers deem as gouging by businesses that piggyback on their sweating backs.

Little new acreage is being turned organic in America. In fact some organic acreage was decertified after the beginning of the 2008 financial debacle. Although the aggregate organic market already totals over $31 billion, most added value is accruing to businesses, not farmers. An economist might conclude that too many farmers are price takers, who need to cooperate in establishing their own processing, handling, and trading arms. In this way they could climb the value chain, to move from production of mere commodities to holding equity in differentiated products.

American humorist Mark Twain (n.d.) advised citizens to buy land because "they're not making it anymore." Certainly landgrabs have never gone out of style. The farmers that Fred Pearce, author of *The Landgrabbers* (2012), documents as victims of foreign and indigenous landgrabs in Africa and other continents are not so different from country people in America. The political economic forces that drove the English Enclosure Movement are not unlike those driving disenfranchisement of rural Americans in the twenty-first century. Although Karl Marx was not terrific at formulating solutions to problems, he was better than most at defining them. Rural dwellers who lose their land often embark on rural-to-urban migration, entering a new class of the urban and periurban proletariat who, with little or no capital, have little to offer but labor at jobs that barely provide a living wage for them. Marx termed landgrabs "primitive accumulation," which geographer David Harvey (2003) today calls "accumulation by dispossession." In the foreword to this book, C. S. Sundaresan (2014: 8, 8–15) finds similar patterns of dispossession in developing countries and India's own Orissa State over "natural resources [and] creation of special business zones."

Concepts of dispossession were also familiar to Dr. Verghese Kurien, known as the "Father of India's White Revolution." Kurien, the 1989 recipient of the World Food Prize, grew up in the British Raj to become a champion of dairy farmers' cooperatives (Scholten 2010). This "Milkman of India," who died in 2012, would have timely advice for American dairy farmers in the world's richest country. While the biotech cotton sector in India is now known for hundreds of thousands of suicides among farmers who have become heavily indebted in attempts to buy Roundup pesticides and other inputs for GM cotton (Bhardwaj 2010), Kurien noted that "where there are cooperatives, there are no suicides." That is because, said Kurien, farmer-members of dairy cooperatives supplying the Amul brand (now starting its own factory in the United States) do not control only production and processing—they also control marketing, where the money is (Scholten and Basu 2012). Farmers around the world have a sense that their livelihoods are more sustainable when they have a hand in adding value in processing and sharing in the profits of marketing. In the United States, checkoffs that side-line organic farmers are protested by them because processing and marketing is controlled by others.

Small farmers in emerging economies and developed countries alike have the most security from monopolistic power and opportunities to share in the value generated by shared learning and innovation, in well-led and well-managed cooperatives. Organizations such as Organic Valley/CROPP could be the U.S. answer to the Gujarat Cooperative Milk Marketing Federation (GCMMF). This Indian cooperative grew from a tiny farmers' strike in 1946 into the Anand Model that carried India past America into the rank of top world milk producer in 1998. Since 1988, the Organic Valley cooperative has grown from a handful of farmer-members to 1,844 in 2014, and it remains a bright spot on the skyline for U.S. organic dairy families and consumers.

Cooperatives are not by themselves guarantors of family farm livelihoods. C. S. Sundaresan (2014; foreword) is one economist who notes that government policy is instrumental in maintaining livelihoods for dairy farmers in India or other countries. If agricultural organizations such as cooperatives or marketing boards are not embedded in government rural policy, they may wither. This was borne out in the United Kingdom when the demise of the Milk Marketing Board cooperative in the 1990s weakened the power relationships for farmers vis-à-vis processors and especially traders. After the dismantling of the MMB, a new voluntary cooperative body called Milk Marque was created to compete with private competitors. Milk Marque got off to a strong start by recruiting 80 percent of the United Kingdom's 29,000 dairy farmers. Unfortunately for them, a coalition of supermarkets filed a complaint with the government's monopolies commission, charging that Milk Marque comprised a monopoly that would harm UK consumers. The UK government bowed to the big supermarkets' complaint and broke up Milk Marque to the point where supermarket power reigned supreme. A company called Dairy Crest, formerly part of MMB and floated on the London Stock Exchange in 1996, seemed able to enlist enough farmers to negotiate fair prices with supermarkets. Ultimately, that was not the case, as too many individual farmers were willing to defect from collective action and make separate agreements with the big supermarkets. Farm gate prices to UK dairy farmers dwindled, and feed prices rose to the point in 2012 when stressed farmers threatened to pour milk on the ground rather than sell at a continued loss to processors.

The lesson of UK farmers for their U.S. counterparts is that a healthy dairy industry requires government policy that is sensitive to family farming. Thus, there are fears in U.S. markets that if Washington, DC, were to rescind the Federal Milk Marketing Orders (FMMOs), the U.S. conventional sector would disintegrate into the beggar-thy-neighbor social Darwinism that agricultural economists call oligopsony (Doyon and Novakovic 1997). The lesson for organic dairy families is that membership in a cooperative strong enough to influence federal legislation is essential, for example, in drafting food labeling legislation to maintain the integrity of organic products.

GMO Labeling: I-522 Denouement in Washington State

U.S. consumers' desire for transparency in food labeling is strong. At the same time, biotech firms have tried to foster the impression—almost a cultural meme—that GMOs are profitable, environmentally sustainable, and inevitable in crops and dairy applications. Despite counterevidence from the Rodale Institute (2011), Cornucopia Institute, and other sources, many Americans continue to accept assurances from biotech firms that they are on the side of science. This understanding is less successful overseas, where people are more likely to see GMO exports as expressions of American political clout. This is so, despite the widespread planting of GMOs in Argentina and Canada. The downsides of GMO canola/rapeseed in Argentina, such as the spawning of superweeds resistant to Roundup/glyphosate and the ensuing need to mix cocktails of old and new pesticides to contain them, are becoming well known (Binimellis et al. 2009). Globally, GMOs are labeled in 64 countries but not the United States, where biotech firms seem determined to block GMO labeling. In California, with a population of 37,253,956, biotech firms and their agribusiness allies spent $25–40 million to block labeling. The funds were collected by entities such as the Grocery Manufacturers Association to defeat Proposition 37 to label GMOs in California in 2012. The GMA overwhelmed spending by Prop. 37 activists, who had a budget of barely $4 million.

In 2013, attention shifted to Washington State as Initiative 522 demanded the labeling of GMO foods, and the biotech lobby again vastly outspent those in support of labeling. In a state with a much smaller population than California's (6,724,540), this amounted to $11 million by the Grocery Manufacturer Association alone and many more millions by corporate entities, as enumerated below. The politics of I-522 activists ratcheted up the naming and shaming of those deemed hypocrites. The Organic Consumers Association (2013d; e) challenged Aurora Organic Dairy to withdraw from the Grocery Manufacturers Association (GMA), which had already directed over $2.2 million against I-522. Earlier in September, health guru Dr. Andrew Weil withdrew his company Weil Lifestyle from the GMA after a petition by more than 25,000 consumers asked him to. Weil eventually announced that he did not concur with the GMA's antilabeling stance, perhaps at some sacrifice to his career.

Surveys consistently show a vast majority of Americans favor food labeling. They also show that organic milk is a gateway for new parents who were previously little concerned with food safety. Thus, it is unsurprising that a group called Moms for Labeling sued the Grocery Manufacturers Association (GMA) for allegedly cloaking its members in anonymity. It was a boon to green consumers on October 16, 2013, when Washington State attorney general Bob Ferguson sued the GMA, charging the GMA had collected and spent over $7 million by that point in the campaign to cloak its donors in anonymity. This violated public disclosure laws. The headline in the *Seattle Post-Intelligencer* (2013) read, "Faced with Lawsuit, Grocery Manufacturers Association Agrees to Disclose Campaign Finances." The attorney general said: "The people of Washington demand transparency in elections." With victory coming so quickly, Ferguson continued, "I'm pleased the GMA board recognized their responsibility to disclose the names of companies who contributed to opposing Initiative 522, and the amount of their contributions." The GMA announced, "In the spirit of continuing cooperation and in an effort to provide Washington voters with full transparency about GMA's funding for the 'No on 522' campaign, the association has voluntarily decided to establish a Washington state political committee and to file reports with the PDC disclosing the source of all funds used in connection with Washington State elections."

Some voters were confused by media messages in the hard-fought campaign. *The Bellingham Herald* (2013) reported local views. One longtime Whatcom County raspberry farmer said GMOs allow conventional farmers to spray less, and he prefers voluntary labeling of non-GMO food to mandatory labeling of GMOs. The co-owner of an orchard between Bellingham and Lynden in Whatcom County said, "We couldn't think of any good reason why somebody should not know that their food is genetically engineered. We have a right to know what we're eating." A Whatcom pioneer in organic farming since the 1970s near the town of Everson who sells in Seattle farmers' markets, supported I-522: "We're not asking for a very big change to happen . . . Labeling GMO products just gives us the choice to decide if we want to eat them or not."

The Cornucopia Institute (2013c) announced, "With a week to go before the November election, the Grocery Manufacturers' Association made two cash donations on Friday totaling about $3.7 million to the *No on 522* campaign making their total donation $11m to defeat I-522." Cornucopia also reported that the top contributor to the I-522 to label GMO foods in Washington State was Dr. Bronner's Magic Soaps (the eco-friendly "hippe soaps" favored by green hikers) giving $1.7 million, and other donors included the Center for Food Safety Action Fund ($350,000), Mercola Health Resources ($300,000), Nature's Path ($170,000), and the Organic Consumers Association ($878,000). Other donors included, unsurprisingly, Annie's ($105,000), Earthbound Farms ($20,000), Food Democracy Now! ($100,000), Hain Celestial ($50,000), Mercola ($300,000), the Organic Valley/CROPP cooperative ($25,000), PCC Natural Markets ($198,344), Stonyfield ($100,000), UNFI ($50,000), Whole Foods ($20,000), and many smaller donors. But Dean Foods—the corporation that had bought market leader Horizon Organic Dairy—was on the other side. They were listed as contributing $120,245 to fight GMO labeling. Green and Black's organic chocolate, another small actor gone corporate, was also an antilabeling donor (in the mid-2000s, the brand was sold to Cadbury Schweppes and then Kraft).

Dr. Mercola's Natural Health Newsletter (2013) also listed donors, observing: "Looks like Pepsi, Coke, and Nestlé are the top funders trying to hide their identity." Top contributors to the "No on I-522" campaign were Monsanto ($4,834,411), DuPont Pioneer ($3,420,159), Bayer CropScience ($591,654), Dow AgroSciences ($591,654), and BASF Plant Science ($500,000). Although financial donations were not listed for them, fighting I-522 with the GMA were further surprises among America's long-trusted brands, including Campbell Soup, Cascadian Farm and Muir Glen organic companies (two early organic pioneers in the state led by Gene Kahn), ConAgra, Del Monte, General Mills, Gerber (makers of baby food), Hershey, Kellogg's, Land O' Lakes cooperative, Morningstar Farms, Nestlé USA, Odwalla, Similac-Abbot Nutrition (makers of Baby Formula), and Welch's. Organic companies belonging to the GMA were in an uncomfortable position. For example, Cascadian Farm is a pioneer organic brand, but as a subsidiary of General Mills, it had limited choice. Membership in the GMA confers certain commercial marketing and political lobbying benefits, but when the attorney general exposed their campaign contributions, these brands risked losing consumer trust. GMA members had to assess the risk of alienating citizens who might be induced to boycott their products, annoyed by corporate support of a political campaign to limit people's right to know if their food contained GMOs.

Some I-522 partisans were infuriated by the biotech industry's media portrayals of the pro-GMO labeling camp as unscientific Luddites. Numerous I-522 supporters maintain hope for the eventual efficacy of biotechnology but demand transparency in labeling and more safety testing for people and the environment.

The Grocery Manufacturers Association continues lobbying to weaken state or national labeling laws. Ronnie Cummins, head of the Organic Consumers Association (OCA 2014a), claimed that due to "the inevitability of mandatory state GMO food labeling laws—laws that will likely, as in Europe, drive GMOs off supermarket shelves—industrial food and biotech corporations are in a panic." Cummins states, "After being forced to spend $70 million, and then barely defeating two citizen ballot initiatives in California and Washington State, Big Food and biotech's front group, the GMA is facing criminal charges in Washington State for illegally laundering over $11 million in campaign donations." As mentioned above, the GMA claims it voluntarily disclosed the identities of donors to its anti-I-522 campaign in Washington in 2013. The GMA appears strong as long as Supreme Court decisions such as Citizens United permit massive anonymous corporate donations to media campaigns. Greens worry while actors they see as shills for biotech companies promote toothless labeling laws that favor industrial processors and traders, not consumers or farmers.

On May 8, 2014, Governor Peter Shumlin signed into law a bill that makes Vermont the first state with a standalone GMO labeling law, meaning that it will go into effect regardless of actions by other states. Notably, the law requires the labeling of all GMO foods and prohibits the use of the terms "natural" and "all natural" on foods containing GMOs. Agribusiness firms have announced plans to sue Vermont, but after signing the bill at a reception that included the state's Ben & Jerry's Ice Cream, Governor Shumlin announced an online campaign to accept funds from all states to support Vermont's law in litigation.

The month before, Vermont's strong labeling law got a boost from an organization identified with conventional farming. On the Odairy discussion list, April 11, 2014, organic pioneer Jim Riddle hailed "strong words" in a statement by National Farmers Union president Roger Johnson (NFU 2014) against House Resolution 4432, "The Safe and Accurate Food Labeling Act of 2014," sponsored by Reps. Mike Pompeo (R-KS) and G. K. Butterfield (D-NC), which contrary to its wording seeks the weak labeling of foods. Because the NFU is such a central actor in American farm politics, the statement is excerpted at length:

> Farmers Union members have clearly stated their position in the policy adopted at our annual meeting in favor of required consumer labeling for foods made from or containing genetically modified organisms (GMOs). The rights of both GMO and non-GMO producers should be respected as appropriate regulatory agencies continue to research and evaluate ethical, environmental, food safety, legal, market and structural issues that impact everyone in the food chain. NFU policy supports conspicuous, mandatory labeling for food products throughout the processing chain to include all ingredients, additives and processes, such as genetically altered or engineered food products . . . Consumers want more information about their food, not less. The prevalence of state-led efforts to label genetically modified organisms (GMOs) only corroborates these findings.

Sting in the Tale

Relations between family-scale organic dairy farms and corporations are asymmetric and not always benign. Greens held their breath when Dean Foods, the name that Suiza Foods took after buying Dean in 2001, second only to Nestlé as a world dairy power, bought Horizon in 2004 (Phil Howard 2014). According to the Dean Foods (c. 2013) website, in 2004 it acquired Horizon Organic Holding Corporation, maker of a full line of organic milk and dairy products, and later that year consolidated Silk, Horizon Organic, and other branded businesses as WhiteWave Foods Company, headquartered in Broomfield, Colorado.

As Samuel Fromartz details in his 2006 book, *Organic, Inc.*, Dean Foods invested $15 million in WhiteWave Silk, an organic soy milk company developed by Steve Demos, beginning in 1998. After Dean took control of White-Wave in 2002, Demos came to regard Dean's management of his artisanal 100 percent organic soymilk company as less than congruent with the Buddhist principles of right craftsmanship that had guided his work. According to The Cornucopia Institute's *Organic Soy Report* (2009a), Demos had made White-Wave soy products, including Silk soymilk, from U.S. organic soybeans, but Dean instructed WhiteWave to match the lower price of Chinese organic soy. They could not do that, so Dean sourced cheaper organic soy from China. Later, Dean sourced most of its beans from *conventional* U.S. sources, using the nontraditional toxic chemical hexane in processing. Demos has called hexane the "dirty little secret of the natural foods industry" (Cornucopia 2009a: 29, 34). Today, perhaps only 6 percent of Silk products are organic. The upshot was that WhiteWave Silk soymilk was conventionalized before all of its organic consumers, or even supermarkets, realized it, because the packaging was virtually unchanged except for substitution of the term *natural* for *organic* in January 2009. Over time Dean-Horizon turned the Rachel's Organic Yogurt line (the Welsh family firm bought by Horizon in 2003, before Dean acquired both in 2004) into conventional products marketed as natural in America (Scholten 2010b). Dean gained organic customers and raised profits by conventionalizing its methods. As Whatcom County farmers say about dairy mergers: "The big fish eat the little fish, and then they eat each other" (LeCompte-Mastenbrook 2004).

Milk farmers—hoping that after decoupling from $12 billion Dean Foods (c.2013), WhiteWave/Horizon would reaffirm its links to family dairy farms—were abashed/[OR DISAPPOINTED] when WhiteWave (Sep. 17, 2014) announced plans to buy So Delicious® Dairy Free Foods for $195 million. The CEO of So Delicious, Chuck Marcy, former CEO of Horizon Organic Dairy, said WhiteWave would "enable us to . . . build on our commitment to bringing dairy-free joy to even more households."

In the complex history of WhiteWave and Horizon, Gregg Engles has been a mainstay. Engles graduated from Yale Law School and clerked for Anthony Kennedy before he joined the Supreme Court. As an entrepreneur, Engles' career has included stints with Suiza, Dean Foods. In 2014 Engles remained CEO of WhiteWave with Blaine McPeak as president of Horizon.

After the Horizon megadairy in Idaho was sold in 2013, Horizon continued to buy milk from it, according to The Cornucopia Institute. WhiteWave/Horizon now had a 46 percent share of the national organic dairy market, but CEO Engles sought more diversity and, in 2014, acquired the nation's largest organic produce grower, Earthbound Farms (*Food Business News* 2014; Cornucopia 2014b,c). Before the acquisition, this brand was valued at $750 million annually at retail. Analysts predicted WhiteWave's annual revenue after buying Earthbound as $2.5 billion. Engles enthused that Horizon now controlled the biggest brands on consumers' two favorite paths into organics: dairy and produce.

WhiteWave/Horizon's next move surprised some observers. Americans love macaroni and cheese. (Kraft cracked the conventional market in the 1960s when its boxes sold for as little as 7/$1. Since 1989, Annie's Organic Macaroni & Cheese, with its bunny logo, sold so well that General Mills offered $820 million for it in September 2014.) In a Cornucopia Institute press release titled "Leading Organic Brand, Horizon, Blasted for Betraying Organics" (Cornucopia 2014a; b). Mark Kastel claimed that by introducing four of six new Horizon Mac & Cheese packaged dinners with "conventional or synthetic ingredients that would not be legally allowed in food labeled as 'organic,'" it had joined a plethora of "corporate agribusinesses that have intentionally blurred the line between products they offer with all certified organic ingredients and others sold in similar packaging but containing materials that would never be accepted for use under the USDA organic seal." Kastel castigated "egregious conventional ingredients" in WhiteWave's nonorganic mac and cheese, such as "milk protein concentrate (MPC), a controversial product, often imported and of dubious quality, that substitutes for fresh milk." John Peck, executive director of Family Farm Defenders, a Wisconsin-based group that lobbies against low-quality imports, was quoted as saying MPCs are "basically a way to drive down dairy prices received by farmers—they are the cheapest way to make dairy products." (Monitoring was increased after 2008 when Chinese milk adulterated with toxic melamine was found in American infant milk replacer. Chinese consumers have since increased demand for safe milk from abroad.)

WhiteWave/Horizon president Blaine McPeak cast the new mixed organic/nonorganic product lines in the best light, claiming they expand consumer choice, in an article by Keith Nunes in *Food Business News* (2014): "Picking the categories was straightforward," said McPeak. "We wanted categories that are sizable and over-index with families with younger kids." Of the six varieties of mac and cheese, two are certified organic, and four are "made with organic," meaning at least 70 percent of ingredients are certified organic. Mark Kastel of The Cornucopia Institute remarked on Odairy (Feb. 22, 2014, "Re: [ODAIRY] 'Additional new Horizon (non-organic) products aimed at "families with young children."'") that WW/Horizon was "using conventional milk protein concentrate (MPC), typically imported, as a way to jazz up the protein content without needing to depend on the cost for organic commodities" that could be sourced from U.S. organic farmers.

Observers skeptical of the marketing of new *conventional* foods under previously established *organic* logos fear devaluing of the organic sector. This incarnation of Gresham's law, like the substitution of paper currency or Bitcoins for gold, worries idealists. WW/Horizon's introduction of nonorganic products in packaging so similar to its organic products may blur the lines between conventional fare and USDA organics. It could strengthen the cynicism of consumers who already ask whether or not there is really any important difference between organic and conventional foods.

That was a question consumers were right to ask in the denouement of a court case involving another big actor in the organic sector, Aurora Organic Dairy. In the aftermath of The Cornucopia Institute legal complaints that resulted in the downsizing of its Colorado farms, Aurora was found guilty of deceptive practices in marketing some milk as organic that did not meet USDA standards. In the settlement Aurora paid about $7.5 million to compensate attorneys' fees and offered up to $30 to consumers who could show proof of purchase. In its *2012 Corporate Citizenship Report* (AOD 2013: 2; see also *Fortune* 2007) Aurora explains:

> Because of our size, certain groups have challenged some of AOD's practices related to organic dairy production. These challenges culminated in a class action lawsuit, which was originally filed in late-2007. Aurora Organic Dairy settled the case in 2013, on behalf of itself and its customers, without admitting any wrongdoing and receiving confirmation from the federal courts that our company at all times maintained valid organic certificates for our dairy products.

To the frustration of organicists, surveys have shown many consumers trust the purity and safety of foods marketed as *natural* more than *organic*, oblivious that natural can be a cover for food containing GMOs, sewage sludge, unnatural ingredients, and traces of chemical inputs and that has been processed with irradiation. The USDA organic certification is a stringent if sometimes contested standard. But quality assurance certifiers were also criticized by the USDA in 2007. After the 2010 Pasture Rule clarifications, inspectors were expected to follow the rules more assiduously.

In early 2014 USDA deputy administrator Miles McEvoy announced the streamlining of additions to the synthetics list that further marginalize citizen and farmer participation in the National Organic Standards Board. Under USDA secretary Vilsack, the inclusion of synthetic ingredients, such as carrageenan and DHA, in dairy organics threatens consumers' ability to differentiate organic foods from their conventional counterparts. Erosion of organic food quality with synthetics picks the pockets of small organic producers and processors who, perhaps by dint of extra labor or specialist place-based knowledge, produce quality foods without resorting to synthetics.

Dairy Farmageddon?

The publication of *Diet for a Small Planet* (1971) by Frances Moore Lappé affected American culture as strongly as *Silent Spring* (1962) by Rachel Carson did a decade earlier. Carson alerted people to the negative environmental externalities of chemical inputs in the food chain, while Lappé quantified the relative costs of different nutrients in the U.S. diet, educating many consumers for the first time on grains and carbohydrates compared to the increasing environmental costs of protein from poultry, swine, and beef fattened on grain. These revelations persuaded many people to ethically audit their diets, question the feeding of most U.S. grown soya to animals, and gave impetus to people considering vegetarianism and veganism.

This should not obscure the enduring role of dairy in Earth's diet. *Hoard's Dairyman* (2013f) writer Ali Enerson argues that humans and livestock will coexist because animals are "efficient human food sources" that eat "by-products or unconsumed residue from human food channels," and grazing land is "typically too rugged, rocky, lacking in nutrients or too dry to grow" human food and "attempting to convert grazing land to cropland is not sustainable and poses great ecological risks." Current high demand for dairy products in a prospering China also suggests continuity on dairying. That said, Enerson claims, "Organic dairy production actually raises carbon emissions 13 percent per unit of milk produced in U.S. systems." Her assertion is disputed by proponents of organic systems using pastures as carbon sinks, claiming grass-fed cows emit less greenhouse gas (GHGs) than conventional cattle (Benbrook Nov. 2012: 18–22). Steiner and Rodale, in their whole-systems analyses discussed in chapter 4, would also question Enerson's analysis. Assuming that conventional dairying will continue reliance on confined-animal feeding operations (CAFOs), it will be interesting to see whether future generations of global consumers gravitate to the position expressed in the St. Paul Declaration (IFOAM 2006), and that of philosophers such as Peter Singer (1975), that CAFO conditions violate contemporary ethics of animal welfare. Years ago, dairy breeders admitted longevity was a concern, but lately they talk of crisis (Biagiotti 2014, forthcoming 2015; see also *Hoard's Dairyman* 2014e). After some digging, this author has obtained global longitudinal data on U.S. dairy cow longevity indicating that the average for conventional cows is barely two lactations; nearly half of cows go to slaughter before they are four years old, around half the age of cows in the 1950s (DHI-Provo 2013). Culling rates for U.S. organic cows are also higher than many consumers would like.

In March 2014, ten months after my visit to the University of New Hampshire organic farm, its faculty advisor Dr. Anita Klein emailed, noting that consultations with cooperative extension specialists yielded contradictory estimates on the relative longevities of conventional and organic cows: "How long a cow remains in the herd varies with respect to lots of factors: size of the farm, whether the herd is undergoing selection for milk yield, etc. In general, the age at which cows are culled has been decreasing over the last several decades; this is another economic problem for the dairy industry because of the investment to raise a calf to a heifer and to her first lactation." Klein cited the same article by veterinarian Paul R. Biagiotti (2014) that this book did earlier. Klein was confident that "the average age of our Organic Dairy Research Farm Jerseys when they are culled is about 4.5 years." This compared to the university's conventional Fairchild Dairy, where Holsteins remain in the herd until about 6.5 years, and its small group of up to eight milking Jerseys last 5–6 years before they are culled. This is anecdotally fascinating; however, as Klein points out, firmer conclusions on longevity require methodologies controlling the numbers, breeds, and management systems of cattle involved.

A formidable study of about half the entire herd of 9 million U.S. dairy cows recently appeared in *Hoard's Dairyman* (2014e) intel report titled "Which Breeds Stay in Herds Longer?" by western editor Dennis Halladay. Statistics compiled by the USDA based on National Dairy Herd Improvement Association (NDHIA) data show that, along with cow care, breed helps determine average longevity. (It may be worth standing this question on its head to ask, Are different demands made upon cows based on their breed?) A higher percent culling figure indicated relatively lower longevity. In declining order, breeds (culling percentage) compared thus: Holsteins (32.7 percent), Red and Whites (32.0 percent), Guernseys (30.8 percent), Brown Swiss (29.8 percent), Shorthorns (28.6 percent), Ayrshires (27.7 percent), crossbreds (over 50 percent: 26.9 percent), crossbreds (50 percent or less: 26.1 percent), and Jerseys (24.9 percent). Jerseys—that hardy, pleasant breed—lasted an average one-third longer than Holsteins in this study. Again, a note of caution: when considering longevity, it is important to understand the context of available replacements (higher numbers of females with use of sexed semen), present stocking density, and beef slaughter prices.

Culling Crisis

Apparently, national U.S. culling rates are being skewed by very intensive conventional megadairies, such as those found in, but not limited to Idaho, populated by high productity Holsteins averaging fewer than two lactations before slaughter, and annual mortality rates around 10 percent. If there is a glut of replacements for high-producing Holsteins (resulting from sexed semen as aforementioned), dairy managers are quicker to make voluntary decisions to cull at this writing in August 2014, because slaughterhouses pay relatively high prices for beef carcasses, since the aggregate total of U.S. dairy is at a historic low of about 9.2 million cows. However, the high mortality rate in Idaho suggests much involuntary culling, when cows suffer lameness, mastitis, or do not conceive on the first breeding attempt.

Conventional and organic veterinarian Paul Biagiotti describes this conventional dairy culling rate crisis in *Progressive Dairyman* (2014; see also chapter 7). When he practiced in Connecticut, years before New England herds averaged about half of the current one hundred cows, clients on Bicentennial Farms (operating since the American Revolution) featured cows with long, productive lives. A recent visit by Biagiotti to one of these farms found a healthy cow 18 years old. This contrasts with his last decade of practice in Idaho, where herds average seven hundred cows, statewide death rates average 10 percent yearly, and Dairy Herd Improvement Association (DHIA) records indicate a 44 percent culling rate. Although cows are not physiologically mature till their third lactation, he noted that, in Idaho, barely 50 percent achieve a second lactation due to poor body condition, lameness, and suspensory ligament damage (blown udders). Canadian megadairies are following this dubious American pattern, despite signs that nature did not design bovines for longevity *and* producing an average of 30,000 pounds of milk per cow annually, as some U.S. herds already do.

The lesson is, when it comes to conventional or organic herd size, bigger is not better for health.

Certified-organic cows generally outlive conventional cows in confinement, but more research is needed to unequivocally establish how and why. Organic consumers assume organic cows enjoy better welfare than their conventional sisters, but fortifying this belief with more empirical evidence could inspire greater loyalty. The rapid enclosure of conventional U.S. dairy cows in CAFOs in the latter half of the twentieth century equated the death of traditional single-family dairy farms, but few citizens still have relatives on such farms.

It is not surprising that the CEO of animal welfare group Compassion in World Farming (CWF), Philip Lymbery, titled his book *Farmageddon* (2014), which condemns conventional megafacilities in global dairy and beef systems. Of course, some megadairies are better than others. But too many emit more pollution than the environment can absorb, and stressful conditions and acidic diets for cows truncate their longevity due to problems in breeding and feeding.

The idea for my own book emerged about 2001, when it became clear many cows on even certified *organic* farms were confined without access to pasture all year except, perhaps, for a month or two as dry cows before birthing another calf. One impetus for the National Organic Program (NOP) in the first place was consumer demand that cows exercise natural behaviors, notably by grazing on pasture. Many greens were angry that that was not the case. Michael Pollan (2001) famously dubbed new "organic" megadairies as part of an "organic-industrial complex", and many consumers went beyond organic to become locavores.

Small family-scale farmers rued megadairies most of all for lowering their bargaining power with the processors that paid them. In *Orion* (2003) Pollan noted how far these industrial permutations were from the original organic dream of the 1960–70s counterculture. Julie Guthman (2004) described the "conventionalization" of organics, with methods similar to conventional farming, but substituting certified organic inputs—often trucked in from far away. This was probably not what Rudolf Steiner or J. I. Rodale had in mind. Consumers objecting to what they saw as factory farming masquerading as organic exercised their political power in a milk boycott. The chapters in this book record the milk boycott of Aurora Organic Dairy, and of Horizon Organic Dairy brands, waged by the Organic Consumers Association with local organizations, such as Puget Consumers Cooperative around Seattle. Also related is litigation initiated by The Cornucopia Institute against Aurora, Horizon, and other farms violating National Organic Program (NOP) regulations on access to pasture. Later it emerged that Aurora deemed this rule ambiguous and interpreted it as allowing cows to be confined during their lactations—though that was not the impression given in bucolic pictures on company websites.

The organic portion of the ten-thousand-cow Vander Eyck Dairy that supplied Dean-Horizon was decertified, before May 2007 when Horizon stated support for the 120 day/30 percent dry matter intake (DMI) minimums and urged the industry to exceed those standards. Prominent dairy farmers and veterinarians considered the standards very easily met, saying they would not consider any operation pasture-based if it did not meet those standards. If it wished, it seems that the newest incarnation of the company, WhiteWave/Horizon, could have a positive bully pulpit to lead the organic dairy industry.

The New York Times (2007) reported court judgments that Aurora Organic Dairy was guilty of 14 willful violations of organic rules. AOD and Horizon were ordered to drastically downsize their herds and raze buildings to enlarge pasture for the cows remaining. Finally in 2010 the USDA clarified its final Pasture Rule to include these minimal standards, adding guidance that cows should graze more when weather permitted.

So far so good. Perhaps consumer politics and grassroots activists had put USDA organics back on the idealist path. A writer might be forgiven for fearing the drama quotient of his latest book had fallen. But as Cornucopia Institute cofounder Will Fantle told me at the April 2013 National Organic Standards Board (NOSB) rules meeting in Portland, Oregon, USDA deputy administrator Miles McEvoy had inspected new megadairies in the Southwest but did not divulge what was found. This was no secret, as McEvoy's trip, however quiet, was mentioned on Cornucopia's website and the Odairy email list, along with realization that miniscule additions to the NOP budget were inadequate for policing the $31-billion organic sector. The National Organic Program staff directory (revised April 2014) listed only eight staffers in the Compliance & Enforcement Division. While USDA-accredited certifying agents (e.g., QAI) are used to monitor compliance with USDA rules, it is perplexing that, after highly publicized disputes over pasturing and stocking density, the USDA does not send its own staff to inspect the farms.

A report for the USDA Agricultural Marketing Service (AMS July 15, 2013: 9, 13–14) by the Office of the Inspector General offers sound advice on how the National Organic Program can retain consumer trust in organic dairying: the NOP needs to clarify its origin of livestock rule, ensure feed brokers are subject to oversight by certifying agents, give agents specific guidance on enforcement, and evaluate the record-keeping requirements on access to pasture (page 14 hints that small farms already keep cows on pasture "the majority of the year").

Valentine Surprise

In mid-February 2014, The Cornucopia Institute dropped a bombshell in a news release (2014d) headlined "Horizon 'Organic' Factory Farm Accused of Improprieties, Again," reading:

> In an open letter published today and addressed to USDA National Organic Program chief Miles McEvoy, The Cornucopia Institute accused the regulatory agency of abdicating its enforcement responsibilities. Cornucopia, an organic industry watchdog, charged that the USDA had allowed Dean Foods and its WhiteWave subsidiary to, allegedly, operate a giant factory farm dairy that has been illegally disadvantaging the nation's family-scale dairy producers. On Feb. 11, Cornucopia Institute filed its third formal legal complaint alleging Dean/WhiteWave's giant industrial dairy, located in Paul, Idaho has continued to operate illegally. "We're hoping that third time's a charm," said Cornucopia's Senior Farm Policy Analyst, Mark Kastel.

If Cornucopia's charge were true, its significance could still be weighed against speculation that the Idaho farm supplied not even 5 percent of Horizon milk, with family-scale farms contributing to its total. In the open letter to McEvoy, Will Fantle mentions reports that the "management claimed they could 'average' the entire herd in an effort to meet the 30 percent minimum dry matter intake." (This is akin to a factory boss honoring a 40-hour workweek by making the most productive workers work 80 hours and the others less.) If true, it would stress the most productive cows and conflict with the spirit of Horizon's own literature. *Horizon Organic Standards of Care* (Horizon no date: 12, 22), a brochure received from that company by this author in late 2013, showed nuanced commitment to balancing forage in the manger and fodder in the pasture: "Because this is so important, our Idaho and Maryland farms produce milk and also serve as learning centers for organic practices." Besides fodder on pasture, Horizon underscored the role of forage: "Because forage is the main portion of our cattle's diet, it is crucial that we ensure its quality, feed value, and organic integrity. In addition to pasture grass, our farms maximize the production of their own forage by growing over 90 percent of their own forage needs onsite."

A *Dairy Reporter* article of February 20, 2014, headlined Horizon's rejection of Cornucopia's claim: "Horizon Organic Regulation Breach Claims 'Wildly False': WhiteWave." WhiteWave/Horizon repudiated "wildly false allegations" that its former Horizon Dairy farm breached USDA National Organic Program (NOP) rules. A WhiteWave spokesperson said Cornucopia had made complaints about its Idaho farm to the USDA twice since 2006, when it was found in compliance with NOP rules, and it fully expected a third such decision.

Mark Kastel responded, "They are welcome to their own opinion but they are not welcome to their own facts." Cornucopia's February 14, 2014, press release repeated their past claim that the USDA NOP had never visited Horizon's Paul, Idaho, dairy in 2007 or more recently. Replying to this author's email query on responses by NOP deputy administrator Miles McEvoy to their complaint, Kastel explained how the head of the enforcement branch acknowledged receipt of their complaint:

I was flabbergasted to learn that they disputed our allegation that the USDA had never investigated the Dean/WhiteWave Paul, Idaho industrial dairy. Their explanation? They had the certifier do the investigation! In this case the certifier was Quality Assurance International (QAI) with a history of major malfeasance. As an example, they certified the Vander Eyk dairy with zero pasture. Our complaints name the certifiers because there is no way for us to determine whether the violations were based on collusion between the certifier and the certified entity.

Kastel added:

In our opinion it is irresponsible for the USDA to have depended on QAI to bless the outfit they are certifying. It would be a different story if the allegation was one of fraud where there was some reason to believe that the certifier was being deceived as well. But we have to assume, since this was a large-scale operation, that like all farms had to file a system operating plan, and that the certifier had to know what was going on.

Cornucopia's Valentine's Day press release (2014d) claimed a strong lobbying presence in Washington has "indemnified" agribusiness dairy giants from enforcement. It drew a parallel with the Wall Street crash of 2008. "Just as we have banks that have become 'too big to fail,'" Cornucopia alleged big organic-industrial firms had "get out of jail free" cards via lobbyists in the capital.

Kastel acknowledged the Idaho operation had finally added some pasture "for the first time" but "also increased the number of times the cows were being milked from twice a day to three and even four times a day." Cornucopia explained the basis of this inference:

> Recent interviews with dairy staff by Cornucopia investigators suggest that, to promote extremely high levels of milk production, the Horizon farm management prevented the cows from being put out on pasture between some of the milkings, and when they were out, made sure their bellies were already full of high-production rations (TMR feed) eaten in the barn.

Kastel conjectured (Cornucopia 2014d): "The cows were either prevented from going out and grazing, or if they did go out on pasture they probably didn't eat much fresh grass but instead lay down and chewed their cud, digesting the ration already eaten in the barn." Cows, consumers, and family farmers are shortchanged when megadairies flout rules or withhold health insurance from their underpaid non-unionized workers. A farmer who milks 45 cows near Wonewoc, Wisconsin says (Cornucopia 2014d):

> Small organic dairies nationwide have struggled with drought, flooding and oppressive heat. Still, we have pastured our cattle as required by the National Organic Program (NOP) . . . We have provided a product that consumers expect when they buy organic and we make it work economically—without cutting corners. If factory farm organic dairies are unwilling or unable to meet the NOP's pasture provisions, then perhaps it is time they are notified that their continued noncompliance to the National Organic Standards has gone on too long and they should seek a non-organic market for their milk.

Goldie Caughlan, a former member of the National Organic Standards Board, who joined the board of The Cornucopia Institute when retiring from her post at Puget Consumers Co-op Seattle noted (Cornucopia 2014b): "Organic consumers are amongst the most loyal in the marketplace. Consumers who patronize all-organic labels, that later blur the lines on the store shelf by adding conventional products under the same brand, are all too often taken advantage of." Consumer trust depends on the monitoring and enforcement of organic rules, and it follows that more transparency might have precluded the February 2014 complaints by The Cornucopia Institute about WhiteWave/Horizon's Idaho supplier.

Watchdogs

Nonprofit organizations, such as The Cornucopia Institute and the Organic Consumers Association, are occasionally accused by organic sector handlers, processors, distributors, and retailers of risking public confidence in the USDA organic label when they expose violations of the National Organic Program. But public trust is eroded by agribusiness forces that contravene the letter and spirit of NOP rules. Large-scale capital investment has been useful in establishing organic products in American supermarkets after most of a century of intensive dairying. But organic pioneers visualized the movement as a progressive one that should improve over time and bear up to increasing scrutiny.

Of course uncontrolled intraorganic carping can frighten consumers while damaging organicists' morale. Respected organic farmers, such as Klaus and Mary-Howell Martens of Lakeview Organic Grain, caution other contributors on the Odairy email list about the "circular firing squad." With good humor, Mary-Howell recently reminded members debating extensions on antibiotics in fruit trees: "We in the organic community are so good at shooting at each other . . . We need you all, alive and well!" (NODPA Odairy Feb. 4, 2014). Wise advice. On the other hand, without watchdogs, someone could rob the farm. Organic agribusinesses wishing to retain consumer loyalty would do well to remember that it was the organic dream of a movement populated by farmers, consumers, and other actors that gave impetus for the Organic Foods Production Act (1990), making their organic industry possible in the first place.

Farmers are also reminded by watchdogs, such as The Cornucopia Institute (2014c), that processors have multiple interests and loyalties. One example is reiterated by Mark Kastel, who claims that processor Horizon cut contracted organic farmers loose during a temporary milk surplus in 2009. Kastel also complains that the cafeteria in WhiteWave/Horizon's modern corporate head-quarters in Colorado features International Delight nondairy coffee creamer made from conventional, not organic, soy.

In the 1970s, USDA secretary Earl Butz mocked organic agriculture as a nice idea while "asserting that 50 million Americans would starve to death if the country were to switch to organic farming techniques" (Youngberg and DeMuth 2013). Butz's statement was political speak in support of President Richard Nixon's Food Power campaign to counteract trade deficits during the Vietnam War by a surge in agricultural exports. Butz's exhortations to small farmers to "Get big, or get out" depopulated rural communities, while setting the stage for the environmentally dubious Gene Revolution led by private firms, such as Bayer, Dow, DuPont, Monsanto, and Syngenta. For decades such corporations captured the media limelight, repeating the assertion that organics cannot feed the world.

Earl Butz was wrong, according to the Rodale Institute, whose aforementioned 30 Year Farming Systems Trial in Pennsylvania supports its bold claim that (*Pace Butz!*) organics *can* feed the world. Rodale (2011: 7, 11) also cites a report from the FAO-UN International Conference on Organic Agriculture and Food Security in Rome, attended by the U.S. branch of Rural Advancement Foundation International (RAFI) and global organizations such as IFOAM. The FAO report (FAO 2007: 4–5, 17) cites two recent models of a hypothetical global food supply grown organically. Some models (Badgley et al. 2007; Halberg et al. 2006) indicate that low and high targets of 2,640 and 4,380 kcal per capita daily, respectively, could be produced for Earth's current population of 7 billion people without increasing the agricultural land base, thus retaining current forests for carbon sequestration. The FAO (ibid.) report relates:

> These results considered the average organic yield ratio of different food categories . . . substituting synthetic fertilizers . . . with nitrogen fixation of leguminous cover crops . . . These models suggest that organic agriculture has the potential to secure a global food supply, just as conventional agriculture today, but with reduced environmental impacts.

Agricultural investment is fundamental to any model of organics maintaining fertility and productivity for 7 billion earthlings, not to mention the two to three billion more expected by the year 2050. The FAO (2007: 3) report specifies that "public investment is essential for agricultural growth" such as infrastructure, research, education, and extension. It also recognized that a world fed by organics would probably require more farm labor than the capital- and technology-prioritizing GMO systems they replaced. This is to be welcomed, as rural communities worldwide need remunerative jobs. Further, previously successful development programs, such as India's Operation Flood 1970–96, not only improved rural nutrition and incomes, but also stimulated the creation of ancillary jobs in the countryside, such as welding shops, retail kiosks, and education, health, and other services (Candler and Kumar 1998; Scholten 2010).

Successful future farm paradigms will also embrace technology. Whether they come under the label of *sustainable intensification* propounded by Jules Pretty and the Royal Society, or the *evergreen revolution* suggested in the evolving approaches of India's Green Revolution pioneer M. S. Swaminathan (see Basu and Scholten 2013; Scholten forthcoming in the *Handbook of Agricultural Globalization*), future approaches utilizing more biological than chemical processes will be complex enough in their integration of livestock with crop rotations that digital computing and monitoring devices will be in all farmers' toolkits.

The USDA (Odairy Feb. 4–5, 2014) admits the need for GMO, conventional, and organic crops to grow in "agricultural coexistence" without adulterating each other's gene pools. But its actions in certifying GMO crops at home and disseminating them abroad increases speculation of government connivance with the biotech industry's hope of a fait accompli, in which it is impossible to return to an Eden-like pre-GMO state.

As the global commons shrinks, it is harder to enable natural, instinctive behaviors, such as the desire of cattle to graze freely. New recognition of ruminants' sentience implies human responsibility to avoid inflicting stress and pain on these creatures (Singer 1975; 2002). That is why people in a line from Rudolf Steiner through contemporary consumers might prefer drinking artificial milk made from organic soybeans to milk from cows confined to stressful lives more brutish and short than their ancestors. Since the 1950s, and certainly the 1980s, CAFO confinement has gradually made traditional

pasture dairying more uneconomic. The 1980s ushered in laissez-faire mar-
ket-oriented USDA policies that, tied to the technology treadmill, depressed
farm gate prices and hastened the exit of roughly a hundred thousand U.S.
farm families from conventional dairying. The result was CAFOs that concentrate points of pollution in the environment, shed farm labor, worsen animal
welfare, and weaken the incomes of families that produce milk with fewer
negative externalities on animals and the environment. If year-round confinement operations are to continue, they should be roomy enough to reduce animal stress, improve health, and increase longevity without recourse to drugs,
such as antibiotics.

Without grazing, organic dairy farming approaches the factory-like industrialization of battery poultry operations. Strong rules on pasture are needed.
In light of IFOAM's 2006 St. Paul Declaration, USDA NOP organic dairying
lost integrity when major actors blocked their cows from expressing natural
instinctive behaviors, such as grazing on pasture. When a longtime dairyman
and conservationist was asked about agribusiness lobbying to weaken organic
access to pasture rules he said (personal communication): "I always thought
pasture was the dividing line between what they can have—and what is ours."

Farmageddon could still strike organics as the result of myopic government
regulation allowing too much appropriation by agribusiness. By May 2014,
organic sectoral conflicts had increased in scope and intensity. While small
farmer versus big farmer controversies continued to smolder on adherence
to pasture rules, conflict ignited along the dairy cold chain pitting organic
farmers against an agribusiness nexus of processors (handlers) and traders
(wholesalers and retailers) over issues such as GMO food labeling, an organic
checkoff, the sunsetting of synthetic ingredients such as carrageenan, and the
statutory power of farmers in the National Organic Standards Board under
the benchmark 1990 Organic Foods Production Act (OCA 2014b). Recent
actions by secretary of agriculture Tom Vilsack and the National Organic
Program's Miles McEvoy are interpreted by critics as moves to weaken the
power of the NOSB to, for example, "sunset" synthetics such as carrageenan
in dairy products and antibiotics in tree fruit. Synthetic materials were formerly allowed on the national organics list for no more than five years by
two-thirds majority vote. In 2013 McEvoy (NOP-AMS Sep. 13, 2013) sent a
memorandum titled "Subject: 'Sunset' Review of the National List of Allowed
and Prohibited Substances (National List)" reversing that arrangement. Now
synthetic substances are virtually left on the list indefinitely, unless they are
expelled by a two-thirds majority. What may have been an innocuous proce-
dural change was feared by idealists to be a harbinger of utter appropriation
and conventionalization of the organic sector by agribusiness.

The original heavyweights behind the Organic Foods Production Act of
1990 reacted. Before the spring 2014 NOSB meeting, April 29–May 2, 2014,
in San Antonio, Texas, Sen. Patrick Leahy (D-VT) and Rep. Peter DeFazio
(D-OR) sent an open letter to USDA secretary Tom Vilsack criticizing the
department's new organic policy in the sunsetting of materials (NOC 2014).
They stated that they believed the recent "sunset policy change made by your

agency . . . to be in conflict with both the letter and intent of the statute." Ominously, the lawmakers added: "We are particularly concerned that such a substantive policy change was made without the benefit of full notice and comment."

Organic politics were sparky in San Antonio. The normally businesslike atmosphere of National Organic Standards Board meetings was disrupted by a protest that hearkened back to the radicalism of the 1960s. Over the preceding winter, the USDA had not only weakened the sunsetting powers of the NOSB, but NOP deputy administrator Miles McEvoy asserted the right to cochair the San Antonio rules meeting with the elected NOSB board chairman Mac Stone, who would normally chair the meeting solo. McEvoy's move seemed to impinge what authority representatives of small farmers and consumers have in the NOSB to prevent total domination of organics by the USDA and agribusiness. Some observers were surprised that McEvoy stamped authority on the NOSB, when the USDA's recent sunset moves had been roundly condemned by members of the public, as well as Senator Leahy and Representative DeFazio.

As McEvoy prepared to speak, Alexis Baden-Mayer, political director of the Organic Consumers Association, held a banner in front of the lectern entitled "Safeguard Organic Standards" (OCA Apr. 30, 2014b). Colleagues joined her in chanting "Don't change sunset!" When the protesters refused to stop, Baden-Mayer was arrested by police, carried from the room, and subsequently refused entry to the remainder of the rules meetings.

Will Fantle of The Cornucopia Institute (2014) reported that as the meeting resumed, NOSB member Jay Feldman, executive director of the group Beyond Pesticides, requested a point of order against the cochairing of the meeting by the USDA's McEvoy. The meeting was halted again. Cornucopia's Will Fantle (2014) reported that McEvoy was seen making cell phone calls (possibly with his boss, USDA secretary Tom Vilsack) and approached Feldman, saying he would cancel the meeting unless Feldman retracted his point of order. After reassembling, Feldman reluctantly did so. Thereupon, McEvoy attempted to explain to the hundreds of attendees why the USDA was exerting more control of the 15-member NOSB board and suggested that after a future training session in Washington, DC, NOSB members would enjoy more transparency in streamlined procedures. It was not the first time that a national official promised that the future is bright. McEvoy did mention that progress was being made on the formulation of an origin of livestock rule, to stop organic factory farms from replacing stressed cows with conventional animals raised on cheaper grain. But this ray of light did not dispel all the disquiet in San Antonio.

From this point, were she present, Alexis Baden-Mayer may have deemed her arrest worthwhile. Exercising what power they could, the NOSB board voted to end the use of streptomycin in organic apples and pears after October 21, 2014—instead of granting a growers' petition to extend its use till 2017. The testy group also sent a measure on the use of methionine in poultry back to committee (which it otherwise might have supported, had it not wished to make a point on USDA actions). OCA head Ronnie Cummins praised Baden-Mayer's action and vowed that its million-member network and allied organizations, such as The Cornucopia Institute and Food Democracy Now!, would continue fighting a corporate takeover of organics (OCA 2014b).

What explains what some grassroots farmers and consumers saw as a power grab by USDA National Organic Program deputy director Miles McEvoy on the NOSB board? Also, could OCA political director Alexis Baden-Mayer's protest be seen as mere political theater to rekindle interest in the group's agenda? Giving the USDA and McEvoy the benefit of the doubt, perhaps they have made a top-down realpolitik assumption that the U.S. will be the most sustainable of all possible worlds when the national list of synthetic nonorganic substances is stretched enough to allow agribusiness to more easily process products that can then be retailed as USDA certified-organic food. In other words, local organics and farmers' markets are well and good, but what really matters is mass marketing by agribusiness retailers, such as Wal-Mart. It is a big gamble. If such a strategy results in turning more acres of the world organic, many observers would justify it in a utilitarian political calculus of the greatest good for the greatest number. But if too many shoppers conclude that the USDA is marginalizing the input of farmers and consumers on the NOSB in order to grease the wheels for processors and retailers, the attraction of the USDA organic label could ebb.

On the other hand, Mark Keating, a former livestock specialist for the NOP who is now a policy consultant for the Organic Farming Research Foundation (OFRF), expressed alarm that some members of the National Organic Standards Board were obsessed with minute quantities of pesticides and other prohibited substances, such as carrageenan, hexane, and streptomycin. It is an interesting point, but when the Consumers Union, Organic Trade Association, public health officials, and members of the public are frightened of, for instance, antibiotic resistance, it is imperative that the NOSB be seen to manage the national list carefully. That said, a refocus on *process* in organic systems (that would be recognized by organic pioneers, including Rudolf Steiner, Eve Balfour, Albert Howard, and J. I. Rodale) and on shorter food chains in regional systems—compared to the present globalized conventional system—would garner support from consumers and organic pioneers. However, materials are much easier to measure and monitor than process. Also, without the support of profit-seeking agribusiness—and tax reforms favoring local organic farming—it is hard to envision a refocus on process in the NOP.

Meanwhile, a host of issues concern small organic farmers in 2014. On the state level, Michigan revoked right-to-farm laws to anyone within an eighth of a mile of other residences, a blow to urban keepers of hens and goats, as well as preexisting farms, in areas including Detroit.

Vermont's recently signed standalone law, requiring labeling of GMO foods and prohibiting "natural" labels on foods containing GMOs, immediately aroused promises of litigation from manufacturers and traders. The online plea of Governor Shumlin to organic partisans across the country to donate money to the state of Vermont's Food Fight Fund may be a development in a new hybrid politics positioning a coalition of organic consumers, farmers, and politicians against agribusiness in the form of the Grocery Manufacturers Association (GMA). How this unfolds in the fall 2014 elections could link to the showdown states of Oregon and Colorado where GMO labeling is on the ballot. Advocacy group Right to Know Colorado GMO (July 25, 2014) claims they "collected more than 125,000 valid signatures, well over the 86,105 needed, and received approval from the Colorado Secretary of State on August 20, 2014" for the measure. Successes in those states would strengthen allied forces in Connecticut, Maine, and other states to label GMO foods. It is possible that the U.S. Supreme Court could invalidate state laws to label GMOs. That could look oddly undemocratic to people in the 65 or so countries around the world where they are labeled.

On the national level, the GMA is lobbying for a weak organic labeling scheme that appears to have the support of President Obama. Weak labeling might allow the inclusion of GMO high-fructose corn syrup (HFCS) in organic labeled foods and fizzy colas. Such industrial appropriation of traditional food systems will be resisted by a coalition of greens, health officials, and nutrition experts.

On January 13, 2014, a Supreme Court decision on *Organic Seed Growers and Trade Association et al. v. Monsanto* quashed pleas by farmers afraid that firms such as Monsanto could sue *them* for patent infringement if trace amounts of GMO DNA are found in their adjacent non-GMO crops (Cornucopia 2014).

Most germane to this book, suspicion remains that the suppliers to the organic dairy sector's biggest corporation violates the USDA National Organic Program final Pasture Rule of 2010. If this proves true, it threatens the integrity of the NOP (Cornucopia 2014). Could agribusiness appropriationism strangle the organic goose? It is already playing with consumer trust. If the NOP lost trust with consumers, what might transpire is predictable. The grassroots movement that followed publication of Rachel Carson's (1962) *Silent Spring* could rise again. Thousands of angry, ecoconscious people would reinforce existing alternative food networks. They would forget the USDA, as many already have, and go "beyond organic"—the phrase coined around 2003 when the program was already bogged down by high certification costs and heavy paperwork for small producers, as it has been. Community supported agriculture (CSA), box schemes, buyers' groups, cooperatives, farmers' markets, slow food, and other groups would continue with renewed vigor, even if just a small percentage of the national system—but with a leavening effect that reminds people of the biological basis of healthy food and the place of animals in agriculture.

Status Quo Ante

Since First Lady Michelle Obama planted an organic garden on the White House lawn with ladybugs and crab meal instead of pesticides, writes Jason Zengerle of *The New Republic* (Apr. 29, 2014), foodies have been underwhelmed by President Barack Obama's policies. Zengerle claims that on issues from antibiotics to GMO labels the "farmer in chief" has "over-promised reforms, underestimated the strength of his opposition, and then flinched." Yet, even if Obama and his USDA secretary Tom Vilsack caved under pressure from the biotechnology industry in deregulating GMO alfalfa and the like, it is worth mentioning that Vilsack held hearings on overcon-solidation in the meat industry in which just four corporations control 84 percent of the beef meat market to the detriment of small farmers' livelihoods and possibly national health (Schlosser 2001; Fromartz 2006; Leonard 2014). Disappointed greens might console themselves that the Obamas do at least enjoy locavore restaurants. New York University nutritionist Marian Nestle says, "I think the White House garden has phenomenal symbolic value." But one longtime Obama supporter, Dave Murphy of the group Food Democracy Now! describes the administration's food policy as "status quo and industrial agribusiness as usual."

Would the country have greener farm and food policies if Mitt Romney had won the White House from Obama in 2012? Hardly. Unlike Romney, Obama has not actually been paid as a consultant for Monsanto strategy. Meanwhile, organic consumers, farmers, processors, retailers, veterinarians, policy makers, and others struggle to keep overlap between the dreams of organic pioneers and current USDA policy. Grazing cows on pasture passes one obvious test of organic integrity. If it is necessary to the welfare and lon-gevity of cows, many people would argue that it should be required for both conventional and organic cows.

Animal behaviorist Temple Grandin says (CSU 2010): "I think using ani-mals for food is an ethical thing to do, but . . . we've got to give those animals a decent life and we've got to give them a painless death. We owe the animal respect."

A step toward respect is illustrated by the experience of this author's friends who are admired farmers. The health of their 150 cows deteriorated after switching from grazing to confinement. Foot problems became so rife, they joked that the specialist who came to treat hooves was on the farm more than the husband who took a part-time truck driving job to supplement income. Exasperated, the couple turned the cows back on pasture. Within two years herd health improved.

Respect—and better incomes—backed by sensible government policies on animal welfare and safe food production are also due families who call those animals by name and manage their farms like living organisms in rural communities.

Bibliography

Note: This bibliography includes a subdivision headed "U.S. Government."

Academics Review (2014). "Organic Marketing Report." Reviewed by Bruce Chassy, David Tribe, Graham Brookes, and Drew Kershen. Anglo-American Association, U.S. 501©3: http://academicsreview.org/wp-content/uploads/2014/04/Academics-Review_Organic-Marketing-Report1.pdf [Accessed Apr. 26, 2014].

Allen, Paul (2011). *Idea Man: A Memoir by the Cofounder of Microsoft.* New York: Portfolio.

Alliance for a Green Revolution in Africa (AGRA) (2013). "FAQs About Seeds, Breeding, and GMOs": http://www.agra.org/resources/faqs-frequently-asked-questions-about-seeds-breeding-and-gmos/ [Accessed Nov. 13, 2013].

American Frozen Food Institute (AFFI) (2013). "2013 AFFI Government Action Summit. Schedule of Events": http://www.affi.org/node/1319 [Accessed Nov. 5, 2013].

American Grassfed Association (AGA) (2011). "Grassfed & Pasture Finished Ruminant Standards." Aug. 2011: http://www.americangrassfed.org/wp-content/uploads/2011/12/AGA-Grassfed-Standards-Fall-2011.pdf [Accessed Aug. 16, 2014].

Anderson, Burton Laurence (1957). *The Scandinavian and Dutch Rural Settlements in the Stillaguamish and Nooksack River Valleys of Western Washington. Seattle*: Diss. University of Washington.

Anderson, Kym (ed.) (2010). *The Political Economy of Agricultural Price Distortions.* Paperback 2013. Cambridge, UK: Cambridge University Press.

Anderson, Kym and Rodney Tyers (1991). *Global Effects of Liberalising Trade in Farm Products.* Trade Policy Research Centre. London: Harvester-Wheatsheaf.

Antaya, N. T., A. F. Brito, K. J. Soder, N. L. Whitehouse, N. E. Guindon, A. B. D. Pereira, and C. C. Muir (2013). "Kelp Meal (*Ascophyllum nodosum*) Did Not Improve Milk Yield or Mitigate Heat Stress but Increased Milk Iodine in Mid Lactation Organic Jersey Cows during the Grazing Season." *Journal of Dairy Science* (E-Suppl. 1) 96:33.

Arsenault, Nicole, Peter Tyedmers, and Alan Fredeen (2009). "Comparing the Environmental Impacts of Pasture-based and Confinement-based Dairy Systems in Nova Scotia (Canada) Using Life Cycle Assessment." *International Journal of Agricultural Sustainability* 7 (1): 19–41.

Atkins, Peter J., Ian Simmons, and Brian K. Roberts (1998). *People, Land & Time.* London: Arnold.

Atkins, Peter and Ian Bowler (2001*). Food in Society: Economy, Culture, Geography.* London: Arnold.

Atkins, Peter (2010). *Liquid Materialities: A History of Milk, Science and the Law.* Surrey, England: Ashgate.

Atlantic, The (2012). "The FDA Did Not Do Enough to Restrict Antibiotics Use in Animals." By Robert S. Lawrence. Apr. 16, 2012.

Aurora Organic Dairy (AOD), Boulder, CO 80302 (www.auroraorganic.com)

—— (2005a). "AOD Taps Vet Juan S. Velez." Press release. Sep. 14.

—— (2005b). "Deal on Organic Dairy Reached." *Greeley Tribune*, by staff, Oct. 20.

—— (2006a). "Animal Health: Natural Breeding."

—— (2006b). "The Role of Organic Pasture in Animal Welfare." Juan Velez, USDA-NOSB Pasture Symposium, Apr.19.

—— (2007). "[Aurora] Dairy Opens Doors to Tour. *Greeley Tribune*, July 1.

—— (2013). *2012 Corporate Citizenship Report*. See pages 2, 14, 16, 19, 20, 21, 22, 28, 30, 32.

Baltimore Sun (2006). "OCA & Cornucopia Expose Horizon's Factory Dairy Farm in Maryland: Shore Farm Accused of Skirting Standards." By Chris Guy, Aug. 28.

Badgley, Catherine, Jeremy Moghtadera, Eileen Quinteroa, Emily Zakema, Jahi Chappella, Katia Avilés-Vázqueza, Andrea Samulona, and Ivette Perfecto (2007). "Organic Agriculture and the Global Food Supply." *Renewable Agriculture and Food System* 22 (02): 86–108.

Balfour, Lady Eve (1977). "Towards a Sustainable Agriculture: The Living Soil." IFOAM Conference in Switzerland in 1977: http://journeytoforever.org/farm_library/balfour_sustag.html [Accessed Dec. 7, 2013].

Bartlett, Rosamund (2008). *Tolstoy: A Russian Life*. New York: Random House.

Basu, Pratyusha, and Bruce A. Scholten (2013). *Technological and Social Dimensions of the Green Revolution: Connecting Pasts and Futures*. Oxon, UK and New York, NY: Routledge (Taylor & Francis).

Bauman, Dale E., and Jude L. Capper (2011). "Efficiency of Dairy Production and Its Carbon Footprint." Derived from Capper et al. *Proceedings of Cornell Nutrition Conference* (2008).

Bellingham Herald, The (2013). "Whatcom County Farmers Debate State GMO Labeling Initiative." By Kie Relyea, Oct. 27: http://www.bellinghamherald.com/2013/10/27/3278591/whatcom-county-farmers-debate.html#storylink=cpy [Accessed Oct. 30, 2013].

Benbrook, Charles M. (2010). "'Shades of Green." User's Manual—Guide and Documentation for a Daily Farm Management System Calculator. Oct. Pullman, WA: The Organic Center.

—— (2012a) *A Deeper Shade of Green: Lessons from Grass-based Organic Dairy Farms*. Aug. Washington, DC: The Organic Center.

—— (2012b) "Impacts of genetically engineered crops on pesticide use in the U.S. - the first sixteen years." *Environmental Sciences Europe* 24:24. [Also, Odairy-OTA Feb. 19, 2014].

—— (2014). "Shades of Green Dairy Farm Calculator Webinar," Mar. 18: http://www.extension.org/pages/31790/shades-of-green-dairy-farm-calculator-webinar [Accessed Mar. 30, 2014].

Benbrook, Charles M., Gillian Butler, Maged A. Latif, Carlo Leifert, and Donald R. Davis (2013). "Organic Production Enhances Milk Nutritional Quality by Shifting Fatty Acid Composition: A United States–Wide, 18-Month Study." *PLoS ONE* 8(12): e82429. doi:10.1371/journal.pone.0082429.

Benbrook, Charles M., Cory Carman, E. Ann Clark, Cindy Daley, Wendy Fulwider, Michael Hansen, Carlo Leifert, Klaas Martens, Laura Paine, Lisa Petkewitz, Guy Jodarski, Francis Thicke, Juan Velez, and Gary Wegner (2010). *A Dairy Farm's*

Footprint: Evaluating the Impacts of Conventional and Organic Farming Systems. Pullman, WA: The Organic Center.

Berkes, Fikret (1999). *Sacred Ecology: Traditional Ecological Knowledge and Resource Management.* Philadelphia: Taylor and Francis.

Bernstein, Eduard (1961). *Evolutionary Socialism: The Classic Statement of Democratic Socialism.* New York: Shocken Books.

Berry, Wendell (1970/1972). *A Continuous Harmony: Essays Cultural and Agricultural.* New York: Harcourt.

Berry, Wendell (1986). *The Unsettling of America: Culture and Agriculture.* San Francisco: Sierra Club.

Biagiotti, Paul R. (2013). "Overcrowding Invites Disease." *Hoard's Dairyman.* Aug. 25: 540.

—— (2014). "Lack of Longevity: Are We in a Culling Crisis?" *Progressive Dairyman.* Feb. 21: http://www.progressivedairy.com/index.php?option=com_content&id=11837:lack-of-longevity-are-we-in-a-culling-crisis&Itemid=71 [Accessed Mar. 15, 2014].

—— (forthcoming 2015). *Practical Organic Dairy Health and Management.* Fort Atkinson, WI: *Hoard's Dairyman.*

Bible (DASV) (2011). The Digital American Standard Version Bible. Project of Dr. Ted Hildebrandt, Gordon College, Wenham, MA 01984. Based on American Standard Version (ASV) of 1901.

Binimelis, Rosa, Walter Pengue, and Iliana Monterroso (2009). "'Transgenic treadmill': Responses to the Emergence and Spread of Glyphosate-resistant Johnsongrass in Argentina." *Geoforum.* 40(4): 623–33.

Biodynamics (2008). "Ehrenfried Pfeiffer, the Threefold Community, and the Birth of Biodynamics in America." By Bill Day, Fall 2008. Milwaukee, WI: The Biodynamic Farming and Gardening Association: https://www.biodynamics.com/threefold-day [Accessed Dec. 7, 2013].

Bonanno, Alessandro (2000). "The Crisis of Representation: The Limits of Liberal Democracy in the Global Era." *Journal of Rural Studies* 16: 305–23.

Born Free USA (2013). "The Destructive Dairy Industry: How Has Milk Production Changed since the 1950s?" http://www.bornfreeusa.org/facts.php?p=373&more=1 [Accessed Oct. 21, 2013].

British Broadcasting Corporation (BBC) (2013a). "Fukushima Leak Is 'Much Worse Than We Were Led to Believe.'" By Matt McGrath: http://www.bbc.co.uk/news/science-environment-23779561 [Accessed Aug. 26, 2013].

—— (2013b). "Second Badger Cull Is Believed to Have Begun in Gloucestershire": http://www.bbc.co.uk/news/uk-england-gloucestershire-23955074 [Accessed Sep. 4, 2013].

—— (2013c). "New Antibiotic That Attacks MRSA Found in Ocean Microbe." By Simon Redfern, July 31.

Business Wire (2006). "Horizon Organic Supports Pasture Requirement: Organic Dairy Leader Urges USDA to Adopt Stricter Regulations." Apr. 21: http://www.thefreelibrary.com/Horizon+Organic %28R%29+ Supports+Pasture+Requirement%3B+Organic+Dairy+Leader...-a0144777421 [Accessed May 1, 2014].

Butler, Gillian, Jacob H Nielsen, Tina Slots, Cris Seal, Mick D. Eyre, Roy Sanderson, and Carlo Leifert (2008). "Fatty Acid and Fat-soluble Antioxidant Concentrations in Milk from High- and Low-input Conventional and Organic Systems: Seasonal Variation." *Journal of the Science of Food and Agriculture* 88:1431–41.

Buttel, Fred (1999). "Agricultural Biotechnology: Its Recent Evolution and Implications for Agrofood Political Economy." *Sociological Research Online*, Vol. 4, no. 3, http://www.socresonline.org.uk/4/3/buttel.html

Buttel, Frederick H. (2000). "The recombinant BGH controversy in the United States: Toward a New Consumption Politics of Food?" *Agriculture and Human Values* 17(1): 5–20.

California Milk Advisory Board (CMAB) (2009). "Two Centuries of Prominence and Personalities. California's Dairy Industry: The Early Years (1769–1900)." June 2009: http://www.californiadairy pressroom.com/Press_Kit/History_of_Dairy_ndustry [Accessed Nov. 30, 2013].

—— (2013). "California: The Nation's Dairy Leader." June 2013: http://www.californiadairypressroom.com/Press_Kit/Nations_Dairy_Leader [Accessed Nov. 30, 2013].

Callenbach, Ernest (1975). *Ecotopia: The Notebooks and Reports of William Weston.* New York: Bantam.

Canada, Government of (2012) "Culling and Replacement Rates in Dairy Herds in Canada." Ottawa, Ontario: www.dairyinfo.gc.ca

Canadian Veterinary Medical Association (1998). *Report of the Canadian Veterinary Medical Association Expert Panel on rbST.* CVMA Expert Panel on rbST. Prepared for Health Canada, November, 1998.

Canada, Parliament of (1999). rBST and the Drug Approval Process: Interim Report— The Standing Senate Committee on Agriculture and Forestry. Chair: L. J. Gustafson, Dep. Chair: E. F. Whelan, P. C., March 1999: http://www.parl.gc.ca/Content/SEN/Committee/361/agri/rep/repintermar99-e.htm [Accessed Nov. 6, 2013].

Candler, Wilfred, and Nalini Kumar (1998). *India: The Dairy Revolution: The Impact of Dairy Development in India and the World Bank's Contribution.* World Bank Operations Evaluation Study (OED). Washington, D.C.: WB.

Capper, Jude L. (2009). "The Environmental Impact of Dairy Production: 1944 Compared with 2007." *Journal of Animal Science* 87 (6): 2160: http://agron-www.agron.iastate.edu/Courses/agron515/Capperetal.pdf [Last accessed Aug. 17, 2014].

Capper, Jude L. (2012). "Can Buying Local Food Really Save the Planet?" *Hoard's Dairyman*, Jan. 25: 43.

Capper, Jude L. (2012). "Comparing the Environmental Impact of Conventional, Natural and Grass-Fed Beef Production Systems." *Animals* 2 (2): 127–43.

Capper, J. L., Castaneda-Gutierrez, E., Cady, R. A., and Bauman, D. E. (2008). "The Environmental Impact of Recombinant Bovine Somatotin (rbST) Use in Dairy Production." *Proceedings of the National Academy of Sciences* 105 (28): 9688–73.

Capper, Jude L., R. A. Cady, and D. E. Bauman (2009). "Efficiency of Dairy Production and Its Carbon Footprint." Author for correspondence: D. E. Bauman, Department of Animal Sciences, Cornell University, Ithaca, NY 14853; email: deb6@cornell.edu. Online paper stored by University of Florida Institute of Food and Agricultural Sciences (IFAS): http://dairy.ifas.ufl.edu/rns/2010/11-Bauman.pdf [Last accessed Aug. 17, 2014]. Extensive portions of this paper were derived from the article by Capper, Cady and Bauman, Proceedings of Cornell Nutrition Conference (2008). The authors acknowledge Roger A. Cady (Elanco Animal Health) for his invaluable contributions to this work.

Carson, Rachel (1962). *Silent Spring.* Boston: Houghton Mifflin.

Center for Food Safety (2007). "Center Says Private Legal Action Will Help Enforce Organic Labeling Requirements and Uphold Organic Standards to Maintain Public Trust." Contact: Joe Mendelson, Oct. 17, press release: www.centerforfoodsafety.org/AuroraPR10_17_07.cfm Press Releases [Accessed Oct. 21, 2007].

Center for Food Safety (2013). "VICTORY!! The Monsanto Protection Act is Finally Dead! Sen. Barbara Mikulski (Dem. MY) Announces Deletion from HJ RES. 59." Sep. 25, 2013: http://salsa3.salsalabs.com/o/1881/p/salsa/web/common/public/content?content_item_KEY=13184 [Accessed Sep. 25, 2013].

Center for Integrated Ag Systems (CIAS) (2005). *Social Implications of Management Intensive Rotational Grazing (MIRG): An Annotated Bibliography*. By M. Mariola, K. Stiles, and S. Lloyd. University of Wisconsin–Madison Center for Integrated Ag Systems. Articles by Joe Mendelsohn and Center for Food Safety (CFS 2006–7).

Centers for Disease Control and Prevention (CDC) (2011). *A Public Health Action Plan to Combat Antimicrobial Resistance*. Interagency Task Force on Antimicrobial Resistance. Co-Chairs: Centers for Disease Control and Prevention, Food and Drug Administration, National Institutes of Health. Participating Agencies: Agency for Healthcare Research and Quality, Centers for Medicare and Medicaid Services, Department of Agriculture, Department of Defense, Department of Veterans Affairs, Environmental Protection Agency, Health Resources and Services Administration, Health and Human Services/Office of the Assistant Secretary for Preparedness and Response.

Centre for Research on Globalization (CRG) (2012). "Monsanto Controls Both the White House and the U.S. Congress." By Josh Sagar. CGR, Montreal, Canada, Oct. 1, 2012: http://www.globalresearch.ca/monsanto-controls-both-the-white-house-and-the-us-congress/5336422 [Accessed Feb. 8, 2014].

Christiansen, Andrew (1995). "Recombinant Bovine Growth Hormone: Alarming Tests, Unfounded Approval: The Story Behind the Rush to Bring rBGH to Market." A project of the Rural Education Action Project (REAP), Vermont, July 1995.

Consumers Union (2014). "Comments of Consumers Union to the National Organic Standards Board." Apr. 8, Docket No. AMS-NOP-14-0006. Signed by Urvashi Rangan, Michael Crupain, and Charlotte Vallaeys.

Chicago Manual of Style (2000). G. J. Duncan and J. Brooks-Gunn (eds.). Chicago, IL: University of Chicago Press.

Collier, Paul (2008). *The Bottom Billion*. Oxford: OUP.

Colloquium of Organic Researchers (2002). *Proceedings of the UK Organic Research 2002: Research in Context*. Mar. 26–28, 2002, University of Aberystwyth, Wales.

Colorado State University (CSU) (2010). "Temple Grandin HBO Movie, Viewing Party Set for Saturday at Colorado State University." CSU Dept. Media Relations, Fort Collins. Feb. 3, 2010: http://www.news.colostate.edu/Release/5014 [Accessed Feb. 7, 2014].

Congressional Research Office (2013). "Farm-to-Food Price Dynamics." By Randy Schnepf, Sep. 27: http://www.fas.org/sgp/crs/misc/R40621.pdf [Accessed May 10, 2014].

Cornucopia Institute (Cornucopia, WI: http://cornucopia.org. 9,000+members in 2013).

—— (2006a). "Maintaining the Integrity of Organic Milk: Showcasing Ethical Family Farm Producers." By Mark Kastel, for USDA-NOSB pasture symposium, Apr. 19.

—— (2006b). "Processors Alliance Final Letter [to USDA]." By anonymous, Oct. 26.

—— (2006c). "Dairy Scorecard." Online resource: www.cornucopia.org/dairysurvey/index.html [Accessed Sep. 13, 2013]. See also "Forward/Update 2008" version: http://www.cornucopia.org/dairysurvey/OrganicDairyReport/Dairy_Report_Update.pdf [Accessed Mar. 30, 2014].

—— (2007a). "USDA Cracking Down on "Organic" Factory Farms—Country's Largest Dairy [Aurora] Likely to Lose Certification." Aug. 14.

—— (2007b). "Organic Consumers Launch Nationwide Lawsuit in 27 States against Aurora Organic Dairy for Violating Organic Standards." Oct. 17.

—— (2009a). "Behind the Bean: The Heroes and Charlatans of the Natural and Organic Soy Foods Industry: The Social, Environmental, and Health Impacts of Soy." With accompanying Organic Soy Scorecard.

—— (2009b). "Organic Milk Scandal: Why Walmart, & Others are Selling 'Organic' Milk at Such Low Prices." By Mark Kastel, Aug. 31, 2009 In Nutrition Digest Vol. 36, No. 4: http://american nutritionassociation.org/newsletter/organic-milk-scandal [Accessed Aug. 17, 2014].

—— (2010). "Scrambled Eggs: Separating Factory Farm Egg Production from Authentic Organic Agriculture: A Report and Scorecard." http://www.cornucopia. org/egg-report/scrambled eggs.pdf [Accessed Mar. 4, 2014].

—— (2012a). "The Organic Watergate: White Paper: Connecting the Dots: Corporate Influence at the USDA's National Organic Program." Wisconsin: www.cornucopia.org.

—— (2012b). "USDA Organic Audit: Procedures for Approving Synthetics Followed, Allegations of Corruption Unexamined." News release, July 24, 2012.

—— (2012c). "Comments to the National Organic Standards Board October 2012 Meeting, Providence, RI. Submitted Sep. 24, 2012": http://www.cornucopia. org/wp-content/uploads/2012/11/Cornucopia-Oct-2012-NOSB-Comment.pdf [Accessed Dec. 5, 2013].

—— (2013a). "State of Fever: Monsanto's GMO Policy Infecting All Levels of Government." By Will Fantle. The Cultivator, newsletter of the Cornucopia Institute: June 26, 2013.

—— (2013b). "Former Chair [Jim Riddle] of National Organic Standards Board Writes Resignation Letter to the Organic Trade Association." Oct. 1, 2013.

—— (2013c). "GMA Injects New Cash vs. GMO Labeling Effort in Washington State!" Oct. 29, 2013: http://www.cornucopia.org/2013/10/food-group-injects-new-cash-vs-labeling-effort#GMOs#YesOn522#RightToKnow#LabelGMOs@ [72977977116:274:The Cornucopia Institute] [Accessed Oct. 29, 2013].

—— (2013d). "Is the USDA a Wholly-owned Subsidiary of Monsanto?" Quoting Food and Water Watch: http://www.cornucopia.org/is-the-usda-a-wholly-owned-subsidiary-of-monsanto [Accessed Nov. 22, 2013].

—— (2014a) "The Supreme Court Denies Family Farmers the Right to Self-Defense from Monsanto Abuse." Jan. 14, 2014: http://www.cornucopia.org/2014/01/supreme-court-denies-family-farmers-right-self-defense-monsanto-abuse/ [Accessed Jan. 14, 2014].

—— (2014b) "Leading Organic Brand, Horizon, Blasted for Betraying Organics: Corporate Parent, WhiteWave, Announces Non-Organic Horizon Products [Mac & Cheese]." Feb. 13, 2014.

—— (2014c) "High Points from Today's WhiteWave Call with Investors." Feb. 13, 2014.

—— (2014d) "News Release: Horizon 'Organic' Factory Farm Accused of Improprieties, Again." See: https://groups.yahoo.com/neo/groups/Odairy/conversations/topics/16500 [Accessed Aug. 26, 2014].

—— (2014e) "Turmoil Shakes National Organic Standards Board Meeting in Texas. Board Decision Making Colored by Restrictions to Authority and Governance Imposed by USDA": http://www.cornucopia.org/2014/05/turmoil-shakes-national-organics-standards-board-meeting-texas [Accessed May 8, 2014].

Croney, C. C., M. Apley, J. L. Capper, J. A. Mench, and S. Priest (2012). "Abstract: Bio-ethics Symposium: The Ethical Food Movement: What Does It Mean for the Role of Science and Scientists in Current Debates about Animal Agriculture?" *Journal of Animal Science* 90 (5): 1570–82.

Cox, Rosie, and Ben Campkin (eds.) (2007). *Dirt: New Geographies of Cleanliness and Contamination*. London: I. B. Tauris. Based on Purity in Food session convened by Rosie Cox and Laura Venn, AAG Denver, Apr. 5–9, 2005.

Daily Mail, The (2013). "What Is the Truth about the MRSA Superbug?" By Rory Clements, Oct. 23: http://www.dailymail.co.uk/health/article-157079/What-truth-MRSA-superbug.html#ixzz2j2k3K3wf [Accessed 25.Oct.13]

Dairy Business (c. 2009). "USDA Survey Provides a Glimpse of Organic Milk and Crop Production: Organic Dairies Represented 3.5% of U.S. Herds, 2.1% of Cows and 1.5% of Milk Volume in 2008." By Dave Natzke. [Accessed May 16, 2013]

Dairy Farmers of Washington (c. 2014). "Facts about the Washington Dairy Industry." 2005–2014 Washington Dairy Products Commission.

Dairy Group, The (2010). *Dairy Cow Housing*. Report prepared for Arla, Morrisons and Dairy Co. Aug. 2012 Somerset: New Agriculture House.

Dairy Herd Improvement (DHI-Provo) (2013) "Rolling Herd Average - % Cows Leaving the Herd." Ladd Muirbrook, Sr. Acct Exec., Dairy Division. Table and chart cover 2004–2013, DFI-Provo Workshop, Oct. 22, 2013. Provo, Utah: DHI Computing Service, Inc.

Dairy Herd Improvement Association (DHIA) (2013). "DHI Glossary." Aug.: http://www.drms.org/PDF/materials/glossary.pdf [Accessed Mar. 15, 2014]

Dairy Reporter (2014). "Horizon Organic Regulation Breach Claims 'Wildly False': WhiteWave." By Mark Astley. Feb. 20. Summary: "WhiteWave Foods Has Slammed 'Wildly False Allegations' That a Former Horizon Dairy Milk Farm Breached US Department of Agriculture (USDA) National Organic Program (NOP) Regulations": http://www.dairyreporter.com/Manufacturers/Horizon-organic-regulation-breach-claims-wildly-false-WhiteWave [Accessed Mar. 4, 2014].

Davis, John H., and Ray A. Goldberg (1957). *A Concept of Agribusiness*. Boston: Harvard University Graduate School of Business Administration.

Dean Foods (2007). "Dean Foods Company (NYSE:DF) Financial and Income Statements 2006." http://finance.google.com/finance?fstype=bi&q=DF [Accessed Aug. 12, 2007].

Dean Foods (c. 2013). "History of Dean Foods Company." With information on Silk, White Wave, Horizon, Rachel's, etc.: http://www.deanfoods.com/our-company/about-us/brief-history.aspx [Accessed Feb. 16, 2014].

DeLaval (2013) Cow Longevity Conference Proceedings, Aug. 28–29, 2013. Tumba, Sweden: DeLaval International AB: http://www.milkproduction.com/Global/PDFs/Cow%20Longevity%20Conference%20Proceedings%20.pdf [Accessed Aug. 20, 2014].

Democratic Senatorial Campaign Committee (DSCC) (2014). "Breaking: New Koch Attack Rampage Pounds Vulnerable Democrats." Feb. 10. *DSCC Rapid Response*; email info@dscc.org.

Diamond, Jared (1997). *Guns, Germs, and Steel: The Fates of Human Societies*. London: W. W. Norton & Co.

Diamond, Jared (2005). *Collapse: How Societies Choose to Fail or Succeed*. New York: Penguin Books.

Digital American Standard Version Bible (DASV) (2011). Digital American Standard Version Bible. Based on ASV of 1901. Project of Ted Hildebrandt. Gordon College,

Wenham, MA: http://faculty.gordon.edu/hu/bi/ted_hildebrandt/DASV/00_DAS-VIndex.htm#TopOfDASVPage [Accessed Sep. 25, 2013].

Drummond, Ian, and Terry Marsden (1999). *The Condition of Sustainability*. London: Routledge.

Doyon, M. A., and Andrew M. Novakovic (1997). "An Application of Experimental Economics to Agricultural Policies: The Case of U.S. Dairy Deregulation on Farm Level Markets." RB 97–11, Dept. of Agr. Res. and Mgrl. Econ, Cornell University.

Dupuis, E. Melanie (2000). "Not in My Body: rBGH and the Rise of Organic Milk." *Agriculture and Human Values* 17:285–95.

—— (2002). *Nature's Perfect Food: How Milk Became America's Drink*. New York: New York University.

Dyer, Christopher (2007). "Conflict in the Landscape: The Enclosure Movement in England, 1220–1349." In *Landscape History*, Vol. 29: http://www.medievalists.net/2010/02/19/conflict-in-the-landscape-the-enclosure-movement-in-england-1220-1349/ [Accessed July 5, 2012].

Ecologist, The (2013). "The GM Lobby and Its 'Seven Sins Against Science': The Pro-GM Lobby Has Sought to Take the 'Scientific High-Ground' by Positioning Itself as the Voice of Reason and Progress, While Painting Its Opponents as Unsophisticated 'Anti-Science' Luddites. In a Scathing Response Peter Melchett Turns the Tables." By Peter Melchett, Jan. 7: file:///I:/Documents%20and%20Settings/Owner/Desktop/PASTURE%20BOOK%2011sep13/GM%20Lobby%20and%20Its%20%27Seven%20Sins%20Against%20Science%27%20_%20Cornucopia%20Institute%207jan13%20ECOLOGIST%20pMELCHETT.htm [Accessed Jan. 8, 2013].

Economist, The (2006). "Natural Foods: The Rise of Big Organic: Wal-Mart Goes Crunchy." June 8.

—— (2007). "Cotton Suicides: The Great Unravelling." Jan. 20.

—— (2010). "Economist Debates: Biotechnology: Statements." Nov. 5: http://www.economist.com/debate/files/view/BASF_artifact0.pdf [Accessed Nov. 7, 2013].

—— (2013a) "Biodiversity: All Creatures Great and Small. Deputy Editor Emma Duncan Argues GMO Intensification Helps Preserve Hainan Gibbons." Sep. 14.

—— (2013b) "How Science Goes Wrong." Oct. 19: 11.

—— (2013c) "Genetically Modified Food: Warning Labels for Safe Stuff: One Way or Another, Labelling of GM Food May Be Coming to America." Nov. 2.

—— (2013d) "Fields of Beaten Gold: Greens Say Climate Change Deniers are Dangerous. So Are Greens Who Oppose GM Crops." Dec. 7: 16.

—— (2013e) "GM Maize, Health and the Séralini Affair: Smelling a Rat." Dec. 1: 86.

Ehrmann, Henry W. (1983). *Politics in France*. Boston: Little Brown & Co.

Elanco (2009). "Recombinant Bovine Somatotropin (rbST): A Safety Assessment, by Eight Authors Led by Richard Raymond, Former Undersecretary for Food Safety at USDA." Non-peer-reviewed paper, presented at 2009 conference in Montreal and publicized by Elanco.

Elden, Stuart (2004). *Understanding Henri Lefebvre*. London: Continuum.

Environmental Working Group (EWG) (c. 2013). "Farm Subsidies: The United States Summary Information": http://farm.ewg.org/region.php [Accessed Mar. 18, 2014].

Ewen, Stanley W., and Arpad Pusztai (1999). "Effect of Diets Containing Genetically Modified Potatoes Expressing Galanthus nivalis Lectin on Rat Small Intestine." *The Lancet* 354 (9187): 1353–54.

Farm Animal Welfare Council (FAWC) (1997). "Five Freedoms." In *FAWK Animal Welfare Council Annual Report* (3797). Advisory body est. by UK Government in 1979: http://www.fawc.org.uk/freedoms.htm [Accessed Dec. 3, 2013].

Farmers' Legal Action Group, Inc. (2008a). "Hushed Up: Confidentiality Clauses in Organic Milk Contracts." By Jill Krueger, Apr. (Support by NODPA). St. Paul, MN: Legal Action Group, Inc.

—— (2008b) "Companion Article: When Your Processor Requires More Than Organic Certification: Additional Requirements in Organic Milk Contracts." St. Paul, MN: Legal Action Group, Inc.

Fine, Ben, Michael Heasman, and Judith Wright (1996). *Consumption in the Age of Affluence: The World of Food*. London: Routledge. [Page 105: appropriationism and substitutionism].

Food and Agricultural Organization of the United Nations (FAO-UN) (2003). "WTO Agreement on Agriculture: The Implementation Experience." ID 127659. [White, Green, Blue, Brown, Yellow Revolutions.] By Anil Sharma. Rome: FAO.

—— (2005). *From the Green Revolution to the Gene Revolution. How Will the Poor Fare?* By Prabhu Pingali and Terri Raney. Rome: FAO.

—— (2006). *Livestock's Long Shadow*. By Henning Steinfeld, Pierre Gerber, Tom Wassenaar, Vincent Castel, Mauricio Rosales, and Cees de Haan. Rome: FAO.

—— (2007). *Organic Agriculture and Food Security*. By Nadia El-Hage Scialabba. International Conference on Organic Agriculture and Food Security, Rome. May 3–5, 2007: http://www.fao.org/organicag/ofs/index_en.htm [Accessed Feb. 13, 2014].

—— (2013). "Dairy Industry Development Programmes: Their Role in Food and Nutrition Security and Poverty Reduction." By Brian T. Dugdill, Anthony Bennett, Joe Phelan, and Bruce A. Scholten. In *Milk and Dairy Products in Human Nutrition*. Rome: FAO. Download free: https://durham.academia.edu/BruceScholten.

FOOD (2007). "Farmers Letter to USDA Acting Sec'y Connor." Sep. 25. Via NODPA.

Food Business News (2014). "WhiteWave Foods Expanding its Horizon." By Keith Nunes. Feb. 20: http://www.foodbusinessnews.net/articles/news_home/Business_News/2014/02/WhiteWave_Foods_expanding_its.aspx?ID=%7BC81BB32F-B073-4CD4-B70A-4D989B1DE99D%7D&cck=1

Food, Inc. (2008). Documentary film directed by Robert Kenner, narrated by Michael Pollan and Eric Schlosser.

Fortune (2007). "An Organic Milk War Turns Sour: Now That His Dairy Company Has Settled Charges That It Violated Organic Food Standards, Aurora President Mark Retzloff Wants to Muzzle Foes." By Marc Gunther, Oct. 3: http://money.cnn.com/2007/10/02/news/companies/aurora_follow.fortune/ [Accessed Oct. 7, 2007].

Foucault, Michel (1975). *Discipline and Punish*. New Edition, Apr. 25, 1991. New York: Penguin.

Fromartz, Sam (2006). *Organic, INC.: Natural Foods and How They Grew*. New York: Harcourt.

Funding Universe (c. 2001). "Horizon Organic Holding Corporation History." Source: *International Directory of Company Histories*, Vol. 37. St. James Press, 2001: http://www.fundinguniverse.com/company-histories/horizon-organic-holding-corporation-history/ [Accessed May 8, 2014].

Gates, Bill (1999). "Conversation List of Greatest Inventions Reflects a Wondrous World." In *The Edge*: http:// www.edge.org/conversation/list-of-greatest-inventions-reflects-a-wondrous-world [Accessed Mar. 19, 2013].

Geller, Bruce, Kimberly Marshall-Batty, Frederick J. Schnell, Mattie M. McKnight, Patrick L. Iversen, and David E. Greenberg (2013). "Gene-Silencing Antisense Oligomers Inhibit Acinetobacter Growth In Vitro and In Vivo." *Journal of*

Infectious Diseases 208 (10): 1553–1560: http://jid.oxfordjournals.org/content/early/2013/09/30/infdis.jit460.abstract

Geisler, C. C., and Thomas A. Lyson (1991). "The Cumulative Impact of Dairy Industry Restructuring." *BioScience* 41 (8): 560–67. In Hinrichs and Lyson (eds.) (2007), *Remaking the North American Food System: Strategies for Sustainability*. Nebraska: University of Nebraska Press.

George, Susan (1977). *How the Other Half Dies: The Real Reasons for World Hunger*. Montclair, NJ: Allenheld, Osmun & Co.

—— (1984). *Ill Fares the Land*. Montclair, NJ: Allenheld, Osmun & Co.

Gilbert, Jess, and Raymond Akor (1988). "Increasing Structural Divergence in U.S. Dairying: California and Wisconsin since 1950." *Rural Sociology* 53 (1): 56–72.

Gilbert, Jess, and Kevin Wehr (2003). "Dairy Industrialization in the First Place: Urbanization, Immigration and Political Economy in Los Angeles County, 1920–1970." *Rural Sociology* 68 (4):467–90.

Gill, Grandon T. (2013). "Case Studies in Agribusiness: An Interview with Ray Goldberg." *Informing Science: the International Journal of an Emerging Transdiscipline*, 16: 203–212: http://www.inform.nu/Articles/Vol16/ISJv16p203-212GillCS02.pdf

Glassman, Jim (2006). "Primitive Accumulation, Accumulation by Dispossession, Accumulation by 'Extra-Economic Means.'" *Progress in Human Geography* 3:608–25.

Goodman, David, Bernardo Sorj, and John Wilkinson (1987). *From Farming to Biotechnology: A Theory of Agro-Industrial Development*. Oxford: Blackwell.

Goodman, David, and E. Melanie DuPuis (2002). "Knowing Food and Growing Food: Beyond the Production-Consumption Debate in the Sociology of Agriculture." *Sociologia Ruralis* 42:5–22.

Granatstein, David: See Washington State University.

Grandin, Temple (1989). "Behavioral Principles of Livestock Handling." *Professional Animal Scientist*, Dec. 1989: 1–11.

—— (n.d.) "Dr.Grandin Speaks out on Animal Welfare Issues: 'I divide animal welfare concerns into two basic categories . . . [#1: Abuse and Neglect, #2: Boredom and Restrictive Environments']." On *Dr. Temple Grandin's Web Page*: http://www.grandin.com/welfare/welfare.issues.html [Accessed Aug. 20, 2014].

Grandin, Temple and Catherine Johnson (2005). *Animals in Translation*. New York, NY: Scribner (Division of Simon and Schuster).

—— (2006). *Thinking in Pictures*. Follows 1995 Vintage edition. New York, NY: Houghton Mifflin Harcourt.

—— (2009). *Animals Make Us Human: Creating the Best Life for Animals*. New York, NY: Houghton Mifflin Harcourt.

Grant, R. (2009). "Stocking Density and Time Budgets." In *Proceedings: Western Dairy Management Conference*. (In *Dairy Group 2010*): 7–17.

Grist (2011). "In a Stunning Reversal, USDA Chief Vilsack Greenlights Monsanto's Alfalfa." By Tom Philpott, Jan. 8: http://grist.org/article/2011-01-27-in-stunning-reversal-usda-chief-vilsack-greenlights-monsantos-al/ [Accessed Jan. 13, 2011].

Growing a Nation (n.d.). "Historical Timeline—Farmers & the Land." http://www.agclassroom.org/gan/timeline/farmers_land.htm [Accessed Sep. 10, 2013].

Guardian, The (1999). "Britain's Scientific Elite Continue to Try to Suppress Dr. Pusztai's Research on Dangers of GE Foods: Pro-GM Food Scientist 'Threatened Editor' GM Food: Special Report." By Laurie Flynn and Michael Sean Gillard, Nov. 1.

—— (2007). "Biofuel Demand to Up Food Prices." By John Vidal, July 5.

—— (2008). "Arpad Pusztai: Biological Divide: The Scientist at the Centre of a Storm over GM Foods 10 Years Ago Tells James Randerson He Is Unrepentant." Jan. 15: http://www.the guardian.com/education/2008/jan/15/academicexperts.higher-educationprofile [Accessed Feb. 16, 2013].

—— (2013a) "Owen Paterson: UK Must Become Global Leader on GM Crops." By John Vidal, June 20.

—— (2013b) "Washington State Alfalfa Crop May Be Contaminated with Genetic Modification: Farmer Reports His Alfalfa Shipment Was Rejected." By Suzanne Goldenberg, Sep. 12.

—— (2013c) "WikiLeaks Publishes Secret Draft Chapter of Trans-Pacific Partnership: Treaty Negotiated in Secret between 12 Nations 'Would Trample over Individual Rights and Free Expression,' Says Julian Assange." By Alex Hern, Nov. 13: http://www.theguardian.com/media/2013/nov/13/wikileaks-trans-pacific-partnership-chapter-secret [Last acchttp://www.theguardian.com/media/2013/nov/13/wikileaks-trans-pacific-partnership-chapter-secretessed Aug. 17, 2014].

—— (2013d) "Will Chimps Soon Have Human Rights? Steven M. Wise Is Campaigning for Chimps' Right to Seek Freedom from Unlawful Detention on the Grounds That They Are Essentially People." By Emine Saner, Dec. 4: http://www.theguardian.com/world/shortcuts/2013/dec/03/chimpanzees-human-rights-us-lawyer?CMP=fb_gu [Accessed Dec. 7, 2013].

Guthman, Julie (2004). *Agrarian Dreams: The Paradox of Organic Farming in California*. Berkeley: UC Press.

Halberg, N. H., F. Alroe, M. T. Knudsen, and E. S. Kristensen (2006). *Global Development of Organic Agriculture: Challenges and Prospects*. Oxfordshire: CABI Publishing.

Hall, Clare, Alistair McVittie, and Dominic Moran (2004). "What Does the Public Want from Agriculture and the Countryside?" *Journal of Rural Studies* 20 (2 April): 211–25.

Hartman Group, The (1997). The Hartman Report: Food and Environment: A Consumer's Perspective, Phase II. Winter. Bellevue, WA: THG. Note: Hartman specializes in organic consumption and culture research; these reports were in newsletters and are available online: http://www.hartman-group.com/about/ [Last accessed Aug. 17, 2014].

—— (2004). Organic Food & Beverage Trends 2004: Lifestyles, Language & Category Adoption. THG.

—— (2006). Organic 2006: Consumer Attitudes & Behaviour, Five Years Later & into the Future. THG.

Harvey, David (1990/2000). *The Condition of Postmodernity*. Oxford: Blackwell.

—— (2003). *The New Imperialism*. Oxford: OUP.

Harvey, David R. (1998) "The U,S, Farm Act: 'Fair' or 'Foul'? An Evolutionary Perspective from East of the Atlantic." *Food Policy* 23 (2), April: 111–121.

Haskell, M. J., L. J. Rennie, V. A. Bowell, M. J. Bell, and A. B. Lawrence (2006). "Housing System, Milk Production, and Zero-Grazing Effects on Lameness and Leg Injury in Dairy Cows." *Journal of Dairy Science* 89: 4259–4266.

Haskell, M. J., S. Brotherstone, A. B. Lawrence and I. M. S. White (2007). "Characterization of the Dairy Farm Environment in Great Britain and the Effect of the Farm Environment on Cow Life Span." *Journal of Dairy Science* 90: 5316–5323.

Health Canada (1998). *rBST (Nutrilac) Canadian Gaps Analysis Report*. By Shiv Chopra, Mark Feeley, Gerard Lambert, and Thea Mueller. Ottawa: Health Canada.

—— (1999). *Report of the Royal College of Physicians and Surgeons of Canada*. (More Mastitis & SCC.) Archived 2013.

Hill, C. T., P. D. Krawczel, H. M. Dann, C. S. Ballard, R. C. Hovey, W. A. Falls, and R. J. Grant (2009). "Effect of Stocking Density on the Behavior of Dairy Cows with Differing Parity and Lameness Status." *Applied Animal Behaviour Science* 117:144–49.

Hinrichs, C. Clare, and Thomas A. Lyson (eds.) (2007). *Remaking the North American Food System: Strategies for Sustainability*. Lincoln: University of Nebraska Press. (Refers to Geisler and Lyson 1991.)

Hoard's Dairyman (2007a). "Ethanol Is Changing our Industry: Corn & Soybean Prices off the Charts." Feb. 25: 129.

—— (2007b). "Safeway Stores in the NW Part of the Country Are the Latest to Say No to Milk from Cows Given BST, According to *The Oregonian*." Hoard's has heard, Mar. 10: 159.

—— (2010). "Organic Dairy Farmers Get Firm Pasture Rule." By Tina Wright. June: 440.

—— (2012a). "Editorial: It's Time to Discontinue Tail Docking." Mar. 25: 189, 194.

—— (2012b). "Managing Sand on the Way Out." May 25: 372.

—— (2012c). "China's Dairy Sector Is Experiencing Growth Pains: The Country's Dairy Farms Face Many of the Same Challenges as U.S. Counterparts. Feed Quality, Genetics and Sanitation Can Be Added to the List." By Kevin Herrick, July: 447.

—— (2013a). "U.S. Dairy Industry Statistics, 2007 to 2012." Mar. 25: 206–7.

—— (2013b). "Nature Thrives Because Ag Keeps Innovating." April 10: 244.

—— (2013c). "Dairy to Serve as Pilot Industry for Sustainability." By Doug Young, Aug. 10: 512.

—— (2013d). "Lameness Can Be Beaten: Getting Cows off Their Feet and into Well-Designed Stalls Can Improve Hoof Health by Leaps and Bounds." By Tara Cheema, Sep. 10: 569.

—— (2013e) "Dairy Farmers, Veterinarians Closely Monitor Antibiotic Use." By Angela King, for Press-Gazette Media. Oct. 22.

—— (2013f). "Animal Agriculture Is Sustainable and Essential to Feeding a Growing World." By Ali Enerson, *Hoard's Dairyman Special Publications Editor* Oct. 28: http://www.hoards.com/blog_animals-humans-coexist [Last accessed Aug. 18, 2014].

—— (2014a). "Why the Yakima Lawsuits Matter to Every Producer." Jan. 10: 7.

—— (2014b). "Organic Dairies Feel the Squeeze, Too." By Tina Wright, Feb. 10: 96.

—— (2014c). "Most Dairy Farms Exit Business since 2007." Mar. 10: 151.

—— (2014d). "National Milk Supply Treads Water in 2013." Mar. 10: 155.

—— (2014e). "Which Breeds Stay in Herds Longer?" By Dennis Halladay, Western Editor, *HD Intel 2014*, Mar. 31: http://hoards.com/intel/140331_art3 [Accessed Mar. 31, 2014].

Holloway, Lewis, Christopher Bear, and Katy Wilkinson (2013). "Robotic Milking Technologies and Renegotiating Situated Ethical Relationships on UK Dairy Farms." *Agriculture and Human Values* 30 (Oct. 2013): 185–199.

Holt, Georgina, and Matt Reed (eds.) (2006). *Sociological Perspectives of Organic Agriculture: From Pioneer to Policy*. London: CABI/OUP.

Horizon Organic Dairy (2004–13). *Producer Post*. Summer and Winter online editions.

Horizon (May 25, 2007). "Horizon Organic® Support NOSB Recommendation of 30% DMI from Grazing—Organic Pioneer Publishes Updated 'Standards of Care.'" Broomfield, CO. Contact: sara.unrue@whitewave.com; www.horizonorganic.com/aboutus/press/2007_5_25.html [Accessed May 25, 2007].

Horizon, WhiteWave (n.d.). "Pasture Management." In Horizon Organic Standards of Care. 25 pp. Excerpt: "Our Idaho and Maryland farms produce milk and also serve as learning centers for organic practices." Received from Horizon before separation from Dean Foods in 2013.

Horizon, WhiteWave (2014). Numerous publications and information available online.

Howard, Philip (2004). "What Do People Want to Know about Their Food? Measuring Central Coast Consumers' Interest in Food Systems Issues." *Research Brief* #5, Winter 2005. Center for Agroecology and Sustainable Food Systems, UC Santa Cruz: http://casfs.ucsc.edu/documents/research-briefs/RB_5_consumer_interest1.pdf [Accessed May 9, 2014].

—— (2005). "What Do People Want to Know about their Food? Measuring Central Coast Consumers' Interest in Food Systems Issues." Jan. 15, 2005. Center for Agroecology and Sustainable Food Systems, UC Santa Cruz.

—— (2014). "Organic Industry Structure: Acquisitions and Alliances Top 100 Food Processors in North America (Who Owns Organic?)." Department of Community Sustainability, Michigan State University. Graphic updated Feb. 2014; incl. Dean-WhiteWave-Horizon spinoff. Free via MSU or http://www.cornucopia.org/wp-content/uploads/2014/02/Organic-chart-feb-2014.jpg *or* http://www.cornucopia.org/wp-content/uploads/2014/02/Organic-chart-feb-2014.jpg.

Howard, Philip H., and Patricia Allen (2006). "Beyond Organic: Consumer Interest in New Labeling Schemes in the Central Coast of California." *International Journal of Consumer Studies* 30: 439–51.

Huffington Post (2012). "Salmon Says: Should You Worry about Radiation in Your Wild Pacific Fish?" By Harriet Sugar-Miller, posted May 30, 2012: http://www.huffingtonpost.ca/harriet-sugarmiller/radiation-pacific-fish_b_1553537.html [Accessed Aug. 26, 2013].

International Food Information Council (IFIC) (2013). *Food Biotechnology: A Communicator's Guide to Improving Understanding*, 3rd ed. David Schmid, CEO: http://www.foodinsight.org/foodbioguide.aspx [Accessed Dec. 13, 2013].

Inc. Magazine (2007). "Wal-Mart Loved Organic Valley's Milk: So Why Cut Off the Flow?" By K. Pattison, July 15.

Independent, The (2013). "Chief Medical Officer Dame Sally Davies: Resistance to Antibiotics Risks Health 'Catastrophe' to Rank with Terrorism and Climate Change." By Michael McCarthy, Mar. 13: http://www.independent.co.uk/news/science/chief-medical-officer-dame-sally-davies-resistance-to-antibiotics-risks-health-catastrophe-to-rank-with-terrorism-and-climate-change-8528442.html [Accessed Mar. 15, 2013].

Indian Country (2014). "Coast Salish Nations Unite to Protect Salish Sea." Feb. 17: http://indiancountrytodaymedianetwork.com/2014/02/17/coast-salish-nations-unite-protect-salish-sea [Accessed Feb. 17, 2014].

International Federation of Organic Agriculture Movements (IFOAM) (2006a–2013a editions) *The World of Organic Agriculture—Statistics and Emerging Trends*. Helga Willer and Minou Yussefi (eds.) Bonn: www.ifoam.org [Accessed often since first edition 2003].

IFOAM (2005, 2006a, corrected 2007). *The IFOAM Basic Standards for Organic Production and Processing Version 2005*. Bonn: IFOAM. http://www.futurepolicy.org/fileadmin/user_upload/Axel/IFOAM_basic_standards.pdf

—— (2006b) "The St. Paul Declaration: On the Occasion of the First IFOAM International Conference on Animals in Organic Production." Aug. 26. University of

Minnesota, Aug. 23–25, 2006: http://infohub.ifoam.org/en/st-paul-declaration [Accessed May 16, 2007].

IFOAM (2006c) "IFOAM Livestock Conference." *The Inspector's Report.* Fall 2006: http://www.ioia.net/images/TIRArchive/V15N4pt2.pdf.

IFOAM (2012). *The IFOAM Norms for Organic Production and Processing, Version.* (2012 Norms are similar to 2005–07 Standards above.) Aug. 2012. Bonn: IFOAM.

IFOAM (2013). The Organic Movement Worldwide: Directory of IFOAM Affiliates 2013. Markus Arbenz, Executive Director. Bonn: IFOAM.

International Fund for Agricultural Development (IFAD) (n.d.). "Livestock and Pastoralists." By Antonio Rota and Sylvia Sperandini: http://www.ifad.org/lrkm/factsheet/pastoralists.pdf [Accessed July 19, 2013].

International Journal of Agricultural Sustainability (2013). "Can Organic and Resource-Conserving Agriculture Improve Livelihoods? A Synthesis." (Cited by OCA, Oct. 2013.) By Mica Bennett and Steven Franzel, Aug., vol. 11, issue 3: 193–215.

International Livestock Research Institute (ILRI) (2011). "The Productivity of 'Nomadic Farming' Over the Long Term." Aug. 11: http://www.ilri.org/ilrinews/index.php/archives/6947 [Accessed Aug. 11, 2011].

International Union for Conservation of Nature (IUCN) (2013?). "Economic Importance of Goods and Services Derived from Dryland Ecosystems in the IGAD Region." IUCN: http://www.ilri.org/ilrinews/index.php/archives/6947 [Accessed July 19, 2013].

Iowa State University (2013). *Organic Dairy Profile.* By Madeline Schultz, revised Jan. 2013 by Diane Huntrods. Agricultural Marketing Resource Center: http://www.agmrc.org/commodities__products/livestock/dairy/organic-dairy-profile/.

Jelinek, Lawrence J. (1979). *Harvest Empire: A History of California Agriculture.* San Francisco: Boyd and Fraser.

Karreman, Hubert J. (2004/2007). *Treating Dairy Cows Naturally: A Handbook for Organic and Sustainable Farmers: Thoughts & Strategies.* 2nd ed. 2007. Austin, TX: Acres USA.

—— (2011). *The Barn Guide to Treating Dairy Cows Naturally.* Austin, TX: Acres USA.

Kastel, Mark (1995). "Down on the Farm: The Real BGH Story: Animal Health Problems, Financial Troubles." Rural Vermont, 1995.

Keeler, John T. S. (1981). "The Corporatist Dynamic of Agricultural Modernisation in the Fifth Republic." In *The Fifth Republic at Twenty*, Wm. G. Andrews and Stanley Hoffman (eds.). Albany: State University of New York Press.

—— (1987). *The Politics of Neocorporatism in France: Farmers, the State and Agricultural Policy-making in the Fifth Republic.* New York: Oxford University Press.

King, Angela (2013). "Dairy Farmers, Veterinarians Closely Monitor Antibiotic Use." For Press-Gazette Media and *Hoard's Dairyman*, Oct. 22, 2013.

King, Barbara J. (2013). *How Animals Grieve.* Chicago: University of Chicago Press. Also NPR, "The Ties That Bind Animals and Humans Alike": http://www.npr.org/blogs /13.7/2013 [Accessed Sep. 1, 2013].

Knapp, J. R., J. L. Firkins, J. M. Aldrich, R. A. Cady, A. N. Hristov, W. P. Weiss, A. D. G. Wright, and M. D. Welch (2011). *Cow of the Future: Research Priorities for Mitigating Enteric Methane Emissions from Dairy.* Working draft July 2011. Innovation Center for U.S. Dairy: http://www.usdairy.com/~/media/usd/public/cowofthefuture whitepaper_7-25-11.pdf.ashx [Accessed Aug. 9, 2013].

Krawczel, P. D., C. T. Hill, H. M. Dann, and R. J. Grant (2008). "Short Communication: Effect of Stocking Density on Indices of Cow Comfort." *Journal of Dairy Science* 91: 1903–7.

Kunstler, James Howard (2007). "Making Other Arrangements: A Wake-Up Call to a Citizenry in the Shadow of Oil Scarcity [*aka* Peak Oil]." *Orion*, Jan./Feb.: www.orionmagazine.org/index.php/articles/ [Accessed Feb. 3, 2007].

Kurien, Verghese (2005). *I Too Had a Dream*. As told to Gouri Salvi. New Delhi: Roli.

Lang, Tim, and Michael Heasman (2004). *Food Wars: The Global Battle for Mouths, Minds and Markets*. London: Earthscan.

Lappé, Frances Moore (1971, 1975, 1982, 1991). *Diet for a Small Planet*. New York City, NY: Ballantine Books.

Latham, Jonathan, and Allison Wilson (2013). "Sponsored Academics Admit Falsely Claiming Safety for Monsanto/Eli Lilly's Bovine Growth Hormone." Organic Consumers Association, June 17, 2013: http://www.organicconsumers.org/articles/article_27787.cfm [Last accessed Aug. 19, 2014].

LeCompte-Mastenbrook, Joyce (2004). "Making Sense of Place: Narratives of Migration, Milk and Modernity in a Northwest Washington Dairy Community [aka Dutch-Whatcom-Stewardship]." Unpublished BA thesis. Anthropology Dept. honors for design and execution of an anthropological research project. University of Washington, Seattle, WA.

Lefebvre, Henri (1974). *La Production de l'Espace* (*Production of Space*). In Elden (2004).

Leonard, Christopher (2014). *The Meat Racket. New York, NY*: Simon & Schuster.

Logsdon, Gene (2010). *Holy Shit: Managing Manure to Save Mankind*. White River, VT: Chelsea Green.

Lockeretz, Willie (2002). "Strategies for Organic Research." Tufts University. In *UK Organic Research: Proceedings of the COR Conference*. Mar. 22–25, 2002, University of Aberystwyth.

Lymbery, Philip, with Elizabeth Oakeshott (2014). *Farmageddon: The True Cost of Cheap Meat*. London: Bloomsbury.

Lynden Heritage Foundation (2007). "Lynden History Timeline." By Troy Luginbill, director/curator: http://lyndenpioneermuseum.com/resources/lyndenTimeline.php [Accessed Jan. 19, 2014].

MacDonald, James, and William McBride (2007). *The Transformation of U.S. Livestock Agriculture Scale, Efficiency, and Risks*. Washington, DC: USDA.

MacMillan, Thomas C. (2002). *Governing Interests: Science, Power and Scale in the rBST Controversy*. PhD thesis, University of Manchester, UK.

Marsden, Terry, A. Flynn, and M. Harrison (2000). *Consuming Interests: The Social Provision of Foods*. London: UC Press.

Marx, Karl (1979). "The Manifesto of the Three Zurichers." In "Circular letter to Bebel, Liebknecht, Bracke, and Others." In Tucker (1978), *The Marx-Engels Reader*.

McDonald's (2013). "Company Profile: McDonald's is Global—and in Your Hometown": http://www.aboutmcdonalds.com/mcd/investors/company_profile.html [Accessed Nov. 24, 2013].

McMinn, Teresa (2013). "New Vet at Rodale." *Reading Eagle*, [Dr. Hubert Karreman], Nov. 6, 2013.

Melchett, Peter. See *The Ecologist*.

Mercola, Joseph (2011). "Beware of These Organic Brands." Cornucopia's Mark Kastel interviewed. *Dr. Mercola's Natural Health Newsletter*. Nov. 19, 2011: http://articles.mercola.com/sites/articles/archive/2011/11/19/mark-kastel-cornucopia-good-food-movement.aspx [Accessed Dec. 5, 2013].

—— (2013). "GMO Labeling Has Its Day in Court." *Dr. Mercola's Natural Health Newsletter*. Oct. 29, 2013: http://articles.mercola.com/sites/articles/archive/2013/10/29/gmo-labeling-campaign.aspx [Accessed Oct. 30, 2013].

Merrill, Jeanne (2005). "Fields of Dreams: A Comparison of the Minnesota and Wisconsin Sustainable Agriculture Movements." M.Sc. thesis, University of Wisconsin–Madison.

Morgan, Dan (1979). *Merchants of Grain*. New York: Viking Press.

Morgan, Kevin, Terry Marsden, and Jonathan Murdoch (2006). *Worlds of Food*. Oxford: OUP.

Murdoch, Jonathan, and Mara Miele (1999). "Back to Nature: Changing Worlds of Production in the Food System." *Sociologia Ruralis* 39: 465–484.

Nairobi Star (2011). "Pastoralists Innovate in the Face of Adversity." May 19, 2011.

Nation, The (2011). "ALEC Exposed: The Koch Connection: Untold Sums of Cash Poured into ALEC by Charles and David Koch." *The Nation*, Aug. 1, 2011.

National Animal Health Monitoring Systems (NAHMS) (2007a). *Dairy 2007 Part II: Changes in the U.S. Dairy Cattle Industry, 1991–2007*, NAHMS, USDA-APHIS, Washington, DC. (Cited by Benbrook 2012).

—— (2007b) *Dairy 2007 Part III: Reference of Dairy Cattle Health and Management Practices in the U.S.*, 2007, NAHMS, USDA-APHIS, Washington, DC.

National Agricultural Safety Database (NASD) (2011). "Diagnosing Silo-Filler's Disease." NASD/Michigan State University Extension: nasdonline.org/document/1037/d000831/diagnosing-silo-filler-039-s-disease.html [Accessed Oct. 7, 2013].

National Farmers Union (NFU) (2014). "Statement on the Safe and Accurate Food Labeling Act." Apr. 11. NFU, 20 F Street NW, Suite 300, Washington, DC.

National Organic Coalition (NOC) (2013). "RE: DOCKET AMS–NOP–13–0049 [phase out antibiotics against fire blight]") Letter by NOC's Liana Hoodes to NOSB's Michelle Arsenault, Sep. 30.

—— (2014). "Coalition Members." http://www.nationalorganiccoalition.org/coalition_members [Accessed Mar. 9, 2014].

—— (n.d.) "Join as a NOC Network Affiliate!" ["Semiannual pre-NOSB meetings"]: www.nationalorganiccoalition.org/_.../NOC_Network_Application_Form [Accessed Mar. 10, 2014].

—— (2014). "Sen. Leahy and Rep. DeFazio Criticize USDA's New Organic Policy in Sunsetting of Materials." Joint Letter by OFPA 1990 authors Leahy and DeFazio to Sec. Vilsack: http://www.nationalorganiccoalition.org/nosb [Accessed May 2, 2014].

National Sustainable Agriculture Coalition (NSAC) (2014). "Funding Once Again Available for Organic Research!!" Mar. 18: http://sustainableagriculture.net/blog/fy14-orei-rfa-announced/ [Accessed Mar. 20, 2014].

Nature (2013a). "Fields of Gold: Research on Transgenic Crops Must Be Done outside Industry if It Is to Fulfill Its Early Promise." 497: 5–6.

Nature (2013b). "Microbiome: Soil Science Comes to Life: Plants May Be Getting a Little Help with Their Tolerance of Drought and Heat." By Roger East. *Nature* 501: S18–S19. Sep. 25: http://www.nature.com/nature/journal/v501/n7468_supp/full/501S18a.html [Accessed Sep. 26, 2013].

Nestle, Marion (2002/3). *Food Politics: How the Food Industry Influences Nutrition and Health*. London: University of California Press.

New York Times, The (2003). "Monsanto Sues Dairy in Maine Over Label's Remarks on Hormones." By David Barboza, July 12.

—— (2007). "[Aurora] Organic Dairy Agrees to Alter Practices." By Andrew Martin, Aug. 30.

—— (2012a). "Has 'Organic' Been Oversized?" By Stephanie Strom, July 8.

—— (2012b). "In India, GM Crops Come at a High Price."

—— (2013a). "Antibiotic-Resistant Infections Lead to 23,000 Deaths a Year, Finds Centers for Disease Control and Prevention." By Sabrina Tavernise, Sep. 16.

—— (2013b). "Misgivings about How a Weed Killer Affects the Soil." By Stephanie Strom, Sep. 19.

—— (2014). "Politics: Farm Bill Reflects Shifting American Menu and a Senator's Persistent Tilling." By Jennifer Steinhauermarch, Mar. 8.

Northeast Organic Farming Association (NOFA) (2006). "Putting Politics out to Pasture: Report on the National Campaign Organic Committee Meeting." By Steve Gilman, NOFA Policy Coordinator. *The Natural Farmer* 2 (69).

Odairy is the email discussion list of the Northeast Organic Dairy Producers Alliance. NODPA Executive Director Ed Maltby moderates discussion and submission of messages on Odairy@yahoogroups.com. Odairy archive is on: www.nodpa.com.

Odairy (2006). "Answers to questions posed by the NOP in the advance notice of proposed rulemaking (ANPR) on pasture." By Ed Maltby. June 9, 2006.

—— (2006). "Processors alliance final letter re: organic dairy systems." USDA Secretary Michael Johanns. Signing processors: George Siemon, CROPP Cooperative/Organic Valley; Nancy Hirshberg, Stonyfield Farm; Kelly Shea, Horizon Organic; Rich Ghilarducci, Humboldt Creamery; Mark Retzloff, Aurora Organic Dairy. Oct. 26.

—— (2007). "Despite the snow, [FOOD] farmers say organic cows need to graze pasture." Mar. 5.

—— (2007). "Ed Maltby applauds Cornucopia re decertification of Vander Eyck Dairy." Spring.

—— (2008). "Coalition of organic groups presses for immediate publication of pasture access and livestock origin rules." June 7.

—— (2013). "Organic farming—environmental benefit, yield cost?" By David Granatstein, WSU, Aug. 27: http://organicfarms.wsu.edu/blog/global-environment/organic-farming-environmental-benefit-yield-cost/ [Accessed Feb. 6, 2014].

—— (2014a). Mary-Howell Martens, Lakeview Organic Grain (Feb 4, 2014). "NOC Urges YES Vote on Final Passage of Farm Bill/methionine & antibiotics [fruit trees]." (Martens warns of "circular firing squad.")

—— (2014b). "Kathleen A. Merrigan: The former USDA deputy secretary will lead new university-wide Sustainability Institute." By Ed Maltby, Feb. 8.

—— (2014c). "Organic Check-off." By Ed Maltby, Mar. 18.

—— (2014d). "Thoughts on NOSB in San Antonio." From Mark Keating. Sent Apr. 27, 2014 10:17 a.m. To: ODAIRY@LISTSERV.NODPA.COM. Excerpt: "As the level of detection movers further and further to the right of the decimal point, we are increasingly losing the concept of the system compared to the components."

—— (2014e). "WhiteWave to remove controversial ingredient [carageenan] from Horizon Organic, Silk products." From New York. (AP), via Edward Maltby. Sent Aug. 21, 2014.

Orden, David, David Blandford, and Timothy Josling (2010). "Determinants of United States Farm Policies." In Anderson (2010): 162–90.

Oregon State University (OSU) (2013). "Beyond Antibiotics: 'PPMOs' Offer New Approach to Bacterial Infection." OSU News and Research Communications. Corvallis, OR. Published online Oct. 15: http://oregonstate.edu/ua/ncs/archives/2013/oct/beyond-antibiotics-%E2%80%9Cppmos%E2%80%9D-offer-new-approach-bacterial-infection [Accessed Oct. 28, 2013; page not found May 8, 2014].

Oregonian (2013). "Genetically Modified Wheat: With Investigation Unresolved and Harvest about to Start, Oregon Farmers Seek Answers." By Eric Mortenson. June 28, 2013.

Organic Center, The: See Benbrook, Charles M.

Organic Consumers Association (OCA) (2006). "OCA & Cornucopia Expose Horizon's Factory Dairy Farm in Maryland." By Chris Guy, *Baltimore Sun*, Aug. 28.

—— (2007a). "Siemon of Organic Valley-CROPP Joins Aurora and Horizon Asking USDA for Pasture Rule Ignoring DMI." Ronnie Cummins, *OCA Web Note*, Mar. 19.

—— (2007b). "Organic Consumers Launch Nation-Wide Lawsuit in 27 States against Aurora Organic Dairy for Violating Organic Standards." Oct. 17.

—— (2013a). "Consumer Alert: Secret Trade Agreements Threaten Food Safety, Subvert Democracy." By Katherine Paul and Ronnie Cummins, June 13.

—— (2013b). "EPA Set to Raise Limits on Glyphosate." OB #384. Also online (OCA) (July 28, 2013): http://www.organicconsumers.org/bytes/ob384.html [Accessed Aug. 8, 2013].

—— (2013c). "How the International Food Information Council Trains Junk Food Companies to Hide the Truth about GMOs." By Alexis Baden-Mayer, Sep.11: http://www.organicconsumers.org/articles/article_28288.cfm [Accessed May 8, 2014].

—— (2013d). "Aurora Organic Dairy Is Supporting Efforts to Defeat GMO Labeling Laws through Its Membership in the GMA." Sep. 19. Finland, MN: ENEWSPF.

—— (2013e). "OCA Calls on Aurora Organic Dairy to End Membership in the GMA.

—— (2013f). "AOD supporting GMA vs. I-522." Sep. 23.

—— (2013g). "Tell the FDA: Don"t Impose Unfair Burdens on Local, Organic & Sustainable Vegetable Farms" [e.g. HACCP.]: http://salsa3.salsalabs.com/o/50865/p/dia/action3/common/public/?action_KEY=12303 [Accessed Nov. 16, 2013].

—— (2013h). "Who Smells a Rat? What Do You Do When Your Scientific Journal Publishes a Study That Monsanto Doesn't Like? You Retract the Study." *Organic Bytes*, Nov. 28: http://www.organicconsumers.org/bytes/ob405.html [Accessed Dec. 9, 2013]

—— (2014a). "Food Fights, Corporate Trade Agreements, and States' Rights: Democracy or Corporatocracy?" Feb. 6: http://www.organicconsumers.org/articles/article_29258.cfm [Accessed Feb. 15, 2014].

—— (2014b). "Statement on Arrest of Organic Consumers Association Political Director at National Organic Meeting in San Antonio: 'Integrity of Organic Standards Is "Non-Negotiable,"' OCA Director Says." OCA, Apr. 30: http://www.organicconsumers.Org/articles/article_29920.cfm [Accessed May 5, 2014].

Organic Trade Association (OTA) (2012). "Organic Food Sales in the U.S. from 2000 to 2011." http://www.stat-ista.com/statistics/196952/organic-food-sales-in-the-us-since-2000/ [Accessed June 6, 2013].

Organic Valley (OV) (n.d.). "Our Story." Web history published early in the digital age: http://www.organicvalley.coop/about-us/overview/our-history/ [Accessed Aug. 20, 2014].

—— (OV) (2013). "First Grass." May 8: http://www.youtube.com/watch?v=SXmW4xqSIEg [Accessed May 8, 2013].

Osman, Loren H. (1985). *W. D. Hoard: A Man for His Time.* Fort Atkinson, WI: W. D. Hoard & Sons Co.

Paull, John (2011). "Biodynamic Agriculture: The Journey from Koberwitz to the World, 1924–1938." *Journal of Organic Systems* 6 (1): 1–15. *Inst. of Social and Cultural Anthrop.*, Univ. Oxford: http://www. organic-systems.org/journal/Vol_6%281%29/pdf/6%281%29-Paull-pp27-41.pdf [Accessed Dec. 7, 2013].

Pearce, Fred (2012). *The Landgrabbers: The New Fight over Who Owns the Earth.* London: Random House.

Perry, Ann (2011). "Putting Dairy Cows Out to Pasture: An Environmental Plus." In *Agricultural Research*, USDA ARS, May 11: http://www.ars.usda.gov/is/AR/2011/may11/cows0511.htm [Accessed Aug. 19, 2019].

Pew Charitable Trusts (2008). "One in 100: Behind Bars in America." Report on prisons: http://www.pewtrusts.org/en/research-and-analysis/reports/2008/02/28/one-in-100-behind-bars-in-america-2008 [Accessed July 14, 2008].

Pew Environment Group (2012a). "Fact Sheet: Reforming Industrial Animal Agriculture." July 18. Contact: Erin Williams. http://www.pewenvironment.org/news-room/fact-sheets/how-corporate-control-squeezes-out-small-farms-8589942044#sthash.w7YoUARm.dpuf [Accessed Nov. 24, 2013].

—— (2012b). "Remarks from Karen Steuer on CAFO Pollution and the Public Outcry for Stronger Regulation." Oct. 11: http://www.pewenvironment.org/news-room/other-resources/remarks-from-karen-steuer-on-cafo-pollution-and-the-public-outcry-for-stronger-regulation-85899422661 [Accessed May 7, 2014].

Pfeiffer, Ehrenfried (1938). *Bio-Dynamic Farming and Gardening*. Published in English, German, Dutch, French, and Italian in New York, NY: Anthroposophical Movement.

Pirog, Rich, Timothy Van Pelt, Kmyar Enshayan, and Ellen Cook (2001). *Food, Fuel and Freeways: An Iowa Perspective on How Far Food Travels, Fuel Usage and Greenhouse Emission*. Ames, IA, Leopold Center for Sustainable Agriculture, Iowa State University: http:///www.leopold.iastate.edu/.Pitesky, Maurice E., Kimberly R. Stackhouse, and Frank M. Mitloehner (2009). "Clearing the Air: Livestock's Contributions to Climate Change." *Advances in Agronomy*, Vol. 103, Burlington: Academic Press, 2009: 1–40: http://animalscience.ucdavis.edu/faculty/mitloehner/publications/2009%20pitesky%20Clearing%20the%20Air.pdf [Accessed June 13, 2014]

Plain Dealer, The (2014). "Former USDA Chief Merrigan Encourages Organic Fans to Get Involved in Federal Policy." By Debbi Snook, Feb. 16: http://www.cleveland.com/cooking/index.ssf/2014/02/former_usda_chief_merrigan_enc.html [Accessed Feb. 18, 2014].

Pollan, Michael (2001). "Behind the Organic-Industrial Complex." *New York Times Magazine*, May 13.

—— (2003). "Getting Beyond Organic." *Orion* magazine.

—— (2006). *The Omnivore's Dilemma*. New York: The Penguin Press.

—— (2013). *Cooked: A Natural History of Transformations*. New York: The Penguin Press.

Portlandia (2011). "Ordering the Chicken." Video satirizing PC consumers. https://www.youtube.com/watch?v=ErRHJlE4PGI [Accessed Feb. 10, 2014].

Prince, J. H. (1977). "The Eye and Vision." In M. J. Swenson (ed.) *Dukes Physiology of Domestic Animals*. New York: Cornell University Press, 696–712.

ProPublica (2011). "Super-PACs and Dark Money: ProPublica's Guide to the New World of Campaign Finance." By Kim Barker and Marian Wang, July 11: http://www.propublica.org/blog/item/super-pacs-propublicas-guide-to-the-new-world-of-campaign-finance [Accessed Feb. 15, 2014].

Pusztai, Arpad (2001). "Genetically Modified Foods: Are They a Risk to Human/Animal Health?" *ActionBioScience*, June: http://www.actionbioscience.org/biotechnology/pusztai.html [Accessed Apr. 24, 2014].

Rachel's Organic Dairy (2005). *Dancing Cows—Born to Graze*. June, Wales: www.rachelsorganic.co.uk/and: http://www.youtube.com/watch?v=BiLpxWeBF8Y [Accessed May 8, 2006].

Red Ice (n.d.). "Monsanto's Government Ties." Swedish radio news program and website with Henrik Palmgren: http://www.redicecreations.com/specialreports/monsanto.html [Accessed Nov. 23, 2013].

Reed, Matt (2001). "Fight the Future! How the Contemporary Campaigns of the UK Organic Movement Have Arisen from their Composting of the Past." *Sociologia Ruralis* 41 (1): 131–46.

—— (2006). "Turf Wars: The Organic Movement's Veto of GM in UK Agriculture." In Holt and Reed (2006).

Richards, Wilf (2014) Abundant Earth (Durham) Ltd. Durham Local Food and Co-operatives UK, and Permaculture Assoc.: http://www.abundantearth.coop/ [Accessed Aug. 20, 2014].

Riddle, Jim (c. 2014). "Jim Riddle: Coordinator, Organic Research Grant Programs, The Ceres Trust." NOSB 2001–06. [Accessed Jan. 31, 2014].

Right to Know Colorado GMO (July 25, 2014) by Janie Gianotsos: http://www.righttoknowcolorado.org/about_us [Accessed Aug. 26, 2014].

Ritzer, George (1993). *The McDonaldization of Society*. Thousand Oaks, CA: Pine Forge.

Robin, Marie-Monique (2008). *The World According to Monsanto: Pollution, Corruption, and the Control of the World's Food Supply*. G. Holoch, translated from French for 2010 edition. New York: The New Press.

Robinson, Paul (2010). "Improving Fertility in the High Yielding Dairy Cow." Newcastle University: Nuffield Farming Scholarships Trust. Dec. 2010: http://www.nuffieldinternational.org/rep_pdf/1295816856 Paul_Robinson_edited_report.pdf.

Rodale Institute (2011). *The Farming Systems Trial: Celebrating 30 Years*. Kutztown, PA: Rodale Institute. http://66.147.244.123/~rodalein/wp-content/uploads/2012/12/FSTbookletFINAL.pdf .

—— (2013). "Mission and History: Our Mission: Through Organic Leadership We Improve the Health and Well-Being of People and the Planet": http://rodaleinstitute.org/search/albert+howard [Accessed Mar. 22, 2013].

Ruegg, Pamela L., M. Gamroth, Y. H. Schukken, K. M. Cicconi-Hogan, R. M. Richert, K. E. Stiglbauer, and N. Lennart (2013). "Impact of Organic Management on Dairy Animal Health Research Completed." University of Wisconsin–Extension: http://milkquality.wisc.edu/whats-new/impact-of-organic-management-research-completed/ [Accessed Feb. 28, 2014].

Rural Vermont (1995). "Down on the Farm: The Real BGH Story Animal Health Problems, Financial Troubles." By Mark Kastel. A Project of the Rural Education Action Project.

Rushen, J. and A.M. de Passlle (2013) "The Importance of Improving Cow Longevity." From University of British Columbia. Article in DeLaval (2013) *Cow Longevity Conference Proceedings*: 3–21: Tumba, Sweden: DeLaval Int'l AB. Excerpt: Canada average culling 30–40%; worse in China, Mexico, Sweden (pp. 3–4).

Sahota, A. (2004). "The Global Market for Organic Food and Drink." Director, *Organic Monitor* (London): www.organicmonitor.com. In: IFOAM (2004).

Sage, Colin (2003). 'Social Embeddedness and Relations of Regard: Alternative 'Good Food' Networks in South-West Ireland."*Journal of Rural Studies* 19: 47–60.

Salatin, Joel (2013). "Opinion/Editorial: A Note from Joel Salatin (Aug. 18, 2013). 'Why do we need more farmers? [Ruralities raise soldiers says USDA Sec. Tom Vilsack].'" Polyface Farm. Posted Aug. 21: http://www.cornucopia.org/2013/08/a-note-from-joel-salatin/ [Accessed Aug. 24, 2013].

Schlosser, Eric (2001). *Fast Food Nation: The Dark Side of the All-American Meal*. New York: Houghton Mifflin.

Scholten, Bruce A.
—— (1989a). "Milk Quotas Help Melt Europe's Butter Mountain." *Hoard's Dairyman*, Jan. 10: 91.
—— (1989b). "BST Burdens EC Decision Makers." *Hoard's Dairyman*, Mar. 10: 183, 194.
—— (1989c). "USA Chancen für Milchquoten: Mit dem US$ überschüsse steigen." *Württembergisches Wochenblatt für Landwirtschaft/Badisches Wochenblatt für Landwirtschaft (WWL/BWL)*, Aug. 26: 10.
—— (1990a). "Animal Welfare a Rising Farmer Concern in Europe." *Hoard's Dairyman*, Feb. 25:190.
—— (1990b). "Pesticide und die Agrar-Umwelt Politik in USA [Pesticides/ag-enviro policy USA]." *VDI-nachrichten*. Düsseldorf: Verein der Deutschen Ingenieren (VDI): Sep. 6: 12.
—— (1990c). "Wird Bush Umwelt-Präsident [Will Bush veto GM crops?]" (*WWL/BWL*), Oct. 20: 8.
—— (1990d). "Europe's Milk Quotas . . . Six Years down the Road." *Hoard's Dairyman*, July: 587.
—— (1995). "Veal Protests Rock UK." *Hoard's Dairyman*, Mar. 25: 217.
—— (1997a). "Brits Bat around Appearance of Countryside." *Hoard's Dairyman*, Mar. 10: 181.
—— (1997b). "Bye-Bye This American Guy." *Dairy Farmer. London.* Nov. 21: 6–7.
—— (1999). "India's Winning the White Revolution: The Success of India's Operation Flood Will Soon Make It the World's Top Producer of Milk." *Hoard's Dairyman*, Apr. 10: 287.
—— (2002). "Organic-Industrial Complex or Herbal Remedy? A Case near Seattle and Vancouver." *UK Organic Research: Proceedings of the Colloquium of Organic Researchers*, Univ. Aberystwyth, Mar. 22–25. Download free: http:durham.academia.edu/BruceScholten.
—— (2006a). "Farmers' Market Movements in the UK & U.S.: Consumers on Rural Spaces in Urban Places." Unpublished paper, Association of American Geographers (AAG) annual meeting, Chicago, Mar. 8–11, 2006, Rural Geography Specialty Group (RGSG) session New Voices in Rural Geography II convened by Peter B. Nelson (Middlebury College) and Randall K. Wilson (Gettysburg College).
—— (2006b). "Firefighters in the UK and the U.S.: Risk Perception of Local and Organic Food." *Scottish Geographical Journal* 122 (2): 130–48.
—— (2006c). "Motorcyclists in the USA & the UK: Risk Perceptions of Local and Organic Food." In Holt and Reed (2006).
—— (2006d). "Polytunnel Perversity & Cow Confinement: Organic Rules Shape Farmscapes." Unpublished paper, RGS-IBG, session by Helen Moggridge and Kate Mahoney. RGS-IBG London, Aug. 31–Sep. 1.
—— (2007a). *Consumer Risk Reflections on Organic and Local Food in Seattle, with Reference to Newcastle upon Tyne*. PhD thesis, Durham University Geography Department (UK). Download free: http:durham.academia.edu/BruceScholten.
—— (2007b). "Dirty Cows: Perceptions of BSE/vCJD." In Rosie Cox and Ben Campkin (eds.), *Dirt: New Geographies of Cleanliness and Contamination*. London: I. B. Tauris.
—— (2007c). "USDA Organic Pasture Wars: Shaping Farmscapes in Washington State and Beyond." Presentation leading to chapter in Proceedings, 6th Quadrennial Meeting of the British, Canadian & American Rural Geographers (2010). Eastern Washington University, Spokane, July 15–20, 2007. Dick Winchell, Doug

Ramsey, Rhonda Koster, and Guy M. Robinson (eds). Brandon, MB, Canada: Brandon University (RDI).

—— (2008). "Lies on the Milk Label: Consumer Boycotts in the USDA Organic Pasture War." Presentation for "The Lie of the Land: Rural Lies, Myths and Realities" for Rural Geography Research Groug (RGRG) session convened by Gareth Enticott (Cardiff University) and Keith Halfacree (Swansea Univ.). Royal Geography Society-Institute of British Geographers (RGS-IBG) meeting (Aug. 27–29, 2008). London.

—— (2010a). "Pasture in the Biofuel Boom: Rescaling of FRG, UK and U.S. Organic Dairy Farms?" In Ingo Mose, Guy Robinson, Doris Schmied, Geoff Wilson (eds.), *Globalization and Rural Transitions in Germany and the UK*. Gottingen: Cuviller Verlag.

—— (2010b). "USDA Organic Pasture Wars: Shaping Farmscapes in Washington State and Beyond." In Dick Winchell, Doug Ramsey, Rhonda Koster, and Guy M. Robinson (eds.), *Geographical Perspectives on Sustainable Rural Change*. Brandon, MB, Canada: Brandon Press RDI.

—— (2010c). *India's White Revolution: Operation Flood, Food Aid and Development*. London: I. B. Tauris.

—— (2011). *Food and Risk in the U.S. and UK: Seattle and Newcastle Academics, Firefighters, Motorcyclists and Others Reflect on Organic and Local Food*. Saarbrücken: Lombard Academic Publishing (LAP).

—— (2013a). U.S. Organic Dairy Politics Survey: http://www.surveymonkey.com/s/83QSVHP.

—— (2013b) "Development through Dairying: An East African Case Study." *Geography Review*. Vol. 27 (2): 26–29. Oxfordshire: Hodder Education.

—— (forthcoming 2014). "The White Revolution and Its Role in Dual Economies." In Guy Robinson and Doris Carson, *Edgar Elgar Handbook of Agricultural Globalization*.

Scholten, B. A., and Brian T. Dugdill (2012). "Avoiding Dairy Aid Traps: The cases of Uganda, India and Bangladesh." In Alpaslan Ozerdem and Rebecca Roberts (eds.), *Challenging Post-Conflict Environments*. Surrey, UK: Ashgate.

Science (1968). "The Tragedy of the Commons." By Garrett Hardin. Dec.

—— (2000). "Health Risks of Genetically Modified Foods: Many Opinions but Few Data." By J. L. Domingo. 288:1748–49.

Scientific American (2011). "Science Sushi: Mythbusting 101: Organic Farming > Conventional Agriculture." By Christie Wilcox. July 18: http://blogs.scientificamerican.com/science-sushi/2011/07/18/myth busting- 101-organic-farming-conventional-agriculture/ [Accessed June 18, 2013].

—— (2012). "How Bacteria in Our Bodies Protect Our Health: Researchers Who Study the Friendly Bacteria That Live inside All of Us Are Starting to Sort out Who Is in Charge—Microbes or People?" By Jennifer Ackerma. May 15: http://www.scientificamerican.com/article.cfm?id=ultimate-social-network-bacteria-protects-health [Accessed Dec. 8, 2013].

Scruton, Roger (1998). *Animal Rights and Wrongs*. London: Metro Books (Demos).

Seattle Post-Intelligencer (2007). "Growth in Organic Dairies Tests Supply of Feed." AP, June 9.

—— (2013). "Faced with Lawsuit, Grocery Manufacturers Association Agrees to Disclose Campaign Finances." By Scott Sunde, Oct. 18.

Séralinia, Gilles-Eric, Emilie Claira, Robin Mesnagea, Steeve Gressa, Nicolas Defargea, Manuela Malatestab, Didier Hennequinc, and Joël Spiroux de Vendômoisa (2012). "Long Term Toxicity of a Roundup Herbicide and a Roundup-Tolerant Genetically

Modified Maize." *Food and Chemical Toxicology* 50 (11): 4221–31. Retracted by Elsevier.

Sewell, Anna (1877). *Black Beauty*. London: Jarrold & Sons.

Short, Brian, Charles Watkins, and John Martin (2007). *The Front Line of Freedom: British farming in the Second World War*. London: The Agricultural History Society.

Sinclair, Upton (1906/1981). *The Jungle*. New York, NY: Bantam.

Singer, Peter (1975/1990). *Animal Liberation*. Second edition 1990. New York: Random House.

——— (2002). *One World: The Ethics of Globalization*. New Haven, CT: Yale University.

Smith, Jeffrey M. (2003). *Seeds of Deception: Exposing Industry and Government Lies about the Safety of Genetically Engineered Foods*. Fairfield, IA: Yes Books, 2nd printing. Smith notes that Monsanto's claim that our stomachs stop IGF-1 is contradicted by the Canadian Gaps Analysis Report.

——— (2009). "You're Appointing Who? Please Obama, Say It's Not So!" *Huffington Post*, July 23: http://www.huffingtonpost.com/jeffrey-smith/youre-appointing-who-plea_b_243810.html [Accessed Nov. 13, 2013].

——— (2010). "Throwing Biotech Lies at Tomatoes—Part 1: Killer Tomatoes." Dec. 31: http://www.huffingtonpost.com/jeffrey-smith/throwing-biotech-lies-at_b_803139.html [Accessed Nov. 13, 2013].

——— (2012). *Genetic Roulette: The Documented Health Risks of Genetically Engineered Foods*. http://geneticroulettemovie.com/ [Accessed Nov. 13, 2013].

Soil Association (2013). "Innovative Farming." Bristol: SA. http://www.soilassociation.org/innovativefarming [Accessed Dec. 8, 2013].

Steiner, Rudolf (1924a). "Report to Members of the Anthroposophical Society after the Agriculture Course, Dornach, Switzerland, June 20, 1924." C. E. Creeger and M. Gardner (trans.). In M. Gardner, "Spiritual Foundations for the Renewal of Agriculture by Rudolf Steiner" (1993: 1–12). Kimberton, PA: Bio-Dynamic Farming and Gardening Association. *Journal of Organic Systems* 6 (1): 2011.

Steiner, R. (1924b). "To All Members: The Meetings at Breslau and Koberwitz; the Waldorf School; the longings of the Youth." *Anthroposophical Movement* 1:17–18.

Steiner, R. (1924c). "To All Members: The Meetings at Koberwitz and Breslau." *Anthroposophical Movement* 1:9–11.

Steinfeld, Henning, Pierre Gerber, Tom Wassenaar, Vincent Castel, Mauricio Rosales, and Cees de Haan (2006). *Livestock's Long Shadow*. See: FAO-UN.

Stevenson, Adlai (1952). "Let's Not Forget the Farmer." Museum of the Moving Image: http://www.livingroom candidate. org/commercials/1952/lets-not-forget-the-farmer [Accessed Nov. 30, 2013].

Stolze, M., A. Piorr, A. Härring, and S. Dabbert (2000). "The Environmental Impacts of Organic Farming in Europe." *Organic Farming in Europe: Economics and Policy* 6. University of Hohenheim, Stuttgart, Germany: https://www.uni-hohenheim.de/i410a/ofeurope/organicfarmingineurope-vol6.pdf.

Sundaresan, C. S. (2014). *Farm Value Chains: For Sustainable Growth and Development*. New Delhi: Regal.

Sustainable Food News (2007a). "'Access to Pasture' Rule Could Be Delayed until 2009." Mar. 30: www.sustainablefoodnews.com/printstory.php?news_id=1570 [Accessed July 7, 2007].

——— (2007b). "UNFI Chief Urges Industry to 'Pressure' OCA over Horizon Boycott." May 5.

——— (2014). "EU Organic Industry Rejects Proposed Rules, Warns of Decline in Organics." May 7, IFOAM EU.

Texmo (2013). "Texmo Pole Buildings." Alvard-Richardson Construction Co., Bell-ingham, WA: http://texmo buildings.com/ [Accessed Nov. 30, 2013].

Tremaine, David G. (1975). *Indian and Pioneer Settlement in the Nooksack Lowland to 1890.* Occasional Paper no. 4, Bellingham, WA: Center for Pacific Northwest Stud-ies, Western Washington State College.

Tucker, Robert C. (ed.) (1978). *The Marx-Engels Reader.* 2nd ed. New York: Norton.

Tuomisto, H. L., I. D. Hodge, P. Riordan, and D. W. Macdonald (2012). "Does Organic Farming Reduce Environmental Impacts? A Meta-Analysis of European Research." *Journal of Environmental Management* 112:309–20.

Twain, Mark (n.d.). "Buy Land, They're Not Making It Anymore": http://www.brainy quote.com/quotes/quotes/m/marktwain380355.html#5510yOm7SIxFT3F2.99. html [Accessed Apr. 30, 2014].

University of New Hampshire Organic Farm (2013). "Organic Dairy Research Farm." Durham, NH: UNH Agricultural Experiment Station: http://colsa.unh.edu/aes/odrf [Accessed May1, 2013].

USA Today (2003). "Consumers May Have a Beef with Cattle Feed." By Elizabeth Weise, June 9.

U.S. Government

Census Bureau (2012). *U.S. Census Bureau, Statistical Abstract of the U.S.: 2012,* "Table 823. Selected Characteristics of Farms by No. American Industry Class. System (NAICS): 2007." ($34,754,031 total conventional/organic U.S. dairy market value in 2007.)

Environmental Protection Agency (EPA) (2012). "Animal Feeding Operations—Overview" [lg med sm sector_table.pdf], U.S. EPA. Last updated on Feb. 16, 2012: http://cfpub1.epa.gov/npdes/home.cfm? program_id=7 [Accessed March 8, 2013].

—— (2013). "Pesticide Tolerances: Glyphosate." U.S. EPA, May 1: http://www. regu-lations.gov/#!documentDetail;D=EPA-HQ-OPP-2012-0132-0009: [Accessed July 15, 2013].

Food and Drug Administration (FDA) (1999). "Report on the FDA's Review of the Safety of Recombinant Bovine Somatotropin." Apr. 23, 2009, update: growth hor-mone in milk in 1989 rat study: http://www.fda. gov/Animal Veterinary/Safety Health/ProductSafetyInformation/ucm130321.htm [Accessed Nov. 6, 2013].

—— (2008). "Animal Health Literacy Pain Measurement Techniques for Food-Pro-ducing Animals Could Lead to Pain Control Drugs." http://www.fda.gov/down-loads/AnimalVeterinary/ResourcesforYou/Animal HealthLiteracy/UCM207088.pdf [Accessed Nov. 6, 2013].

—— (2009). "Report on the Food and Drug Administration's Review of the Safety of Recombinant Bovine Somatotropin." Report updated Apr. 23, 2009, to clarify quan-tities of growth hormone found in milk and those used in 1989 rat study. Response to Vermont Public Interest Group request: http://www.fda. gov/animalveterinary/safetyhealth/productsafetyinformation/ucm130321.htm [Accessed Nov. 6, 2013].

—— (2013). "Meet Michael R. Taylor, JD, Deputy Commissioner for Foods and Vet-erinary Medicine." Page last updated Jan. 30, 2013: http://www.fda.gov/aboutfda/centersoffices/officeoffoods/ucm196721.htm [Accessed Nov. 20, 2013].

House of Representatives (1999). "Review of the Environmental Protection Agen-cy's new agricultural and silvicultural regulatory programs." J. Charles Fox, Asst. Admin., EPA (p. 15). Agric. Comm., House of Rep., 106th Congress: http://

commdocs.house.gov/committees/ag/hag10640.000/hag10640_0f.htm [Accessed Nov. 24, 2013].

USDA United States Department of Agriculture

―――― (1996 [2011]). *Understanding Cooperatives: Farmer Cooperative Statistics.* Cooperative Information Report 45, Section 1: Cooperative Information Report 45, Section 13. Dec. 1996, revised June 2011: http://www.rurdev.usda.gov/support-documents/CIR45-13.pdf

―――― (1999 and 2007). "Sustainable Agriculture: Definitions and Terms." http://www.nal.usda.gov/afsic/pubs/terms/srb9902.shtml

―――― (2000). "Change in Animal Units for Confined Milk Cows: From 1982 to 1997." Natural Resources Conservation Service. Oct. 2000, based on 1997 Census of Agriculture, NASS: http://www.nrcs.usda.gov/Internet/FSE_DOCUMENTS/nrcs143_012576.pdf [Accessed Sep. 10, 2013].

―――― (2001a). *Confined Animal Production and Manure Nutrients/AIB-771.* June. ERS/USDA *AIB-771* (3), http://www.ers.usda.gov/publications/aib771/aib771c.pdf [Accessed July 14, 2007].

―――― (2001b). *U.S. Organic Farming in 2000–2001.* See section on "Certified Organic Livestock and Pasture" Economic Research Service AIB-780: 22–24.

―――― (2001c). "Rural America." By Doris J. Newton and Robert Hoppe, Agricultural Structure Branch, Resource Econ. Div./ERS 16(1).

―――― (2002a). "Recent Growth Patterns in the U.S. Organic Foods Market." By C. Dimitri and C. Greene, USDA RED/ERS, Market and Trade. Info Bulletin 777, Hartman & FMI find two-thirds bought organics.

―――― (2002b). "Veneman Marks Implementation of USDA National Organic Standards." Rel. 0453.02, Oct. 21.

―――― (2002c). "National Organic Program (NOP): Organic Production and Handling Standards: Livestock Standards: www.ams.usda.gov/nop/FactSheets/Prod-HandE.html [Accessed Sep. 24, 2007].

―――― (2004). "Livestock Health Care Practice Standard, Origin of Dairy Livestock Antibiotics." By R. H. Matthews, NOP manager.

―――― (2006). "Ag Subsidies Go Mainly to Commodity Export Crops such as Wheat."

―――― (2007a). "Q&A's on Aurora Consent Agreement. Agricultural Marketing Service" (AMS includes NOP): www.ams.usda.gov/nop/newsroom/aurora/auroraqas.pdf Accessed Sep. 5, 2007]

―――― (2007b). *Profits, Costs, and the Changing Structure of Dairy Farming.* See "Changes in the Size and Location of U.S. Dairy Farms." By James MacDonald, Erik O'Donoghue, William McBride, Richard Nehring, Carmen Sandretto and Roberto Mosheim (2007). Economic Research Report No. (ERR-47). Sep. 2007: http://www.ers.usda.gov/publications/err-economic-research-report/err47.aspx#.Ubnj PNhZ55s [Accessed June 13, 2013].

―――― (2008). *Organic Production Survey.* Volume 3 • Special Studies • Part 2, AC-07-SS-2.

―――― (2009). *Characteristics, Costs, and Issues for Organic Dairy Farming.* By William D. McBride and Catherine Greene. ERS 82, USDA. Last updated June 19, 2012.

―――― (2010). Federal Register Notice: Final Rule National Organic Program (NOP) Access to Pasture (Livestock) (PDF). Rayne Pegg, Administrator, AMS. Feb. 9, 2010.

―――― (2013). "Table 2: U.S. certified organic farmland acreage, livestock numbers, and farm operations, 1992–2011." http://www.ers.usda.gov/data-products/organic-production.aspx#.Ux3mGc7wp5s [Last updated Sep. 27, 2013].

—— AMS (2006). "Notice: Additional Clarification: Harvey Final Regulation [whole herd conversion]": http://www.ams.usda.gov/AMSv1.0/getfile?dDocName=STEL DEV3049635&acct=nopgeninfo.

—— AMS (2013a). *National Organic Program: Organic Standards.* Apr. 4: http://www.ams.usda.gov/AMSv1.0/ams. [Last Modified Apr. 4, 2013].

—— AMS (2013b). *National Organic Program: Organic Milk Operations.* July 15, USDA AMS, Office of the Inspector General, Audit Report 01601-0002-32 July 2013.

—— AMS (2014). "Instruction: Accreditation Policies and Procedures: NOP 2000." Effective date Feb. 28: http://www.ams.usda.gov/AMSv1.0/getfile?dDocName=ST ELPRDC5087104 [Accessed May 12, 2014].

—— National Organic Program (NOP) (2013). "NOP Report Sound and Sensible." NOSB meeting, Portland, OR, Apr. 9, 2013. By Miles V. McEvoy, deputy administrator, USDA National Organic Program. PowerPoint presentation: http://www.ams.usda.gov/AMSv1.0/getfile?dDocName=STELPR DC5103640 [Accessed Feb. 10, 2014].

—— NOP-AMS (Sep. 13, 2013). "'Sunset' Review of the National List of Allowed and Prohibited Substances (National List)." By NOP dep. administrator Miles McEvoy.

—— National Organic Program (NOP) (2013). *National Organic Standards Board (NOSB) Executive Subcommittee (ES) Meeting Notes* [multiple meetings; apples/pears antibiotics: p10/46]: http://www.ams.usda.gov/AMSv1.0/getfile? dDocName=STELPRDC5102397 [Accessed Mar. 9, 2013].

—— ERS (2012a). "Cattle Feedlots." Last updated May 26, 2012: http://www.ers.usda.gov/topics/animal-products/cattle-beef/background.aspx#. U2okL3aGqih).

—— ERS (2012b). "Cattle & Beef." Updated May 26, 2012: http://www.ers.usda.gov/topics/animal-products/cattle-beef/background.aspx#.UjAKOH9Z55s [Accessed May 29, 2012].

—— ERS (2013). "Organic Production." By Catherine Greene, USDA, Economic Research Service, Oct.

—— ERS (2014). "Organic Agriculture." By Catherine Greene, USDA, Apr.

—— ERS (c. 2001d) "USDA Milk Production: 1950–2000, App. table 1." http://www.ers.usda.gov/publications/sb-statistical-bulletin/sb978.aspx#.Ux3NZM7wp5sAc-cessed [Last updated: May 27, 2012]

—— ERS (2013). "Estimated U.S. Sales of Organic and Total Fluid Milk Products, Monthly and Annual, 2006-2011": www.ers.usda.gov/.../Organic...Organic_Milk_Sales.../amsmilksales [Accessed July 8, 2014].

—— ERS (2013). *USDA ERS—Rural Poverty & Well-being: Geography of Poverty:* www.ers .usda.gov/topics/rural...poverty.../geography-of-poverty.aspx [Accessed June 20, 2013].

—— NAL (2013). *NAL Agricultural Thesaurus,* National Agricultural Library: http://agclass.nal.usda.gov/ [Accessed Oct. 19, 2013].

Union of Concerned Scientists (2009). "Failure to Yield: Evaluating the Performance of Genetically Engineered Crops." By Doug Gurian-Sherman, UCS. Cambridge, MA: UCS Publications.

University of Newcastle School of Agriculture, Food and Rural Development (2006). "Nafferton Farm" (Prof. Phillip Cain, mgr.): www.ncl.ac.uk/afrd/about/facilities/nafferton.htm [Accessed May 31, 2006].

Universitry of Michigan (Winter 2014). "Recent Family Economics Study (FES) News: Wealth and the Great Recession." *The FES News*. Ann Arbor: Institute of Social Research, 4–5.

Utne Reader (2006). "Horizon Ads Promise to Add 2500 Acres Organic Pasture." Dec.

—— (2009). "50 Visionaries Who Are Changing Your World." (Mark Kastel, Cornucopia Institute.) By Staff, Nov.–Dec. 2009.

Vaarst, Mette (2001). "Mastitis in Danish Organic Dairy Farming, DIAS, Denmark." *Proceedings of the British Mastitis Conference 2001*. Garstang: Institute for Animal Health/Milk Development Co., 1–12.

Van der Ploeg, Jan Douwe (2009). *The New Peasantries: Struggles for Autonomy and Sustainability in an Era of Empire and Globalization*. London: Earthscan: http://www.jandouwevanderploeg. com/EN/doc/Struggles_for_Autonomy.pdf [Last accessed Aug. 19, 2014].

Washington Post, The (2013). "The Threat from Antibiotic Use on the Farm." By Donald Kennedy, Aug. 22: http://articles.washingtonpost.com/2013-08-22/opinions/41435819_1_food-animal-production-antibiotics-fda [Accessed Sep. 3, 2013].

—— (2013). "Report: Feeding Antibiotics to Livestock Is Bad for Humans, but Congress Won't Stop it." By Melinda Henneberger, Oct. 23.

Washington State University (2011). "Organic Fire Blight Control and the NOSB." By David Granatstein, WSU, Sep.: http://www.tfrec.wsu.edu/pdfs/P2398.pdf [Accessed Feb. 2, 2014].

—— (c. 2013). "Eggert Family Organic Farm." Pullman, WA: http://css.wsu.edu/organicfarm/eggert-family-organic-farm/ [Accessed Jan. 27, 2014].

—— (2013). "Organic Farming—Environmental Benefit, Yield Cost?" By David Granatstein, Aug. 27: http://organicfarms.wsu.edu/blog/global-environment/organic-farming [Accessed Feb. 1, 2014].

—— (2014). "Fire Blight Control in Organic Apples and Pears." WSU Tree Fruit Research and Extension Center: http://www.tfrec.wsu.edu/pages/organic/fireblight [Accessed Feb. 1, 2014].

Washington Tilth Producers (2012). "Organic Dairy Farming in Washington State": http://tilthproducers.org/programs/washington-organic-week/organic-agriculture/organic-dairy-farming-in-washington-state/ [Accessed Aug. 24, 2013].

Whatmore, Sarah (2002). *Hybrid Geographies: Natures, Cultures, Spaces*. London: Sage.

White Wave (2014). "The WhiteWave Foods Company Announces Agreement to Acquire So Delicious® Dairy Free." Sep. 17: http://www.whitewave.com/news [Accessed Sep. 22, 2014].

Wilkinson, Richard G., and Kate Pickett (2009). *The Spirit Level: Why more equal societies almost always do better*. London: Allen Lane.

Willis, Ken, and Guy Garrod (1992). "Assessing the Value of Future Landscapes." *Landscape and Urban Planning* Vol. 23 (1): 17–32.

Wilson, David (2006). *A Response to Globalisation from the Anglican Communion with Special Reference to Communication, Social Justice and Culture*. Durham University, UK: http://etheses.dur.ac.uk/2343/ [Accessed Sep. 5, 2013].

Wilson, Geoff A. (2007). *Multifunctional Agriculture: A Transition Theory Perspective*. Wallingford, Oxfordshire, UK: CABI.

Wired (2003). "Sour Grapes Over Milk Labeling." By Kristen Philipkoski, Sep. 16: http://www.wired.com/medtech/health/news/2003/09/60132?currentPage=all [Accessed Nov. 20, 2013].

Youngberg, Garth, and Suzanne P. DeMuth (2013). "Organic Agriculture in the United States: A 30-Year Retrospective." *Renewable Agriculture and Food Systems, First-View.* Vol. 28 (4): 1–35. Cambridge, UK: Cambridge University Press.

Zengerle, Jason (Apr. 29, 2014) "How Barack Obama Sold Out the Kale Crowd: The Story of How the Foodiest President Let Down His Foodie Base, Apr. 29." *The New Republic:* http://www.newrepublic.com/article/117504/obama-failed-foodies [Accessed Aug. 28, 2014].

Index

Printed and bound in Great Britain by
CPI Group (UK) Ltd, Croydon, CR0 4YY